Oxford Lecture Series in
Mathematics and its Applications 19

*Series Editors*
John Ball Dominic Welsh

OXFORD LECTURE SERIES
IN MATHEMATICS AND ITS APPLICATIONS

1. J. C. Baez (ed.): *Knots and quantum gravity*
2. I. Fonseca and W. Gangbo: *Degree theory in analysis and applications*
3. P. L. Lions: *Mathematical topics in fluid mechanics, Vol. 1: Incompressible models*
4. J. E. Beasley (ed.): *Advances in linear and integer programming*
5. L. W. Beineke and R. J. Wilson (eds): *Graph connections: Relationships between graph theory and other areas of mathematics*
6. I. Anderson: *Combinatorial designs and tournaments*
7. G. David and S. W. Semmes: *Fractured fractals and broken dreams*
8. Oliver Pretzel: *Codes and algebraic curves*
9. M. Karpinski and W. Rytter: *Fast parallel algorithms for graph matching problems*
10. P. L. Lions: *Mathematical topics in fluid mechanics, Vol. 2: Compressible models*
11. W. T. Tutte: *Graph theory as I have known it*
12. Andrea Braides and Anneliese Defranceschi: *Homogenization of multiple integrals*
13. Thierry Cazenave and Alain Haraux: *An introduction to semilinear evolution equations*
14. J. Y. Chemin: *Perfect incompressible fluids*
15. Giuseppe Buttazzo, Mariano Giaquinta and Stefan Hildebrandt: *One-dimensional variational problems: an introduction*
16. Alexander I. Bobenko and Ruedi Seiler: *Discrete integrable geometry and physics*
17. Doina Cioranescu and Patrizia Donato: *An introduction to homogenization*
18. E. J. Janse van Rensburg: *The statistical mechanics of interacting walks, polygons, animals and vesicles*
19. S. Kuksin: *Analysis of Hamiltonian PDEs*

# Analysis of Hamiltonian PDEs

Sergei B. Kuksin

*Professor of Mathematics*
*Heriot-Watt University, Edinburgh*
*and*
*Steklov Mathematical Institute, Moscow*

OXFORD
UNIVERSITY PRESS

OXFORD
UNIVERSITY PRESS

Great Clarendon Street, Oxford OX2 6DP

Oxford University Press is a department of the University of Oxford.
It furthers the University's objective of excellence in research, scholarship, and education by publishing worldwide in

Oxford New York

Athens Auckland Bangkok Bogotá Buenos Aires Calcutta
Cape Town Chennai Dar es Salaam Delhi Florence Hong Kong Istanbul
Karachi Kuala Lumpur Madrid Melbourne Mexico City Mumbai
Nairobi Paris São Paulo Singapore Taipei Tokyo Toronto Warsaw
with associated companies in Berlin Ibadan

Oxford is a registered trade mark of Oxford University Press
in the UK and in certain other countries

Published in the United States
by Oxford University Press Inc., New York



First published 2000



A catalogue record for this book is available from the British Library

Library of Congress Cataloging in Publication Data

ISBN 0 19 850395 4 (Hbk)

Typeset by Newgen Imaging Systems (P) Ltd., Chennai, India
Printed in Great Britain
on acid-free paper by
T.J. International Ltd., Padstow

# Preface

The book was written to present a proof of the following KAM theorem: most space-periodic finite-gap solutions of a Lax-integrable Hamiltonian partial differential equation (PDE) persist under a small Hamiltonian perturbation of the equation as time-quasiperiodic solutions of the perturbed equation. In order to prove the theorem we develop a theory of Hamiltonian PDEs (Chapter 1) and give short presentations of abstract Lax-integrable equations (Chapter 2) as well as of classical Lax-integrable PDEs (Chapters 3 and 4). Next, in Chapters 5–7 we develop normal forms for Lax-integrable PDEs in the vicinity of manifolds, formed by the finite-gap solutions. Finally, we prove the main theorem applying an abstract KAM theorem (Chapters 8 and 10 of Part II) to equations, written in the normal form. Our presentation is rather complete; the only non-trivial result which is given without a proof is the celebrated Its–Matveev theta formula for finite-gap solutions of a Lax-integrable PDE. The above-mentioned normal form results, and the abstract KAM theorem, are important effective tools to study non-linear PDEs, apart from the persistence of finite-gap solutions (e.g. see Kuksin 1993, Bobenko and Kuksin 1995*a*, and Kuksin and Pöschel 1996 for some other KAM results).

We have restricted ourselves to the so-called 'finite volume case'. That is, we are concerned with equations for functions (or vector-functions) $u(t, x)$, where the space-variable $x$ belongs to a bounded domain and the equations are supplemented by appropriate boundary conditions. The reason is that in the infinite-volume case time-quasiperiodic solutions are very exceptional and disappear under general perturbations of the equation; see Sigal (1993). Accordingly, all preliminary results on Hamiltonian PDEs and infinite-dimensional Hamiltonian systems are designed to treat PDEs in finite volume.

The book is devoted to global aspects of the 'KAM for PDEs' theory and it does not include the two local theories, namely perturbations of linear equations and small oscillations in non-linear equations. References for these two topics can be found in Section 8.5 and their discussion in Kuksin (1998*a*). For detailed treatment of the first topic see Kuksin (1993).

The book is aimed at the reader with a 'standard' mathematical background. However, some knowledge of basic symplectic geometry, non-linear PDEs, Sobolev spaces, and interpolation would simplify reading. As possible references for these four subjects, we suggest Arnold (1974) and Hofer and Zehnder (1994) (for the first), Lions (1969) (for the second), and Reed and Simon (1975) (for the last two). No knowledge of KAM techniques is assumed. To help the reader to understand a rather technical proof of the abstract KAM theorem, we give an Addendum where the same techniques and ideas are used in much easier finite-dimensional situations to prove the classical theorem of A.N. Kolmogorov (which originated the whole of KAM theory).

This book finalizes my research on the topic 'KAM for PDEs', started with the papers Kuksin (1987, 1988). It was written piece by piece in my home institutes and during visits to FIM (ETH, Zürich), IHES (Bures sur Yvette), IAS (Princeton), and the University of Arizona (Tucson). I sincerely thank these institutions for their hospitality, excellent working conditions, and for typing some parts of the manuscript.

While working on the book (and on the whole KAM topic), I have benefited greatly from discussions and collaboration with many colleagues. I am much obliged to all of them. I am especially grateful to Jürgen Moser for the many discussions we had during my two-year stay at FIM and for his support of my KAM research. It was Professor Moser who encouraged me to complete my research in the form finally presented in this book.

*Edinburgh* S.B.K.
*January* 2000

# Contents

*To the memory of Jürgen Moser*

# Notation

**Sets.** Everywhere in the book 'domain' means 'non-empty open set'. An over-line signifies the closure of a set.

If $Y$ is a Banach space, $y \in Y$ and $\delta > 0$, then by $\mathcal{O}_\delta(y, Y)$ we denote the open $\delta$-neighbourhood of $y$; if $y = 0$, then we abbreviate $\mathcal{O}_\delta(0, Y)$ to $\mathcal{O}_\delta(Y)$. If $F$ is a subset of a metric space, then $F + \delta$ is the $\delta$-neighbourhood of $F$; that is, $F+\delta = \{m \mid \operatorname{dist}(m, F) < \delta\}$ (so $\mathcal{O}_\delta(y, Y) = \{y\}+\delta$). By $\mathbb{T}^n$ we denote the $n$-torus $\mathbb{T}^n = \mathbb{R}^n/(2\pi\mathbb{Z}^n)$ and abbreviate $\mathbb{T}^1 = S^1 = S$. By $U(\delta)$ we denote its complex $\delta$-neighbourhood

$$U(\delta) = \{q \in \mathbb{C}^n/(2\pi\mathbb{Z}^n) \mid |\operatorname{Im} q| < \delta\} \supset \mathbb{T}^n.$$

For a Hilbert scale $\{Y_s\}$ and its complexification $\{Y_s^c\}$ we denote

$$\mathcal{Y}_s = \mathbb{R}^n \times \mathbb{T}^n \times Y_s\,, \qquad \mathcal{Y}_s^c = \mathbb{C}^n \times (\mathbb{C}^n/2\pi\mathbb{Z}^n) \times Y_s^c\,.$$

**Sets of indexes.** By $\mathbb{Z}_{\geq 0}$ and $\mathbb{Z}_0$ we denote the sets of non-negative and non-zero integers. For any $n \geq 1$ and any integer $n$-vector $V = (V_1, \ldots, V_n)$, $V_j > 0$, we set

$$\mathbb{N}_V = \{m \in \mathbb{N} \mid m \neq V_j \ \ \forall j\},$$
$$\mathbb{Z}_V = \{m \in \mathbb{Z}_0 \mid m \neq \pm V_j \ \ \forall j\} = \mathbb{N}_V \cup -\mathbb{N}_V.$$

If $V$ is the vector $V^n = (1, \ldots, n)$, then we abbreviate $\mathbb{N}_{V^n}$ to $\mathbb{N}_n$ and $\mathbb{Z}_{V^n}$ to $\mathbb{Z}_n$.

**Infinity.** Everywhere in the book an inequality $s \geq a$ is understood as $s > a$ if $a = -\infty$. Similarly, $s \leq b$ is understood as $s < b$ if $b = \infty$. Accordingly, a segment $[a, b]$ is understood as $(a, b]$ if $a = -\infty$, etc.

**Sequences.** In KAM proofs we use positive sequences $\{\varepsilon_m\}$, $\{\delta_m\}$, and $\{e(m)\}$. They are defined in Part II, Chapter 10, Section 10.2.

**Measures.** $\operatorname{mes}_m$ stands for the $m$-dimensional Lebesgue measure and $\operatorname{mes}_m^{\mathcal{H}}$ stands for the $m$-dimensional Hausdorff measure.

**Linear maps.** All linear operators between Banach spaces are assumed to be bounded. For a linear operator $L$ between Hilbert spaces we denote by $L^*$ the conjugated operator; if the spaces are complex, then $L^*$ is conjugated with respect to complex bilinear scalar products; see Section 1.1. By $\bar{L}$ we denote the operator $-L^{-1}$ (provided that it is well defined). If $L$ is a linear map from a Hilbert scale $\{X_s\}$ to a scale $\{Y_s\}$, then $\|L\|_{a,b}$ stands for its operator norm as a map $X_a \to Y_b$.

**Lipschitz maps.** Let $M$ and $N$ be two metric spaces and $f$, $f_1$, and $f_2$ be maps $M \to N$. We write

$$\mathrm{dist}\,(f_1, f_2) = \sup_{m\in M} \mathrm{dist}_N(f_1(m), f_2(m))\,,$$

$$\mathrm{Lip}\, f = \sup_{m_1\neq m_2} \frac{\mathrm{dist}_N(f(m_1), f(m_2))}{\mathrm{dist}_M(m_1, m_2)}\,.$$

If the metric space $N$ is an Abelian group and $\mathrm{dist}_N(n_1, n_2) = \mathrm{dist}_N(0, n_1 - n_2)$ for any $n_1, n_2 \in N$, [1] then we write $\|f\|_N^M = \mathrm{dist}\,(0, f)$ (0 signifies the map that sends all of $M$ to the zero in $N$) and

$$\|f\|_N^{M,\mathrm{Lip}} = \max(\mathrm{Lip}\, f, \|f\|_N^M)\,.$$

Our final notations are technical and are used in KAM proofs only: if $O$ is a domain in a metric space $B$ and $f$ is a map from $O \times M$ to $N$, then we write

$$\|f\|_N^{O,M} = \sup_{b\in O} \|f(b,\cdot)\|_N^{M,\mathrm{Lip}}\,;$$

if $N = \mathbb{C}^n$, then we abbreviate $\|f\|_{\mathbb{C}^n}^{O,M}$ to $|f|^{O,M}$.

**Differentiable maps.** For a smooth map $f : X \to Y$ we denote by $f_*(x)$ the linearized maps $T_xX \longrightarrow T_{f(x)}Y$ and by $f^*(x)$ the adjoint maps $\left(f_*(x)\right)^* : \left(T_{f(x)}Y\right)^* \longrightarrow \left(T_xX\right)^*$. We call a smooth map $f : X \supset O \to Y$ a diffeomorphism if it is a diffeomorphism of the domain $O$ on the range $f(O)$.

**Vector fields.** If $V(t, x)$ is a non-autonomous vector field, then $S_t^\tau$ stands for its flow map which sends $x(t)$ to $x(\tau)$ ($x(\cdot)$ is a solution for the equation $\dot{x} = V(x)$); if the vector field $V$ is autonomous, then we write $S_t^\tau = S^{\tau-t}$. By $S_{t**}^\tau$ we denote flow maps of the linearized equation, so $S_{t**}^\tau = S_{t*}^\tau$ if the flow maps $S_t^\tau$ are smooth. By $V_H$ we denote the Hamiltonian vector field with a Hamiltonian $H$.

[1] Example: $N$ is a Banach space, or a torus, or the former times the latter.

# I UNPERTURBED EQUATIONS

# 1
# Some analysis in Hilbert spaces and scales

## 1.1 Differentiable and analytic maps

Throughout the book differentiability of maps between Hilbert (or Banach) spaces is understood in the sense of Fréchet. Since the category of $C^r$-smooth Fréchet maps with $r \geq 2$ is rather cumbersome and since only analytic objects arise in our main constructions, we mostly restrict ourselves to two extreme cases: with few exceptions the maps will be either $C^1$-smooth or analytic. Below we fix corresponding notations and briefly recall some properties of $C^1$-smooth and analytic maps.

Let $X$ and $Y$ be Hilbert spaces and $O$ be a domain in $X$. A continuous map $f : O \to Y$ is called continuously differentiable, or $C^1$-smooth (in the sense of Fréchet), if there exists a bounded linear map $f_*(x) : X \to Y$ which continuously depends on $x \in O$, such that $f(x + x_1) - f(x) = f_*(x)x_1 + o(\|x_1\|_X)$ provided that $x, x + x_1 \in O$. We call $f_*(x)$ a *derivative* of $f$ or its *tangent map*. By $f^*(x)$ we denote the adjoint map $f^*(x) = (f_*(x))^* : Y \to X$.

For $C^k$-smooth maps with $k \geq 2$, see Cartan (1993*a*), Lang (1972).

If $f : X \supset O \to \mathbb{R}$ is a $C^1$-smooth map, then $f_*(x) \in X^*$. Identifying $X^*$ with $X$ by the Riesz theorem, we denote an element of $X$ corresponding to $f_*(x)$ as $\nabla f(x)$ and call it a *gradient* of $f$ at $x$. Thus we obtain a *gradient map* $\nabla f : O \to X$. If this map is $C^1$-smooth (that is, if $f$ is $C^2$-smooth), then the tangent map $\nabla f(x)_* : X \to X$ is a symmetric (hence, a self-adjoint) linear operator:

$$\langle \nabla f(x)_* \xi, \eta \rangle_X = \langle \nabla f(x)_* \eta, \xi \rangle_X, \quad \forall \xi, \eta.$$

Indeed, the l.h.s. equals

$$\frac{\partial^2}{\partial \alpha \, \partial \beta} f(x + \alpha \eta + \beta \xi) \,|_{\alpha = \beta = 0}$$

and the r.h.s. equals

$$\frac{\partial^2}{\partial \beta \, \partial \alpha} f(x + \alpha \eta + \beta \xi) \,|_{\alpha = \beta = 0},$$

so they coincide.

For a real Hilbert space $X$ we denote by $X^c$ its *complexification*, $X^c = X \otimes_{\mathbb{R}} \mathbb{C}$. That is, $X^c = X \oplus X$ and multiplication by $i = \sqrt{-1}$ is given by the formula $i(x_1, x_2) = (-x_2, x_1)$. We extend the inner product $\langle \cdot, \cdot \rangle_X$ of the space $X$ to a complex-bilinear pairing $X^c \times X^c \to \mathbb{C}$, so $\|u\|^2 = \langle u, \bar{u}\rangle_X$ for any $u \in X^c$. We denote this pairing by $\langle\cdot,\cdot\rangle_X$ or $\langle\cdot,\cdot\rangle_{X^c}$.

Similarly, if $Z$ is a complex Hilbert space, then $\langle\cdot,\cdot\rangle = \langle\cdot,\cdot\rangle_Z$ denotes a pairing which is a complex-bilinear symmetric quadratic form such that $\|\mathfrak{z}\|^2 = \langle \mathfrak{z}, \bar{\mathfrak{z}}\rangle$; we shall call this pairing a (complex) scalar product. Accordingly, if $Z_1$ and $Z_2$ are complex Hilbert spaces and $L : Z_1 \to Z_2$ is a linear operator, then $L^*$ is a linear operator $Z_2 \to Z_1$, conjugated to $L$ with respect to the corresponding complex-bilinear pairings $\langle\cdot,\cdot\rangle_1$ and $\langle\cdot,\cdot\rangle_2$.

**Examples.** If $X$ is an $L_2$-space or a Sobolev space of real-valued functions, then $X^c$ is a corresponding space of complex functions. If $X$ is an abstract separable Hilbert space and $\{\phi_j\}$ is its Hilbert basis, then $X = \{\sum x_j\phi_j \mid x_j\text{'s are real and } \sum |x_j|^2 < \infty\}$, while $X^c = \{\sum z_j\phi_j \mid z_j\text{'s are complex and } \sum |z_j|^2 < \infty\}$. □

Let $X^c$ and $Y^c$ be complex Hilbert spaces and $O^c$ be a domain in $X^c$. A map $f : O^c \to Y^c$ is called (Fréchet-) analytic if it is $C^1$-smooth in the sense of real analysis (when we treat $X^c$ and $Y^c$ as real spaces) and the tangent maps $f_*(x)$ are complex-linear. Locally near any point in $O^c$ such a map can be represented as a normally convergent series of homogeneous maps (see Vishik and Fursikov 1980 and Pöschel and Trubowitz 1987).

For real Hilbert spaces $X$ and $Y$ and a domain $O \subset X$, a map $F : O \to Y$ is analytic if it can be extended to a complex-analytic map $F : O^c \to Y^c$, where $O^c$ is a complex neighbourhood of $O$ in $X^c$. (The extension $F : O^c \to Y^c$ is uniquely defined if the domain $O^c$ is connected and $O$ is non-empty.)

A map $F : X \supset O \to Y$ is called $\delta$-*analytic* ($\delta$ is a positive real number) if it extends to a bounded analytic map $(O + \delta) \to Y^c$ ($O + \delta$ is the $\delta$-neighbourhood of $O$ in $X^c$).

We note that compositions of analytic maps, as well as their linear combinations, are analytic. Moreover, any analytic map is $C^k$-smooth for every $k$.

There is an important *criterion of analyticity*: a map $f : X^c \supset O^c \to Y^c$ is analytic if and only if it is locally bounded[1] and *weakly analytic*; that is, for any $y \in Y^c$ and any affine complex line $\Lambda \subset X^c$ the complex function $\Lambda \cap O^c \to \mathbb{C}$, $\lambda \mapsto \langle F(\lambda), y\rangle_Y$ is analytic in the sense of one complex variable. Moreover, it is sufficient to check, the analyticity of these functions for a countable system $y = y_1, y_2, \ldots$ of vectors in $Y$ such that the linear envelope of this system is dense in $Y$ (see Pöschel and Trubowitz 1987).

If $O^c$, $X^c$, and $Y^c$ are as above and $Y_1^c$ is a closed subspace of $Y^c$, then a map $f : O^c \to Y_1^c$ is analytic if and only if it is analytic as a map $O^c \to Y^c$ and $f(O^c) \subset Y_1^c$. This trivial consequence of the definition is useful to check the analyticity of some maps, given by non-linear differential operators.

The *Cauchy estimate* states that if a map $F : X^c \supset O^c \to Y^c$ admits a bounded analytic extension to $O^c + \delta$, then for any $u \in O^c$ one has

$$\|F_*(u)\|_{X,Y} \le \delta^{-1} \sup_{u' \in O^c + \delta} \|F(u')\|_Y .$$

[1] That is, any point $x \in O^c$ has a neighbourhood, where $f$ is bounded. In particular, any continuous map is locally bounded.

(The estimate readily follows from its one-dimensional version applied to the holomorphic functions $\mathcal{O}_\delta(\mathbb{C}) \ni \lambda \mapsto \langle F(u+\lambda x), y\rangle_Y$, where $\|x\|_X = \|y\|_Y = 1$.) In particular, this estimate applies to $\delta$-analytic maps between subsets of real Hilbert spaces.

If $F : X^c \supset O^c \to Y^c$ is an analytic map and for some point $x \in O^c$ the tangent map $F_*(x)$ is an isomorphism, then by the *inverse function theorem* in a sufficiently small neighbourhood of $x$ the map $F$ can be analytically inverted. The same is true for real analytic maps. See Vishik and Fursikov (1980) and Pöschel and Trubowitz (1987).

For *Banach spaces* everything is much the same with one extra difficulty: there is no canonical way to give a norm to the complexification $X^c$ of a real Banach space $X$. This difficulty should not worry us since all Banach spaces used in this book are natural and one can immediately guess the right norms. For example, if $X$ is the space of bounded linear operators $Y_1 \to Y_2$ where $Y_1$ and $Y_2$ are Hilbert spaces, then $X^c$ is the complex space of linear over reals operators $Y_1 \to Y_2^c$ with the natural norms, etc.

## 1.2 Scales of Hilbert spaces and interpolation

Let $X_0$ be a Hilbert space (real or complex) with a scalar product $\langle\cdot,\cdot\rangle$ and a Hilbert basis $\{\phi_k \mid k \in \tilde{\mathbb{Z}}\}$, where $\tilde{\mathbb{Z}}$ is a countable set which is an even subset of some $\mathbb{Z}^n$ (so $-\tilde{\mathbb{Z}} = \tilde{\mathbb{Z}}$). Let us take a positive sequence $\{\vartheta_j \mid j \in \tilde{\mathbb{Z}}\}$ such that $\vartheta_j = \vartheta_{-j}$ and $\vartheta_k \to \infty$ as $|k| \to \infty$. For any real number $s$ we define $X_s$ as a Hilbert space with the Hilbert basis $\{\phi_k \vartheta_k^{-s} \mid k \in \tilde{\mathbb{Z}}\}$. By $\|\cdot\|_s$ and $\langle\cdot,\cdot\rangle_s = \langle\cdot,\cdot\rangle_{X_s}$ we denote the norm and the scalar product in $X_s$. Then

$$\|u\|_s^2 = \sum |u_k|^2 \vartheta_k^{2s} \quad \text{if } u = \sum u_k \phi_k,$$

and $\|u\|_s^2 = \langle u, u\rangle_s$ (in the complex case, $\|u\|_s^2 = \langle u, \bar{u}\rangle_s$); clearly $\langle\cdot,\cdot\rangle_0 = \langle\cdot,\cdot\rangle$. The totality $\{X_s\}$ is called a *Hilbert scale*, and the basis $\{\phi_k\}$ is called *a basis of the scale*.

We do not distinguish between Hilbert scales formed by the same spaces with equivalent norms.

A Hilbert scale may be *continuous* or *discrete*: the parameter $s$ may be real or an integer. Below we state all results for real scales with $s \in \mathbb{R}$, but they admit trivial reformulations for the discrete case.

A Hilbert scale possesses the following obvious properties:

1. $X_s$ is compactly embedded to $X_r$ if $s > r$ and is dense there.
2. The spaces $X_s$ and $X_{-s}$ are conjugated with respect to the scalar product $\langle\cdot,\cdot\rangle$: for any $u \in X_s \cap X_0$ we have

$$\|u\|_s = \sup\{|\langle u, u'\rangle| \mid u' \in X_{-s} \cap X_0,\ \|u'\|_{-s} = 1\}.$$

3. For $-\infty < a < b < \infty$ and $0 \le \theta \le 1$ the space $X_c$, $c = (1-\theta)a + \theta b$, *interpolates* the spaces $X_a$ and $X_b$: in the notation of Lions *et al.* (1968), $X_c = [X_a, X_b]_\theta$. In particular, for any $u \in X_b$ the following *interpolation inequality* holds:

$$\|u\|_c \le \|u\|_a^{1-\theta} \|u\|_b^{\theta}.$$

The inequality immediately follows from the Hölder one. Indeed, if $u = \sum u_k \phi_k$, then

$$\|u\|_c^2 = \sum |u_k|^{2\theta} \vartheta_k^{2\theta b} |u_k|^{2(1-\theta)} \vartheta_k^{2(1-\theta)a}$$
$$\leq \Big(\sum |u_k|^2 \vartheta_k^{2b}\Big)^{\theta} \Big(\sum |u_k|^2 \vartheta_k^{2a}\Big)^{1-\theta} = \|u\|_b^{2\theta} \|u\|_a^{2(1-\theta)}.$$

For more on the interpolation theory, see Lions and Magenes (1968) and Reed and Simon (1975).

By property 2, the scalar product $\langle \cdot, \cdot \rangle$ extends to a bilinear pairing $X_s \times X_{-s} \to \mathbb{R}$. Abusing language, we call this pairing an $X_0$-scalar product. We say that we 'multiply in $X_0$ vectors $u_s \in X_s$ and $u_{-s} \in X_{-s}$', etc. For the complexified scale $\{X_s^c\}$ we denote by $\langle \cdot, \cdot \rangle$ complex-bilinear pairings $X_s \times X_{-s} \to \mathbb{C}$.

For any space $X_s$ (real or complex) we identify its adjoint $(X_s)^*$ with the space $X_{-s}$.

We denote by $X_{-\infty}$ and $X_\infty$ the linear spaces $X_{-\infty} = \bigcup X_s$, and $X_\infty = \bigcap X_s$ and give them no norms. The space $X_\infty$ is dense in each $X_s$ since it contains all finite linear combinations of the basis vector $\phi_k$. Vectors from the space $X_\infty$ are called *smooth*.

If $\{\vartheta_k\}$ and $\{\vartheta_k'\}$ are two positive sequences as above such that all the ratios $\vartheta_k/\vartheta_k'$ are uniformly bounded from below and from above, then for any $s$ the two corresponding spaces $X_s$ coincide (and their norms are equivalent). In particular, if $k \in \mathbb{Z}^n \setminus 0$ and $0 < \vartheta_k = C|k|^m + o(|k|^m)$ with some $m > 0$, then the sequence $\vartheta_k' = |k|^m$ defines the same scale $\{X_s\}$. Moreover, if $\tilde{\vartheta}_k = |k|$ and $\{\tilde{X}_s\}$ is the corresponding scale, then $\tilde{X}_{ms} = X_s$ for all $s$. We state this result as

**Proposition 1.1.** *If two Hilbert scales $\{X_s\}$ and $\{\tilde{X}_j\}$ correspond to the same original Hilbert space $X = X_0 = \tilde{X}_0$, to the same basis $\{\phi_k\}$, and to sequences $\{\vartheta_k \mid k \in \mathbb{Z}\,(or\ k \in \mathbb{Z}_0)\}$ and $\{\tilde{\vartheta}_k\}$ such that $0 < \vartheta_k = c|k|^m + o(|k|^m)$ and $\tilde{\vartheta}_k = \langle k \rangle \equiv \max(|k|, 1)$, then for any $s$ the identity map defines an isomorphism of the spaces $X_s$ and $\tilde{X}_{ms}$.*

Scales $\{X_s\}$ of *Sobolev spaces* that arise naturally in PDEs (see Reed and Simon 1975; Lions and Magenes 1968; and Examples 1.1 and 1.2 below) correspond to the case when $X_0$ is a space of square-summable functions and $\{\vartheta_k\}$ has a power growth in $|k|$. After linearly stretching the index $s$, these scales equal some scales with $\vartheta_k = \langle k \rangle$.

**Example 1.1 (scale of Sobolev spaces).** Let us take for $X_0$ the $L_2$-space of $2\pi$-periodic functions given the trigonometric basis $\{\varphi_k \mid k \in \mathbb{Z}\}$, where

$$\varphi_0 = \frac{1}{\sqrt{2\pi}}; \qquad \varphi_k = \frac{1}{\sqrt{\pi}} \cos kx; \qquad \varphi_{-k} = -\frac{1}{\sqrt{\pi}} \sin kx, \qquad \text{for } k = 1, 2, \ldots \tag{1.1}$$

(the minus sign is introduced for later purposes). We choose $\vartheta_k = \langle k \rangle = \max(|k|, 1)$. Then the space $X_s$ equals the Sobolev space of $2\pi$-periodic functions $H^s = H^s(S^1, \mathbb{R})$ and $S^1 = \mathbb{R}/2\pi\mathbb{Z}$. In particular, for $s \in \mathbb{N}$ the space $X_s$ has the form

$$X_s = \left\{ u(x) \in X_0 \mid \frac{\partial^k u}{\partial x^k} \in X_0 \quad \text{for} \quad k \leq s \right\},$$

where $\partial^k/\partial x^k$ stands for a derivative in the sense of distributions. Indeed, $H^s$ with $s \in \mathbb{N}$ is a Hilbert space with the scalar product $\langle u, v \rangle_s = \int (uv + u^{(s)} v^{(s)})\, dx$. For the functions $\varphi_k$ defined above we have: $\langle \varphi_k, \varphi_l \rangle_s = (1 + |k|^{2s})\delta_{k,l}$. So the functions $\{(1+|k|^{2s})^{-1/2}\varphi_k\}$ form a Hilbert basis of the Sobolev space $H^s$. Hence, $H^s = X_s$ since $1 < (1 + |k|^{2s})^{1/2}/\vartheta_k^s < 2$ for all $k$.

The space $X_\infty$ is formed by smooth periodic functions; so for the Sobolev scale, smooth vectors are just smooth functions.

Complexification $X_s^c$ of a space $X_s = H^s$ is the space $H^s(S^1; \mathbb{C})$ of complex Sobolev functions.

The operator $-\Delta = -\partial^2/\partial x^2$ sends each $\varphi_k$ to $k^2\varphi_k$ and defines an unbounded self-adjoint operator in $X_0$ with the domain of definition $X_2 = H^2$. For $s > 0$, the operator $(-\Delta)^{s/2}$ treated as an unbounded operator in $X_0$ has the domain of definition $H^s$. So the Sobolev spaces $H^s$ can be defined as domains of definition of some degrees of the minus Laplacian. For more on this way of constructing Hilbert scales, see Lions and Magenes (1968).

By $H_0^s = H_0^s(S^1, \mathbb{R})$ we denote a sub-scale of $\{H^s\}$, formed by functions with zero mean value. For a basis of this scale we take $\{\varphi_k \mid k \in \mathbb{Z}_0\}$, where the functions $\varphi_k$ are the same as above. □

**Example 1.2.** Let $X_0$ be the $L_2$-space of complex valued functions on the torus $\mathbb{T}^n = \mathbb{R}^n/(2\pi\mathbb{Z})^n$, treated as a real Hilbert space with the scalar product

$$\langle u(x), v(x) \rangle = \operatorname{Re} \int \bar{u} v \, dx,$$

and given the basis $\{\phi_k = (2\pi)^{-n/2} e^{ik\cdot x} \mid k \in \mathbb{Z}^n\}$. We choose $\theta_0 = 1$ and $\theta_k = |k|$ for $k \neq 0$. Then $X_s$ is the Sobolev space $X_s = H^s(\mathbb{T}^n; \mathbb{C}) \simeq H^s(\mathbb{T}^n; \mathbb{R}^2)$. □

Given two scales $\{X_s\}$ and $\{Y_s\}$ and a linear map $L : X_\infty \to Y_{-\infty}$, we denote by $\|L\|_{s_1,s_2} \le \infty$ its norm as a map $X_{s_1} \to Y_{s_2}$. We say that the map $L$ defines a *morphism of order $d$ of the scales $\{X_s\}$ and $\{Y_s\}$* for $s \in [s_0, s_1]$, if $\|L\|_{s,s-d} < \infty$ for each $s \in [s_0, s_1]$ with some fixed $-\infty \le s_0 \le s_1 \le +\infty$.[2] If in addition the inverse map $L^{-1}$ exists and defines a morphism of order $-d$ of the scales $\{Y_s\}$ and $\{X_s\}$ for $s \in [s_0 + d, s_1 + d]$, then we say that $L$ defines an *isomorphism of order $d$* of the two scales for $s \in [s_0, s_1]$. If $\{Y_s\} = \{X_s\}$, then an isomorphism $L$ is called an *automorphism*. We shall drop the specification 'for $s \in [s_0, s_1]$' and shall write $\operatorname{ord} L = d$ , if the segment $[s_0, s_1]$ is fixed for a moment, or can be easily recovered, or is irrelevant.

A morphism of a Hilbert scale to itself of a negative order $-\Delta < 0$ is called a *$\Delta$-smoothing morphism*.

In particular, a bounded linear operator $L : X_{s_0} \to Y_{s_0 - d}$ can be regarded as a morphism of order $d$ for $s \in [s_0, s_0]$.

We note that an order $d$ of a linear morphism is not uniquely defined since any $d' > d$ is an order of the morphism as well.

**Example.** Multiplication by a $C^r$-smooth periodic function defines a zero-order morphism of the Sobolev scale $\{H^s(S^1, \mathbb{R})\}$ for $-r \le s \le r$. In general, it does not define a zero-order morphism of this scale for $s_0 \le s \le s_1$, where $s_0 < -r$ or $s_1 > r$. □

---

[2] If $s_0 = -\infty$, then $s > s_0$. Similarly, $s < \infty$ if $s_1 = \infty$.

If $L : X_s \to Y_{s-d}$ is a morphism of order $d$ for $s \in [s_0, s_1]$, then the adjoint maps $L^* : (Y_{s-d})^* = Y_{-s+d} \to (X_s)^* = X_{-s}$ form a morphism of the scales $\{Y_s\}$ and $\{X_s\}$ of the same order $d$ for $s \in [-s_1 + d, -s_0 + d]$. We call it the *adjoint morphism*.

A morphism $L$ of a Hilbert scale $\{X_s\}$, complex or real, is called *symmetric* (*anti-symmetric*) if $\langle Lu, v\rangle = \langle u, Lv\rangle$ ($\langle Lu, v\rangle = -\langle u, Lv\rangle$) for all smooth vectors $u$ and $v$ (recall that in the complex case $\langle\cdot,\cdot\rangle$ stands for a complex-bilinear pairing). That is, $L = L^*$ ($L = -L^*$) on the space $X_\infty$.

In particular, a linear operator $L : X_{s_0} \to Y_{s_0-d}$ is called symmetric (anti-symmetric) if $L = L^*$ (respectively $L = -L^*$) on the space $X_\infty$.

If $L$ is a symmetric morphism of $\{X_s\}$ of order $d$ for $s \in [s_0, d - s_0]$, where $s_0 \geq -\infty$, then $L^*$ also is a morphism of order $d$ for $s \in [s_0, d - s_0]$. Since $L^* = L$ on $X_\infty$, then by continuity $L = L^*$ as the scale's morphisms. We call such a morphism *self-adjoint*. *Anti-self-adjoint* morphisms are defined similarly.

**Example.** The operator $-\Delta$ defines a self-adjoint morphism of order 2 of the Sobolev scale $\{H^s\}$. The operator $\partial/\partial x$ defines an anti-self-adjoint morphism of order 1. The same operators define a self-adjoint and an anti-self-adjoint automorphism of the scale $\{H_0^s\}$. □

Linear maps from one Hilbert scale to another obey the Interpolation Theorem:

**Theorem 1.1** (*see* Adams 1975; Liones and Magenes 1968; Reed and Simon 1975). *Let $\{X_s\}$ and $\{Y_s\}$ be two real Hilbert scales and $L : X_\infty \to Y_{-\infty}$ be a linear map such that $\|L\|_{a_1,b_1} = C_1$ and $\|L\|_{a_2,b_2} = C_2$. Then for any $\theta \in [0, 1]$ we have $\|L\|_{a,b} \leq C_\theta$, where $a = a_\theta = \theta a_1 + (1 - \theta)a_2$, $b = b_\theta = \theta b_1 + (1 - \theta)b_2$, and $C_\theta = C_1^\theta C_2^{1-\theta}$. This result, with $C_\theta$ replaced by $4C_\theta$, remains true for complex Hilbert scales.*

In particular, if under the theorem's assumptions $a_1 - b_1 = a_2 - b_2 =: d$, then $L$ extends to a morphism of order $d$ of the scales $\{X_s\}$, $\{Y_s\}$ for $s \in [a_1, a_2]$.

**Amplifications.** *(1) Let $L = L_u$, where $u$ is a vector from a domain in some complex Hilbert space. Let $L_u$ depend analytically on $u$ as an operator $X_{a_1} \to Y_{b_1}$ as well as an operator $X_{a_2} \to Y_{b_2}$ and norms of these operators are bounded uniformly in $u$ ($\{X_s\}$ and $\{Y_s\}$ are complex Hilbert scales). Then for any $0 \leq \theta \leq 1$, $L_u$ depends analytically on $u$ as an operator $X_{a_\theta} \to Y_{b_\theta}$ and a norm of this operator is bounded uniformly in $u$ and $\theta$.*

*(2) An obvious $C^1$-version of this result holds if the operator depends on a parameter from a domain in a real Hilbert space (e.g. from an interval of the real line).*

*Proof.* (1) For any $\theta$ the operator we discuss is weakly analytic in $u$; because of the theorem its norm is uniformly bounded. Hence, the operator is analytic by the criterion of analyticity.

(2) The result readily follows from the definition of $C^1$-differentiability. □

**Corollary.** *Let a bounded linear operator $L : X_a \to X_b$ be symmetric (or anti-symmetric) in a real or complex Hilbert scale $\{X_s\}$. Then $L$ extends to a self-adjoint (or anti-self-adjoint) morphism of order $a-b$ of the scale $\{X_s\}$ for $s \in [-b, a]$ (or $\in [a, -b]$ if $-b > a$). Moreover, if the scale is complex, and if the operator $L = L_u : X_a \to X_b$*

*analytically depends on a parameter $u$ from a complex domain and is bounded uniformly in $u$, then all operators $L : X_s \to X_{s-a+b}$ with $s$ as above are analytic in $u$ and are uniformly bounded.*

*Proof.* Since $\|L\|_{a,b} = \|L^*\|_{-b,-a} = \|L\|_{-b,-a}$, then the first assertion follows from the Interpolation Theorem. The second one results from the Amplification. □

Both the amplification and the corollary admit obvious reformulations for linear operators that depend on a paramcter $u$ continuously.

Let $-\infty < a \le b \le \infty$ and $O_s \subset X_s, s \in [a,b]$, be a system of domains *compatible* in the following sense: $O_{s_1} \cap O_{s_2} = O_{s_2}$ if $s_1 \le s_2$. Let $F : O_a \to Y_{a-d}$ be an analytic (or $C^k$-smooth) map such that its restriction to the domains $O_s$ with $a \le s \le b$ define analytic (or $C^k$-smooth) maps $F : O_s \to Y_{s-d}$ . Then we say that $F$ is an *analytic (or $C^k$-smooth) morphism of the scale $\{X_s\}$ of order $d$* for $a \le s \le b$.

**Example 1.1,** continuation. The spaces $H^s$ with $s > 1/2$ are Banach algebras: $\|uv\|_s \le C_s \|u\|_s \|v\|_s$ (see Adams 1975 or Appendix in Kuksin and Pöschel 1996). Therefore for any segment $[a,b]$, $1/2 < a \le b \le \infty$, the map $u(x) \mapsto F(u(x))$, where $F$ is a polynomial, defines an analytic map $H^s \to H^s$ of order zero for $s \in [a,b]$. If $g(x)$ is any fixed function, then the map $u(x) \mapsto F(u(x)) + g(x)$ defines an analytic morphism of the Sobolev scale of order zero for $s \in [a,b]$ if and only if $g \in H^b$. The same is true for a map defined by an analytic function $F$. More generally, this is true for the map $u(x) \mapsto F(u(x), x)$, where $F(u,x)$ is a $C^b$-smooth function of $u$ and $x$, which is $\delta$-analytic in $u$ with some $x$-independent $\delta > 0$. Indeed, for simplicity let $s$ be an integer number $\ge 1$. Elementary calculations show that $\|F(u,x)\|_s \le C(K)$ if $\|u\|_s \le K$ and $\|\text{Im}\, u\|_s \le \delta/2$; i.e. the map is bounded on bounded subsets of a complex neighbourhood of the space $H^s$. If $u(x)$ and $v(x)$ are complex Sobolev functions such that $|\text{Im}\, u(x)| < \delta/2$, then for any function $\phi_j(x)$ from the trigonometric basis (1.1), the function $\lambda \mapsto \langle F(u(x) + \lambda v(x)), \varphi_j \rangle$ is analytic in $\lambda$ from some neighbourhood of the origin in $\mathbb{C}$. So the map $H^s \to H^s, u(x) \mapsto F(u,x)$, is analytic by the criterion of analyticity. □

Given a real Hilbert scale $\{X_s\}$ and a $C^k$-smooth function $H : X_d \supset O_d \to \mathbb{R}$, $k \ge 1$, we consider its *gradient map* with respect to the scalar product $\langle \cdot, \cdot \rangle$:

$$\nabla H : O_d \to X_{-d}, \qquad \langle \nabla H(u), v \rangle = H_*(u)v \quad \forall v \in X_d\,.$$

Let us assume that $k \ge 2$ and that for every $u \in O_d$ the linearized gradient map is a linear map of order $d_H \le 2d$, i.e. $\nabla H(u)_* : X_d \to X_{d-d_H}$. Since $\nabla H_*$ is a symmetric linear operator, i.e. $\langle \nabla H(u)_* v_1, v_2 \rangle = \langle \nabla H(u)_* v_2, v_1 \rangle$ for smooth vectors $v_1$ and $v_2$, then by the corollary from Theorem 1.1, $\nabla H(u)_*$ defines a bounded self-adjoint linear morphism of the scale $\{X_d\}$ of order $d_H$ for $s \in [d_H - d, d]$.

If the domain $O_d$ belongs to a system of compatible domains $O_s$ $(a \le s \le b)$ and the gradient map $\nabla H$ defines a $C^{k-1}$-smooth morphism of order $d_H$ in this system of domains, then we say that

$$\text{ord}\, \nabla H = d_H\,.$$

**Example.** If $A$ is a self-adjoint morphism of a scale $X_s$ of order $d$ and $h(u) = \frac{1}{2}\langle Au, u\rangle$, then $h$ is a smooth functional on $X_s$ with $s \geq d/2$. Now $\nabla h(u) = Au$, so ord $\nabla h = d$ for $s \in (-\infty, \infty)$. □

## 1.3 Differential forms

For $d \geq 0$ and a domain $O$ in a Hilbert space $X_d$ from a Hilbert scale $\{X_s\}$ we identify tangent spaces $T_\mathfrak{x}O$ with $X_d$ and treat differential $k$-forms on $O$ as continuous functions

$$O \times \underbrace{(X_d \times \cdots \times X_d)}_{k} \longrightarrow \mathbb{R},$$

which are polylinear and skew-symmetric in the last $k$ arguments (see more in Cartan 1976*b* and Lang 1972). We write 1-forms as $a(\mathfrak{x})\, d\mathfrak{x}$, where $a : O \to X_{-d}$ and

$$a(\mathfrak{x})\, d\mathfrak{x}[\xi] \stackrel{\text{def}}{=} \langle a(\mathfrak{x}), \xi\rangle \quad \text{for } \xi \in X_d.$$

Moreover, we write 2-forms as $A(\mathfrak{x})\, d\mathfrak{x} \wedge d\mathfrak{x}$, where

$$A(\mathfrak{x})\, d\mathfrak{x} \wedge d\mathfrak{x}[\xi, \eta] \stackrel{\text{def}}{=} \langle A(\mathfrak{x})\xi, \eta\rangle \quad \text{for } \xi, \eta \in X_d,$$

and $A(\mathfrak{x}) : X_d \to X_{-d}$ is a bounded anti-self-adjoint operator.

**Example.** Let $X_0$ be the Euclidean space $\mathbb{R}^n = \{(x_1, \ldots, x_n)\}$ and $A(x)$ be a linear operator in $X_0 = \mathbb{R}^n$ with an anti-symmetric matrix $(A_{ij})$. Then

$$A(x)\, dx \wedge dx = -\sum_{i<j} A_{ij}(x)\, dx_i \wedge dx_j.$$

Indeed, $A\, dx \wedge dx[\xi, \eta] = \sum_{i \neq j} A_{ij}\xi_j\eta_i = -\sum_{i \neq j} A_{ij}\xi_i\eta_j$ and

$$\sum_{i<j} A_{ij}(x)\, dx_i \wedge dx_j[\xi, \eta] = \sum_{i<j} A_{ij}\xi_i\eta_j - \sum_{i<j} A_{ij}\eta_i\xi_j = \sum_{i \neq j} A_{ij}\xi_i\eta_j. \qquad \square$$

A $k$-form $\omega_k$ on $O \subset X_d$ is called $C^k$-smooth or analytic if the corresponding map from $O$ to the linear space of skew-symmetric polylinear functions

$$\underbrace{(X_d \times \cdots \times X_d)}_{k} \longrightarrow \mathbb{R}$$

is $C^k$-smooth or analytic.[3]

[3] The space of polylinear functions is given the following Banach norm: a norm of a function equals its supremum over the polysphere $\{\|\mathfrak{x}\|_d = 1\} \times \cdots \times \{\|\mathfrak{x}\|_d = 1\}$. Thus, for $k = 1$ we get the (Hilbert) norm of the space $X_{-d}$ and for $k = 2$ we get a norm equivalent to the uniform norm in the space of bounded linear operators $X_d \to X_{-d}$. The complexification of the space under discussion is a space of polylinear complex functions.

To define the differential $d\omega_k$ of a $C^1$-smooth $k$-form $\omega_k$ we use the Cartan formula:

$$d\,\omega_k(\mathfrak{x})[\xi_1,\ldots,\xi_{k+1}] = \sum_{i=1}^{k+1}(-1)^{i-1}\frac{\partial}{\partial\xi_i}\omega_k(\mathfrak{x})[\xi_1,\ldots,\hat{\xi}_i,\ldots,\xi_{k+1}]. \qquad (1.2)$$

Here the vectors $\xi_j \in T_{\mathfrak{x}}O \simeq X_d$ are extended to constant vector fields on $O$. So the r.h.s. of (1.2) is well defined (and the commutator terms in the r.h.s. of the classical Cartan formula—see, for example, Godbillon 1969 and Lang 1972—vanish).

This definition agrees well with the finite-dimensional situation, as stated in the following obvious lemma:

**Lemma 1.1.** *Let $\omega_k$ be a k-form on a domain $O \subset X_d$, $L$ be a finite-dimensional affine subspace of $X_d$, and $L^O = L \cap O$. Then $d\omega_k\,|_{L^O} = d(\omega\,|_{L^O})$.*

*Proof.* Both forms are given by the same formula (1.2). □

**Example 1.3.** (1) The differential of a $C^1$-function $f$ on $O \subset X_d$ (= a zero-form) equals $df = \nabla f(\mathfrak{x})\,d\mathfrak{x}$. (2) The differential of a 1-form $a(\mathfrak{x})\,d\mathfrak{x}$, $a : O \to X_{-d}$, equals $d(a(\mathfrak{x})\,d\mathfrak{x}) = (a(\mathfrak{x})_* - a(\mathfrak{x})^*)\,d\mathfrak{x}\wedge d\mathfrak{x}$. Indeed, the operator $A(\mathfrak{x}) = a(\mathfrak{x})_* - a(\mathfrak{x})^* : X_d \to X_{-d}$ is bounded anti-self-adjoint and

$$d\big(a(\mathfrak{x})\,d\mathfrak{x}\big)[\xi,\eta] = \langle a(\mathfrak{x})_*\xi,\eta\rangle - \langle a(\mathfrak{x})_*\eta,\xi\rangle = \langle A(\mathfrak{x})\xi,\eta\rangle. \qquad \square$$

Let $\omega_t$ be any $C^1$-smooth closed $k$-form on a domain $O \subset X_d$, $C^2$-smoothly depending on a parameter $t \in [0,1]$. Let $V(t;\mathfrak{x})$ be a non-autonomous $C^1$-smooth Lipschitz vector field on $O$. We consider the equation

$$\dot{\mathfrak{x}}(t) = V(t;\mathfrak{x}), \quad \mathfrak{x}(t) \in O,$$

and denote by $S^t_{t_0}$ its flow maps, i.e. $S^t_{t_0}\mathfrak{x}(t_0) = \mathfrak{x}(t)$. These maps are $C^1$-smooth, see Cartan (1967*a*). We assume that a sub-domain $Q \subset O$ is such that $S^t_0 Q \subset O$ for $0 \le t \le 1$ and abbreviate $S^t_0$ to $\varphi^t$.

**Lemma 1.2 (Cartan's identity).** *For $0 \le t \le 1$ we have*

$$\frac{d}{dt}\varphi^{t*}\omega_t = \varphi^{t*}\frac{\partial\omega_t}{\partial t} + d\varphi^{t*}(V\rfloor\omega_t) = \varphi^{t*}\left(\frac{\partial\omega_t}{\partial t} + d(V\rfloor\omega_t)\right)$$

*everywhere in $Q$.*[4]

*Proof.* Since $\varphi^t$ is a $C^1$-smooth map which $C^1$-smoothly depends on $t$ (see Cartan 1967*a*), then both parts of the relation we have to prove are well-defined $k$-forms on $Q$.

We abbreviate $\|\cdot\|_d$ to $\|\cdot\|$ and $X_d$ to $X$. For $N \ge 1$ we denote by $X^{(N)}$ the linear envelope in $X_d$ of the basis vectors $\varphi_j$ with $|j| \le N$ and denote $O_N = O \cap X^{(N)}$.

[4] Here $V\rfloor\omega$ stands for the form $(\xi_1,\ldots,\xi_{k-1}) \mapsto \omega[V,\xi_1,\ldots,\xi_{k-1}]$.

By continuity, to prove the identity it is sufficient to check that for arbitrary $\mathfrak{x} \in O_N$ and $\xi_1, \dots, \xi_k \in X^{(N)}$ we have

$$\frac{d}{dt}(\omega_t(\mathfrak{x}(t))[\xi_1(t), \dots, \xi_k(t)])$$
$$= \frac{\partial \omega_t}{\partial t}(\mathfrak{x}(t))[\xi_1(t), \dots, \xi_k(t)] + d\beta_t(\mathfrak{x}(t))[\xi_1(t), \dots, \xi_k(t)], \qquad (*)$$

where $\mathfrak{x}(t) = \varphi^t(\mathfrak{x})$, $\xi_j(t) = \varphi^t(\mathfrak{x})_* \xi_j$, and $\beta_t = V(t, \mathfrak{x}) \rfloor \omega_t$.

For any $M \geq N$ let us denote by $\pi_M$ the natural projector $X \to X^{(M)}$ and denote $V_M = \pi_M \circ V$. We treat $V_M$ as a map from $O$ to $X^{(M)}$ or as a vector field on $O_M$. For $M \geq N$ let us consider the equation

$$\dot{\mathfrak{x}}_M = V_M(t, \mathfrak{x}_M), \quad \mathfrak{x}_M \in O_M,$$

denote by $\mathfrak{x}_M(t)$ its solution such that $\mathfrak{x}_M(0) = \mathfrak{x}$ (we note that $\mathfrak{x} \in O_N \subset O_M$), and denote by $\varphi_M^t$ the corresponding flow maps, so $\mathfrak{x}_M(t) = \varphi_M^t(\mathfrak{x})$. Since $\|\mathfrak{x}(t) - \pi_M(\mathfrak{x}(t))\| \to 0$ as $M \to \infty$ uniformly on the curve $\mathfrak{x}(t) = \varphi^t(\mathfrak{x})$ with $0 \leq t \leq 1$ and since

$$\|\dot{\mathfrak{x}}_M(t) - \pi_M \dot{\mathfrak{x}}(t)\| \leq \|V(t, \mathfrak{x}_M) - V(t, \mathfrak{x})\|$$
$$\leq C\|\mathfrak{x}_M - \mathfrak{x}\| = C\left(\|\mathfrak{x}_M - \pi_M \mathfrak{x}\| + \|(1 - \pi_M)\mathfrak{x}\|\right),$$

then by the Gronwall lemma we have

$$\|\mathfrak{x}_M(t) - \mathfrak{x}(t)\| = o(1) \quad \text{as } M \to \infty \text{ for } 0 \leq t \leq 1.$$

In particular, it proves that the maps $\varphi_M^t$ with $0 \leq t \leq 1$ and sufficiently large $M$ are well defined in the vicinity of $\mathfrak{x}$ in $X^{(m)}$.

Quite similarly, $\left\| \left(\varphi_M^t(\mathfrak{x})_* - \varphi^t(\mathfrak{x})_*\right) \xi \right\| = o(1)\|\xi\|$ for $0 \leq t \leq 1$ and $\xi \in X^{(N)}$.

Now $(*)$ follows by transition to the limit as $M$ goes to infinity since for $\varphi^t$ replaced by $\varphi_M^t$ (and $\mathfrak{x}(t)$ replaced by $\mathfrak{x}_M(t)$), $(*)$ becomes the classical Cartan identity for the flow $\varphi_M^t$ and the closed $k$-form $\omega_t \mid_{X^{(M)}}$ on the finite-dimensional space $X^{(M)}$ (see, for example, Guillemin and Sternberg 1977, and Hofer and Zehnder 1994). □

In what follows we shall also work with $k$-forms in sub-domains of the direct products $Z_d$,

$$Z_d = X \times Y_d, \quad Z_d \ni z = (x, y),$$

where $X$ is a finite-dimensional Euclidean space and $Y_d$ is a space from a Hilbert scale $\{Y_s\}$.[5] We write the linear operators $\mathfrak{A}$ in $Z_d$ in the block form

$$\mathfrak{A} = \begin{pmatrix} \mathfrak{A}_{XX} & \mathfrak{A}_{XY} \\ \mathfrak{A}_{YX} & \mathfrak{A}_{YY} \end{pmatrix},$$

where $\mathfrak{A}_{XY} : Y_d \to X, \mathfrak{A}_{YX} : X \to Y_d$ and $\mathfrak{A}_{XX} : X \to X, \mathfrak{A}_{YY} : Y_d \to Y_d$ are bounded linear operators. The operator $\mathfrak{A}$ is anti-self-adjoint (with respect to the scalar product

[5] Obviously, the spaces $\{Z_s\}$ also form a Hilbert scale.

in $X \times Y_0$) if $\mathfrak{A}_{XY} = -\mathfrak{A}^*_{YX}$ and $\mathfrak{A}_{XX}$, $\mathfrak{A}_{YY}$ are anti-self-adjoint operators. Accordingly we write the 2-form $\mathfrak{A}\, dz \wedge dz$ as

$$\mathfrak{A}(z)\, dz \wedge dz = \mathfrak{A}_{XX}(x, y)\, dx \wedge dx + \mathfrak{A}_{XY}(x, y)\, dy \wedge dx \\ + \mathfrak{A}_{YX}(x, y)\, dx \wedge dy + \mathfrak{A}_{YY}(x, y)\, dy \wedge dy.$$

We note that in our notation

$$\mathfrak{A}_{YX}(x, y)\, dx \wedge dy[(\delta x_1, \delta y_1), (\delta x_1, \delta y_2)] = \langle \mathfrak{A}_{YX}\, \delta x_1, \delta y_2 \rangle_Y = -\langle \delta x_1, \mathfrak{A}_{XY}\, \delta y_2 \rangle_X.$$

For sub-domains of the manifolds $\mathcal{Y}_d$, where

$$\mathcal{Y}_d = \mathbb{R}^n \times \mathbb{T}^n \times Y_d = \{(p, q, y)\}, \quad \mathbb{T}^n = \mathbb{R}^n / 2\pi \mathbb{Z}^n, \tag{1.3}$$

we use natural versions of the notation given above. We note that $\mathcal{Y}_d$ is a metric Abelian group and $\mathrm{dist}_{\mathcal{Y}_d}(\mathfrak{h}_1, \mathfrak{h}_2) = \mathrm{dist}_{\mathcal{Y}_d}(\mathfrak{h}_1 - \mathfrak{h}_2, 0)$ for any $\mathfrak{h}_1$ or $\mathfrak{h}_2$ in $\mathcal{Y}_d$. Moreover, the Hilbert space $Z^d = \mathbb{R}^{2n} \times Y_d$ covers $\mathcal{Y}_d$ by the natural projection $\pi$,

$$\pi : Z_d = \mathbb{R}^{2n} \times Y_d \to \mathbb{R}^n \times \mathbb{T}^n \times Y_d,$$

which is a local isometry.

The Poincaré lemma states that 'locally' each closed form is exact. The proof is constructive and is well applicable to infinite-dimensional problems (see Cartan 1967*b* and Lang 1972). We shall need a version of the lemma for a closed 2-form defined in a neighbourhood $O \subset \mathcal{Y}_d$ of the set $P \times \mathbb{T}^n \times \{0\}$, where $P$ is a sub-domain of $\mathbb{R}^n$, such that fibres of the natural fibration $O \to \mathbb{R}^n \times \mathbb{T}^n$ are convex. Below we state the result, denoting by $w$ points from $\mathbb{R}^n \times \mathbb{T}^n$:

**Lemma 1.3.** *If $\omega_2(w, y)$ is a closed 2-form in $O$ and $\omega_2(w, 0) = 0$, then $\omega_2 = d\omega_1$, where*

$$\omega_1(w, y)(\delta w, \delta y) = \int_0^1 \omega_2(w, ty)[(0, y), (\delta w, t\, \delta y)]\, dt.$$

*In particular, if*

$$\omega_2 = A_{WW}(w, y)\, dw \wedge dw + A_{WY}(w, y)\, dy \wedge dw - A^*_{WY}(w, y)\, dw \wedge dy,$$

*then $\omega_1 = a(w, y)\, dw$, where $a(w, y) = \left( \int_0^1 A_{WY}(w, ty)\, dt \right) y$.*

This result follows from its finite-dimensional version (see Arnold 1974, Arnold and Givental 1989, and Weinstein 1977) and Lemma 1.1: For any $(w, y) \in O$ and $\xi_1, \xi_2 \in T_{(w,y)}\mathcal{Y}_d \simeq \mathbb{R}^{2n} \times Y_d = Z_d$ we denote by $Q$ a sufficiently small neighbourhood of $(w, y)$ in $\mathcal{Y}_d$ and treat $Q$ as a domain in $Z_d$. Now we take for $L$ the affine 3-space through $(w, y)$ in the directions $(0, y)$, $\xi_1$, and $\xi_2$ and get that $d\omega_1(w, y)[\xi_1, \xi_2] = \omega_2(w, y)[\xi_1, \xi_2]$.

## 1.4 Symplectic structures and Hamiltonian equations

In a domain $O_d \in X_d$ with $d \geq 0$ let us take a closed 2-form $\alpha_2 = \bar{J}(\mathfrak{x})\, d\mathfrak{x} \wedge d\mathfrak{x}$ such that the anti-self-adjoint operator $\bar{J}(\mathfrak{x}) : X_d \to X_{-d}$ $C^1$-smoothly depends on $\mathfrak{x} \in O_d$ and defines a linear isomorphism

$$\bar{J}(\mathfrak{x}) : X_d \underset{\sim}{\longrightarrow} X_{d+d_J}, \quad d_J \geq 0.$$

The form $\alpha_2$ supplies $O_d$ with a symplectic structure. This structure is termed analytic (or $C^k$-smooth, $k \geq 1$) if the form $\alpha_2$ is one. This happens if the anti-self-adjoint operator $\bar{J}$ analytically ($C^k$-smoothly) depends on $\mathfrak{x} \in X_d$.

To a $C^1$-smooth function $h$ on $O_d$ the symplectic structure as above corresponds the Hamiltonian vector field $V_h$, defined by the usual (see Arnold 1974) relation:

$$\alpha_2[V_h, \xi] = -dh(\xi) \quad \text{for all } \xi \in TO_d.$$

For any $\mathfrak{x} \in O_d$ we have $\langle \bar{J}(\mathfrak{x})V_h(\mathfrak{x}), \xi \rangle = -\langle \nabla h(\mathfrak{x}), \xi \rangle$ for each $\xi \in X_d$. Thus,

$$V_h(\mathfrak{x}) = J(\mathfrak{x})\nabla h(\mathfrak{x}), \quad \text{where } J = (-\bar{J})^{-1}. \tag{1.4}$$

The operators $\bar{J}(\mathfrak{x})$ and $J(\mathfrak{x})$ are called *operators of the symplectic and the Poisson*[6] *structures*, respectively.

The operator $J$ defines an anti-self-adjoint automorphism of the scale of order $d_J$,

$$J(\mathfrak{x}) : X_{s+d_J} \underset{\sim}{\longrightarrow} X_s, \quad -d - d_J \leq s \leq d, \tag{1.5}$$

which $C^1$-smoothly depends on $\mathfrak{x} \in O_d$; the maps (1.5) analytically depend on $\mathfrak{x}$ if the symplectic structure is analytic (see the corollary to Theorem 1.1).

Since the functional $h$ is $C^1$-smooth, then the gradient map $\nabla h : O_d \to X_{-d}$ is continuous. Using (1.5) we get that the vector field $V_h$ defines a continuous map $O_d \to X_{-d-d_J}$. Usually we shall impose an additional restriction and assume that the vector field $V_h$ is smoother than that and $\operatorname{ord} V_h = d_1 < 2d + d_J$.

To stress that a domain $O_d \subset X_d$ is given a symplectic structure as above we shall write it as a pair $(O_d, \alpha_2)$. If the form $\alpha_2$ is defined on the whole space $X_s$ for each $s \geq s_0$ with some fixed $s_0$ and is there $C^1$-smooth, then we shall say that $(\{X_s\}, \alpha_2)$ is a *symplectic Hilbert scale*.

A basis $\{\phi_j(\mathfrak{x}) \mid j \in \mathbb{Z}_n\}$ of a tangent space $T_\mathfrak{x} O_d = X_d$ is called *symplectic* if

$$\alpha_2[\phi_j(\mathfrak{x}), \phi_k(\mathfrak{x})] = \nu_j(\mathfrak{x})\delta_{j,-k} \tag{1.6}$$

for any $j \in \mathbb{N}_n$ and each $k \in \mathbb{Z}_n$, with some positive real numbers $\nu_j(\mathfrak{x})$, $j \in \mathbb{N}_n$.

For any $C^1$-smooth function $h$ on $O_d \times \mathbb{R}$ a Hamiltonian equation with the Hamiltonian $h(\mathfrak{x}, t)$ is the equation

$$\dot{\mathfrak{x}}(t) = J(\mathfrak{x})\nabla h(\mathfrak{x}, t) =: V_h(\mathfrak{x}, t). \tag{1.7}$$

If $\operatorname{ord} V_h = 0$ and the vector field $V_h$ is $C^1$-smooth and Lipschitz in $O_d$, then the initial-value problem for eqn (1.7) is well posed: for any given initial condition $\mathfrak{x}(0) \in O_d$ it

---

[6] This name is justified by Definition 1.3 below.

has a unique solution defined while it stays in $O_d$. This solution $C^1$-smoothly depends on the initial condition. If the map $V_h : O_d \times \mathbb{R} \to X_d$ is $\delta$-analytic in $\mathfrak{x} \in O_d$ ($\delta$ is $t$-independent), then the map $\mathfrak{x}(0) \to \mathfrak{x}(t)$ is analytic. For all these facts see Cartan (1967*a*) and Lang (1972). The analyticity is not discussed in these references but it directly follows from the arguments that prove the differentiability since in the analytic case all the derivatives are complex-linear.

A partial differential equation, supplemented by appropriate boundary conditions, is called a *Hamiltonian PDE* if, under a suitable choice of a symplectic Hilbert scale $(\{X_s\}, \alpha_2)$, a domain $O_d \subset X_d$ and a Hamiltonian $h$, the equation can be written in the form (1.7). In this case the vector field $V_h$ is unbounded, ord $V_h = d_1 > 0$:

$$V_h : O_d \times \mathbb{R} \to X_{d-d_1}. \tag{1.8}$$

Usually the domain $O_d$ belongs to a system of compatible domains $O_s$, $s \geq d_0$, and the map $V_h$ defines an analytic morphism of order $d_1$ for $s \geq d_0$.

For a vector field $V_h$ as in (1.8) with $d_1 > 0$, different classes of solutions for (1.7) can be considered. For this book we choose the following definition: a continuous curve $\mathfrak{x} : [0, T] \to O_d$ is a *solution of (1.7) in a space* $X_d$ if it defines a $C^1$-smooth map $[0, T] \to X_{d-d_1}$ and both parts of (1.7) coincide as curves in $X_{d-d_1}$. A solution $\mathfrak{x}(t)$ is called *smooth* if it defines a smooth curve in each space $X_l$.

If a solution $\mathfrak{x}(t)$, $t \geq \tau$, of (1.7) with $\mathfrak{x}(\tau) = \mathfrak{x}_\tau$ exists and is unique, we write

$$\mathfrak{x}(t) = S^t_\tau \mathfrak{x}_\tau \quad \text{or} \quad \mathfrak{x}(t) = S^{t-\tau}\mathfrak{x}_\tau \quad \text{if the equation is autonomous.}$$

The operators $S^t_\tau$ and $S^{t-\tau}$ are called *flow maps* of the equation. In fact, it would be more correct to call these operators 'local flow maps' since their domains of definition might depend on $t$ and $\tau$. With some abuse of language we drop the specification 'local' but in each concrete case we check if the flow map is defined on a set we need.

If (1.7) is a Hamiltonian PDE, then this definition of its solution is close to the definition of a classical solution of the corresponding PDE (if $\{X_s\}$ is a scale of Sobolev functions and $d$ is sufficiently large compared with $d_1$, then the solutions defined above are classical solutions of the PDE; see examples below).

For eqn (1.7) with $d_1 > 0$ there is no general existence theorem for a solution of the corresponding initial-value problem that would guarantee existence of the flow maps. Proving the existence is an art we do not touch in this book.

**Example 1.4 (semilinear equation).** Let (1.7) be a *semilinear equation*

$$\dot{\mathfrak{x}} = V(\mathfrak{x}), \qquad V = B + V^0,$$

where $B$ is a linear operator, bounded or unbounded. It is assumed that the operator $B$ generates a continuous group of linear transformations of the space $X_d$,

$$\|e^{tB}\|_{d,d} \leq C_1 e^{C_2|t|},$$

and the non-linearity $V^0$ is Lipschitz uniformly on bounded subsets of $X_d$.

**Proposition 1.2.** *If (1.7) is a semilinear equation as above (i.e. $V_h = B + V^0$, where $\operatorname{ord} V^0 = 0$), then for any $C$ its flow maps $S^t : \mathcal{O}_C(X_d) \to X_d$ are well defined for $|t| \le T$, where $T = T(C) > 0$; if in addition the map $V^0 : X_d \to X_d$ is $C^1$-smooth (analytic), then the flow maps are $C^1$-smooth (analytic). If every solution for (1.7) in $X_d$ for every $t$ satisfies an a priori estimate $\|x(t)\|_d \le g(t, x(0)) < \infty$, where $g$ is a continuous function on $\mathbb{R} \times X_d$, then all flow maps $S^t : X_d \to X_d$ are well defined and as smooth as above.*

This result admits an obvious reformulation for the case when the vector field $V$ is defined on a sub-domain of $X_d$. For all these results, see Pazy (1983) and Kuksin (1993).

Some important Hamiltonian PDEs are semilinear. For example, the nonlinear Schrödinger equation,

$$\dot{u}(t,x) = i(\Delta u + f(|u|^2)u), \quad x \in \mathbb{T}^n,$$

where $f$ is a smooth real-valued function (see Kuksin 1993). However, the semilinearity assumption is very restrictive since it fails for many important Hamiltonian PDEs (e.g. for the KdV). □

**Example 1.5 (non-linear string).** Space-periodic oscillations of a non-linear string which obey a non-linear Hooke's law and do not move as a whole, are described by the following (strongly) non-linear string equation:

$$u_{tt} = u_{xx} + \frac{\partial}{\partial x} f\left(\frac{\partial u}{\partial x}\right), \qquad u(t,x) \equiv u(t, x+2\pi), \qquad \int_0^{2\pi} u(t,x)\,dx \equiv 0,$$

where $f(v)$ is an analytic function of the form $f(v) = \text{const} + av^2 + \cdots$ $(a \ne 0)$ at zero. We can write this equation as a system of two first-order equations:

$$\dot{u} = v, \qquad \dot{v} = u_{xx} + \frac{\partial}{\partial x} f\left(\frac{\partial u}{\partial x}\right).$$

Denoting $w = (u, v)$, we get for $w$ the equation

$$\dot{w} = Aw + F(w), \tag{1.9}$$

where $A(u,v) = (v, u_{xx})$ and $F$ is the non-linear term. In the scale $\{Z_s = H_0^s \times H_0^s\}$ the map $A$ becomes a linear morphism of order 2 and $F$ becomes an analytic (for $s \ge 2$) map of the same order. Equation (1.9) has the Hamiltonian form (1.7) with $J(u,v) = (v, -u)$ and $h(u,v) = \int \left(\frac{1}{2}|v|^2 + \frac{1}{2}|u_x|^2 + f(u_x)\right) dx$.

The non-linear string equation possesses some rather unpleasant properties: due to P. Lax (see Lax 1964 and Klainerman 1983), the only $C^2$-smooth solution of the equation that exists for all $t$ is the zero solution. In particular, eqn (1.9) has no non-trivial time-quasiperiodic (see Appendix 1 below) solutions in $Z_s, s \ge 3$. For $f = 0$ all solutions of the corresponding linear equation are quasiperiodic or almost periodic in time. Thus, arbitrarily small non-linear $f$ kills all non-zero time-quasiperiodic solutions of the linear equation. The reason for this lack of persistence is that eqn (1.9) is strongly non-linear: $\operatorname{ord} A = \operatorname{ord} F$. □

Our main concern in this book are *quasilinear Hamiltonian equations*, i.e. equations (1.7) that have the form (1.9) with ord $A >$ ord $F$ ($A$ is the linear part of an equation); possibly ord $F > 0$, i.e. the equation may be non-semilinear (so the non-linearity in Example 1.5 is too strong and in Example 1.4 it is non-necessarily weak). We call a Hamiltonian $h$ *quasilinear* if it defines a quasilinear Hamiltonian equation.

Let $Q \subset O_d$ be a sub-domain such that the flow maps $S_\tau^t : Q \to O_d$ are well defined and are $C^1$-smooth for $T_1 \leq \tau, t \leq T_2$, where $-\infty \leq T_1 < T_2 \leq \infty$ (here and in similar situations below, $t > T_1$ if $T_1 = -\infty$ and $t < T_2$ if $T_2 = \infty$). Then differentiating a solution $\mathfrak{x}(t)$ of (1.7) in the initial condition we get that the curve $\zeta(t) := S_{\tau *}^t(\mathfrak{x}(\tau))\zeta$ satisfies the linearized equation

$$\dot{\zeta}(t) = V_{h*}(\mathfrak{x}(t), t)\zeta(t), \quad \zeta(\tau) = \zeta. \tag{1.10}$$

The assumption that the map $S_\tau^t$ is $C^1$-smooth is very restrictive since to check the smoothness of flow maps for many important equations (even for the KdV!) is a non-trivial task. To get rid of it we give the following definition.

**Definition 1.2.** Let $\mathfrak{x}(t)$, $t \in \mathbb{R}$, be a solution of eqn (1.7). If for each $\zeta \in X_d$ and each $\theta$ the linearized equation

$$\dot{\zeta}(t) = V_{h*}(\mathfrak{x}(t), t)\zeta(t), \quad \zeta(\theta) = \zeta,$$

has a unique solution $\zeta(t) \in X_d$ defined for all $t$ and such that $\|\zeta(t)\|_d \leq C\|\zeta\|_d$ uniformly in $\theta$ and $t$ from a compact segment, then we write $\zeta(t) = S_{\theta **}^t(\mathfrak{x})\zeta$ and say that the *flow* $\{S_{\theta **}^t(\mathfrak{x})\}$ *of the linearized equation (1.10) is well defined* in $X_d$.

Sometimes we shall use an obvious version of this definition for the case when the solution $\mathfrak{x}(t)$ (and the linearized equation) are defined only on a finite segment of the real line.

We note that under the assumptions of this definition the maps $S_{\theta **}^t$ and $S_{t **}^\theta$ are inverses of each other.

The property described in Definition 1.2 characterizes the flow only in the 'infinitesimal vicinity' of a solution of (1.7). It suits well our goal to study special families of solutions rather than the whole flow of the equation. If the flow maps $S_\tau^t$ are $C^1$-smooth, then $S_{\tau *}^t(\mathfrak{x}) = S_{\tau **}^t(\mathfrak{x})$, but the map in the r.h.s. of this relation can be well defined while the map in the l.h.s. is not.

**Example 1.6 (equations of the Korteweg–de Vries type).** Let us take for $\{X_s\}$ the scale of Sobolev spaces $H_0^s$ as in Example 1.1. We define a Poisson structure by means of the operator $J = \partial/\partial x$, so $d_J = 1$ and $-\bar{J} = J^{-1}$ is the operator $(\partial/\partial x)^{-1}$ of integrating with zero mean value. We get the symplectic Hilbert scale $(\{H_0^s\}, -(\partial/\partial x)^{-1}\, du \wedge du)$. We stick to the discrete scale $\{s \in \mathbb{Z}\}$: it is sufficient since the orders of all involved operators are integer. The trigonometric basis $\{\varphi_j \mid j \in \mathbb{Z}_0\}$ introduced in Example 1.1 (see (1.1)) is symplectic since for $j \geq 1$ and any $k$ we have

$$\alpha_2[\varphi_j, \varphi_k] = \langle \bar{J}\pi^{-1/2} \cos jx, \varphi_k(x)\rangle = j^{-1}\langle -\pi^{-1/2} \sin jx, \varphi_k(x)\rangle = j^{-1}\delta_{j,-k}.$$

For a Hamiltonian $h$ we take $h(u) = \int_0^{2\pi}(-\frac{1}{8}u'(x)^2 + f(u))\,dx$ with some analytic function $f(u)$. Then $\nabla h(u) = \frac{1}{4}u'' + f'(u)$ [7] and $V_h(u) = \frac{1}{4}u''' + \frac{\partial}{\partial x}f'(u)$. Thus the Hamiltonian equation takes the form

$$\dot{u}(t,x) = \frac{1}{4}u''' + \frac{\partial}{\partial x}f'(u) \tag{1.11}$$

(for $f(u) = \frac{1}{4}u^3$ we get the KdV equation; the factor 1/4 is introduced to make the formulas which integrate the KdV more elegant). Since Sobolev spaces $H^s$ with $s \geq 1$ are Banach algebras, for $s \geq 1$ the maps $H_0^s \to \mathbb{R}$, $u(x) \mapsto \int f(u)\,dx$, and $H_0^s \to H^s$, $u(x) \mapsto f'(u(x))$ are analytic (see Example 1.1). Now the map $H_0^s \ni u \mapsto V_h(u) \in H_0^{s-3}$ is analytic for $s \geq 1$. That is, the vector field $V_h$ defines an analytic morphism of order 3 for $s \geq 1$.

Being supplemented by an initial condition $u(0,x) = u_0(x) \in H_0^s$ with $s \geq 3$, eqn (1.11) has a unique solution in $H_0^s$. This solution exists for $|t| < T(\|u_0\|_s)$ ($T$ is a continuous positive function) and the flow maps $S^t$ ($|t| < T$) are $C^1$-smooth. This is a non-trivial result see, for example, Kato (1975).

On the contrary, if $u(t,x)$, $0 \leq t \leq T$, is a smooth solution of (1.11), then the linearized equation

$$\dot{v} = \frac{1}{4}v''' + \frac{\partial}{\partial x}(f''(u)v), \quad v(0,x) = v_0 \in H_0^s, \tag{1.12}$$

has a unique solution in $H_0^s$ for any $s \geq 0$, by the following *trivial* reasons.

To prove *uniqueness* we have to check that a solution $v(t,x)$ with $v(0,x) = 0$ vanishes identically. We denote $\partial^k = (\partial/\partial x)^k$, $k \in \mathbb{Z}$, treating $\partial^k$ as an isomorphism of the scale $\{H_0^s\}$, and multiply the equation by $\partial^{-4}v$ in $H_0^0$, i.e. in $L_2(S^1, dx)$. We get

$$\frac{1}{2}\frac{d}{dt}\|v(t)\|_{-2}^2 = -\frac{1}{4}\int v\partial^{-1}v\,dx - \int f''(u)v\partial^{-3}v\,dx.$$

The first term in the r.h.s. vanishes. Integrating by parts several times, one finds that the second term equals

$$\int\left(\left(\frac{1}{2}\partial^2 - \partial\right)f''(u)\right)(\partial^{-2}v)^2\,dx + \frac{1}{2}\int \partial^3 f''(u)(\partial^{-3}v)^2\,dx.$$

Since $f''(u)$ is a smooth function, this implies the inequality

$$\frac{1}{2}\frac{d}{dt}\|v(t)\|_{-2}^2 \leq C\|v\|_{-2}^2 + C_1\|v\|_{-3}^2 \leq C_2\|v\|_{-2}^2,$$

so $v \equiv 0$ by Gronwall.

To prove *existence* we start with an a priori estimate for a smooth solution $v(t,x)$. Multiplying (1.12) by $v(t,\cdot)$ in $H_0^s$ we get that $d/dt\|v\|_s^2 \leq C(u)\|v\|_s^2$. Hence,

$$\|v(t,\cdot)\|_s \leq e^{C_1 t}\|v_0\|_s \quad \text{for } 0 \leq t \leq T, \tag{1.13}$$

where $C_1 = C(u)/2$. Now we can use the Galerkin method and this estimate to construct a solution $v(t,x)$ of the equation in $H_0^s$, provided that $v_0 \in H_0^\infty$. In this way we get linear

[7] Since $dh(u)v = \int -\frac{1}{4}u'(x)v'(x) + f'(u(x))v(x)\,dx = \langle \frac{1}{4}u''(x) + f'(u(x)), v(x)\rangle_{L_2}$.

flow maps $S^t_{0**}$, defined on $H_0^\infty$, and such that $\|S^t_{0**}\|_{s,s} \le e^{C_1 t}$. By continuity we extend these maps to the whole $H_0^s$. When $t \to 0$, the operators $S^t_{0**}$ remain uniformly bounded because of (1.13) and converge to identity on the dense subset $H_0^\infty \subset H_0^s$. Therefore, $S^t_{0**} \to \text{id}$ in the strong operator topology of the operators in $H_0^s$ (see, for example, Kato 1966) and the curves $v(t,\cdot) = S^t_{0**} v_0$ are continuous in $H_0^s$ for any $v_0 \in H_0^s$. Since they satisfy the equation, they are $C^1$-smooth in $H^{s-3}$ (these arguments are obvious for a smooth vector $v_0$, while a vector $v_0 \in H_0^s$ has to be approximated by smooth ones).

For any $v_0 \in H_0^s$ we constructed a unique solution of the linearized equation (1.12) in $H_0^s$. Thus, the flow maps $S^t_{\tau**}(u(\tau))$ of the linearized equation are well defined 'gratis'. □

We shall often work with equations in a sub-domain $\mathcal{O}_d$ of the manifold $\mathcal{Y}_d$ ($d \ge 0$) as in (1.3), given a symplectic structure by means of a 2-form $(dp \wedge dq) \oplus (\Upsilon(y)\, dy \wedge dy)$, where $dp \wedge dq$ is the classical symplectic form on $\mathbb{R}^n \times \mathbb{T}^n$ and $\Upsilon(y)\, dy \wedge dy$ is a closed 2-form in a domain in $Y_d$. For this symplectic structure, the Hamiltonian system with a $C^1$-smooth Hamiltonian $H(p,q,y)$ is the system

$$\dot p = -\nabla_q H, \qquad \dot q = \nabla_p H, \qquad \dot y = \Upsilon \nabla_y H.$$

Solutions of these equations are defined in the same way as solutions for (1.7).

## 1.5 Symplectic transformations

Let $\{X_s\}$ and $\{Y_s\}$ be two real Hilbert scales and $d, \tilde d \ge 0$. Let $O \subset X_d$ and $Q \subset Y_{\tilde d}$ be two domains given continuous symplectic structures by 2-forms $\alpha_2 = \bar J(\mathfrak{x})\, d\mathfrak{x} \wedge d\mathfrak{x}$ and $\beta_2 = \bar\Upsilon(y) dy \wedge dy$ as in Section 1.4. A $C^1$-smooth map $\Phi : Q \to O$ is called a *symplectic map* (or a *symplectic transformation*) if $\Phi^*\alpha_2 = \beta_2$. That is, if for any $y \in Q$ with $\Phi(y) = \mathfrak{x} \in O$ we have

$$\langle \bar J(\mathfrak{x}) \Phi_*(y)\xi, \Phi_*(y)\eta \rangle_{X_0} = \langle \bar\Upsilon(y)\xi, \eta\rangle_{Y_0}$$

for all $\xi, \eta \in Y_{\tilde d}$, or

$$\Phi^*(y) \circ \bar J(\mathfrak{x}) \circ \Phi_*(y) = \bar\Upsilon(y). \tag{1.14}$$

A symplectic map $\Phi$ is an immersion since by (1.14) its tangent maps are embeddings.

If a symplectic map $\Phi$ is such that the tangent maps $\Phi_*(y)$ define isomorphisms of the spaces $Y_{\tilde d}$ and $X_d$, then $\Phi$ is called a *symplectomorphism*. Obviously, a $C^1$-diffeomorphism $\Phi : Q \to O$ is a symplectomorphism if and only if each tangent map $\Phi_*(y)$, $y \in Q$, sends a symplectic basis of the space $T_y Q$ to a symplectic basis of the space $T_{\Phi(y)} O$ and $\nu_j(y) = \nu_j(\Phi(y))$ for all $j$ and $y$ (see (1.6)).

We shall need an obvious version of the definitions above for the case when $O^c$ and $Q^c$ are *complex* domains in complex spaces $X^c_d$ and $Y^c_{\tilde d}$ and the operators $\bar J(\mathfrak{x})$ and $\bar\Upsilon(y)$ are anti-self-adjoint with respect to the complex-bilinear scalar products $\langle\cdot,\cdot\rangle_{X_0^c}$ and $\langle\cdot,\cdot\rangle_{Y_0^c}$. Namely, a $C^1$-smooth map $\Phi_1 : (Q^c, \alpha_2) \to (O^c, \beta_2)$ is symplectic if $\langle \bar J(\mathfrak{x}) \Phi_{1*}(y)\xi, \Phi_{1*}(y)\eta\rangle_{X_0^c} \equiv \langle \bar\Upsilon(y)\xi, \eta\rangle_{Y_0^c}$.

Analytic symplectic forms $\alpha_2 = \bar{J}(\mathfrak{x})\,d\mathfrak{x}\wedge d\mathfrak{x}$ and $\beta_2 = \bar{\Upsilon}(y)\,dy\wedge dy$ on domains $O \subset X_d$ and $Q \subset Y_d$ analytically extend to some complex neighbourhoods $O^c \subset X_d^c$ and $Q^c \subset Y_d^c$. There they define complex symplectic structures as above. Any analytic symplectomorphism $\Phi : (Q, \alpha_2) \to (O, \beta_2)$ analytically extends to a sufficiently small complex neighbourhoods of $Q$, where it defines a complex symplectomorphism. Below we often use this symplectic analytic extension in the case when the forms $\alpha_2$ and $\beta_2$ are constant coefficients.

From now on for the sake of simplicity we restrict ourselves to the case we need below:

$$d = \tilde{d} \geq 0, \quad \operatorname{ord} \bar{J}(\mathfrak{x}) = \operatorname{ord} \bar{\Upsilon}(y) = -d_J, \quad \forall \mathfrak{x}, y.$$

**Proposition 1.3.** *Let us assume that $\bar{J}(\mathfrak{x}) = \bar{J}$ and $\bar{\Upsilon}(y) = \bar{\Upsilon}$ are constant isomorphisms of the corresponding scales of order $-d_J$. Then:*

*(1) If $\Phi : (Q^c, \beta_2) \to (O^c, \alpha_2)$ is an analytic symplectomorphism of complex domains such that $\|\Phi_*(y)\|_{d,d}, \|(\Phi_*(y))^{-1}\|_{d,d} \leq C$ for every $y \in Q^c$, then for every $y \in Q^c$ and every $\theta \in [-d - d_J, d]$ we have $\|\Phi_*(y)\|_{\theta,\theta}, \|(\Phi_*(y))^{-1}\|_{\theta,\theta} \leq C_1$. The maps $\Phi_* : Y_\theta \to X_\theta$ and their inverses analytically depend on $y \in Q^c$.*

*(2) If $\Phi : (Q, \beta_2) \to (O, \alpha_2)$ is a $C^1$-symplectomorphism, then a $C^1$-version of this result holds true.*

*Proof.* (1) By (1.14) we have $\Phi^* = -\bar{\Upsilon} \circ \Phi_*^{-1} \circ J$. So $\|\Phi^*(y)\|_{d+d_J,d+d_J} \leq C'$ for every $y$. Hence, $\|\Phi_*(y)\|_{-d-d_J,-d-d_J} \leq C'$ and the estimate for $\|\Phi_*\|_{\theta,\theta}$ follows by interpolation. The estimates for $\Phi_*^{-1}$ follow from the identity $(\Phi^*)^{-1} = -\bar{J} \circ \Phi_* \circ \Upsilon$ which implies that $\Phi_*^{-1} = \left((\Phi^*)^{-1}\right)^* = (-\bar{J} \circ \Phi_* \circ \Upsilon)^*$ is a zero-order morphism for $s \in [-d - d_J, d]$.

The maps $\Phi_*$ and $(\Phi_*)^{-1}$ are analytic in $y$ by Amplification 1 to Theorem 1.1.

(2) If the map $\Phi$ is a $C^1$-symplectomorphism, then the assertion follows from the same estimates as above and Amplification 2 to Theorem 1.1. □

Literally the same arguments prove the following result:

**Proposition 1.3′.** *Let the symplectic spaces $(X_d^c, \alpha_2)$ and $(Y_d^c, \beta_2)$ be as above and $\Psi(w) : X_d^c \to Y_d^c$ be a linear symplectomorphism, analytic in $w$ from some complex domain and bounded uniformly in $w$. Then for any $\theta \in [-d - d_J, d]$ the map $\Psi(w)$ defines a symplectomorphism $X_\theta^c \to Y_\theta^c$, analytic in $w$ and bounded uniformly in $w$ and $\theta$. An obvious $C^1$-version of this result also holds true.*

Now let us consider the case when $(\{X_s\}, \alpha_2) = (\{Y_s\}, \beta_2)$ and $\Phi$ is a symplectomorphism $\Phi : (Q^c, \alpha_2) \to (X_d^c, \alpha_2)$, $Q^c \subset X_d^c$:

**Proposition 1.4.** *(1) Let an analytic symplectomorphism $\Phi$ satisfy the assumptions of item (1) of Proposition 1.3 and have the form $\Phi = \mathrm{id} + \Xi$, where the map $\Xi$ is $\Delta$-smoothing ($\Delta \geq 0$) and $\|\Xi_*(x)\|_{d,d+\Delta} \leq C$ for all $x \in Q^c$. Then for every $s \in [-d - \Delta - d_J, d + \Delta]$ the linearized map $\Xi_*(x)$ is analytic in $x$ as a map $X_s \to X_{s+\Delta}$.*

*(2) If $Q \subset X_d$ and $\Phi : Q \to X_d$ is a $C^1$-symplectomorphism, then a $C^1$-version of this result holds.*

*Proof.* (1) Owing to Proposition 1.3, the maps $\Phi_*(x)$, $x \in Q^c$, define zero-order automorphisms of the scale $\{X_s\}$ for $s \in [-d - \Delta - d_J, d + \Delta]$, which are analytic in $x$.

Substituting $\Phi = \text{id} + \Xi$ in (1.14) we get that $\Xi^*(x)\bar{J} + (\text{id} + \Xi^*(x))\bar{J}\Xi_*(x) = 0$. Hence, $\Xi^*(x) = \Phi^*(x)\bar{J}\Xi_*(x)J$. By assumptions, $\|\bar{J}\Xi_*(x)J\|_{d+d_J, d+\Delta+d_J} \le C'$. Since adjoint maps $\Phi^*(x)$ define zero-order automorphisms for $s \in [-d - \Delta, d + \Delta + d_J]$, the maps $\Xi^*(x) : Y_{d+d_J} \to Y_{d+\Delta+d_J}$ are analytic in $x \in Q^c$; so the maps $\Xi_*(x)$ : $Y_{-d-\Delta-d_J} \to Y_{d-d_J}$ also are. Using the assumptions as well as the analyticity criterion we get that $\Xi_*(x)$ defines an analytic in $x$ map $Y_d \to Y_{d+\Delta}$. Interpolating these two results and using Amplification 1 we get the statement.

(2) This assertion follows from Amplification 2. □

This proposition admits an obvious reformulation for parameter-depending symplectomorphisms, similar to Proposition 1.3′. We do not state this result but use it later on.

As in the finite-dimensional case, symplectic maps transform Hamiltonian equations to Hamiltonians. Let $\Phi : Q \to O$ be a $C^1$-smooth symplectomorphism such that

$$\Phi_*(y) : Y_s \to X_s \text{ is a linear map, continuous in } y \in Q, \text{ for any } |s| \le d. \quad (1.15)$$

If $\bar{J}(x) = \bar{J}$ and $\bar{\Upsilon}(y) = \bar{\Upsilon}$ are constant isomorphisms of zero-order, then assumption (1.15) is satisfied due to Proposition 1.3, item (2).

**Theorem 1.2.** *Let the domains $O \subset X_d$ and $Q \subset Y_d$, $d \ge 0$, be given symplectic structures by the 2-forms $\alpha_2$ and $\beta_2$ as above with* $\text{ord}\,\bar{J} = \text{ord}\,\bar{\Upsilon} = -d_J$. *Let the vector field $V_h = J\nabla h$ of eqn (1.7) define a $C^1$-smooth map $V_h : O \times \mathbb{R} \to X_{d-d_1}$ of order $d_1 \le 2d$ and let $\Phi : Q \to O$ be a symplectic map satisfying (1.15), such that the vector field $V_h$ in $O$ is tangent to $\Phi(Q)$ in the following sense:*

$$V_h(\Phi(y)) = \Phi_*(y)\xi \quad \textit{for any } y \in Q \textit{ with an appropriate } \xi = \xi(y) \in Y_{d-d_1}. \quad (1.16)$$

*Then the map $\Phi$ transforms solutions of the Hamiltonian equation*

$$\dot{y} = \Upsilon(y)\nabla_y H(y, t), \quad H = h \circ \Phi, \quad \Upsilon = (-\bar{\Upsilon})^{-1}, \quad (1.17)$$

*to solutions of (1.7).*

We note right away that the assumption (1.16) becomes empty if $\Phi$ is a symplectomorphism (to be more specific, (1.16) now follows from (1.15) since $d - d_1 \ge -d$).

*Proof.* Let $y(t)$ be a solution of (1.17). By (1.15) the curve $\mathfrak{x}(t) = \Phi(y(t))$ is $C^1$-smooth in $Y_{d-d_1}$ and is continuous in $Y_d$. It remains to check that it satisfies (1.7). Since $\dot{\mathfrak{x}} = \Phi_*(y)\dot{y}$ and $\nabla_y H = \Phi^*(y)\nabla_{\mathfrak{x}} h$, then

$$\dot{\mathfrak{x}} = \Phi_*(y)\Upsilon(y)\Phi^*(y)\nabla_{\mathfrak{x}} h = -\Phi_*(y)\Upsilon(y)\Phi^*(y)\bar{J}(\mathfrak{x})V_h(\mathfrak{x}).$$

By (1.16), $V_h(\mathfrak{x}) = \Phi_*(y)\xi$. So the r.h.s. is $-\Phi_*(y)\Upsilon(y)\Phi^*(y)\bar{J}(\mathfrak{x})\Phi_*(y)\xi$. By (1.14) it equals

$$-\Phi_*(y)\Upsilon(\mathfrak{x})\bar{\Upsilon}(\mathfrak{x})\xi = \Phi_*(y)\xi = V_h(\mathfrak{x}).$$

Thus, $\mathfrak{x}(t)$ satisfies eqn (1.7). □

**Corollary.** *Let $O \subset (X_d, \alpha_2)$ and a Hamiltonian vector field $V_h$ be as in the theorem and $\Phi : O \to X_d$ be a $C^1$-smooth map, satisfying (1.15) (with $Y_s \equiv X_s$), such that $\Phi^*\alpha_2 = K\alpha_2$ for some $K \neq 0$. Then the map $\Phi$ transforms solutions of the equation*

$$\dot{x} = K^{-1}J(x)\nabla H(x,t)$$

*to solutions of (1.7).*

In particular, if $K = 1$, then $\Phi$ preserves the set of solutions of eqn (1.7). If $K > 0$ and the Hamiltonian $H$ is autonomous, then $\Phi$ preserves the set of solutions up to time scaling.

*Proof.* The result follows from the theorem with $\{Y_s\} = \{X_s\}$ and $\beta_2 = K\alpha_2$ (so $\bar{\Upsilon} = K\bar{J}$ and $\Upsilon = K^{-1}J$). □

To apply Theorem 1.2 we have to be able to construct a sufficient number of symplectic transformations. An important way to construct symplectomorphisms of domains in $(O \subset X_d, \alpha_2)$ is to obtain them as flow maps $S_t^\tau$ of an additional non-autonomous Hamiltonian equation

$$\dot{\mathfrak{x}} = J(\mathfrak{x})\nabla_{\mathfrak{x}} f(t,\mathfrak{x}) = V_f(t,\mathfrak{x}), \quad \mathfrak{x} \in O, \tag{1.18}$$

where the Hamiltonian $f$ is such that the vector field $V_f$ is Lipschitz:

**Theorem 1.3.** *Let $f$ be a $C^1$-smooth function on $\mathbb{R} \times O$, $O \subset X_d$, such that the map $V_f : \mathbb{R} \times O \to X_d$ is Lipschitz in $(t,\mathfrak{x})$ and $C^1$-smooth in $\mathfrak{x}$. Let $O_1$ be a subdomain of $O$. Then the flow maps $S_t^\tau : (O_1, \alpha_2) \to (O, \alpha_2)$ are symplectomorphisms, provided that they map $O_1$ to $O$. Moreover, $\|S_{t*}^\tau(x)\|_{d,d} \leq \exp(|\tau - t|C_*)$, where $C_* = \sup_{t,x} \|V_{f*}(t,x)\|_{d,d}$. If the vector field $V_f$ is analytic, then the flow maps are analytic as well.*

*Proof.* The flow maps are $C^1$-smooth diffeomorphisms in the smooth case and are analytic in the analytic case; see Section 1.4. The estimates on the linearized flow maps hold since the curves $\tau \to S_{t*}^\tau(x)\xi$ satisfy the linearized equation (cf. (1.10)).

It remains to check that the linearized maps $S_{t*}^\tau$ are symplectic. This follows from below Theorem 1.3′, where a more general result is proven (cf. Definition 1.2 and its discussion). □

Let us assume that the form $\alpha_2$ is a constant coefficient: $\alpha_2 = \langle \bar{J}d\mathfrak{x}, d\mathfrak{x}\rangle$, where $\bar{J}$ is an isomorphism of the scale of order $-d_J$. Proposition 1.3 applies to flow maps $S_t^\tau$ since they are $C^1$-smooth (or analytic) as well as their inverses (the flow maps $S_\tau^t$). So for any $y$ the maps $S_{t*}^\tau(y)$ define zero-order morphisms of the scale for $s \in [-d - d_J, d]$. Let us also assume that the vector field $V_f$ is $\Delta$-smoothing:

$$\|V_{f*}(t,\mathfrak{x})\|_{d,d+\Delta} \leq C', \quad \forall \mathfrak{x} \in O, \forall t$$

with some $\Delta \geq 0$. Since

$$S_t^\tau(\mathfrak{x}) = \mathfrak{x} + \int_t^\tau V_f(\theta, S_t^\theta(x))\, d\theta,$$

then

$$S^{\tau}_{t*}(\mathfrak{x}) = \mathrm{id} + \int_t^{\tau} V_{f*}(\theta, S^{\theta}_t(x)) S^{\theta}_{t*}(x)\, d\theta.$$

Since the maps $V_{f*}$ are $\Delta$-smoothing and the maps $S^{\tau}_{t*}$ satisfy the estimate of Theorem 1.3, the flow maps $S^{\tau}_t$ are symplectomorphisms, close to the identity up to $\Delta$-smoothing maps:

$$\|S^{\tau}_{t*}(\mathfrak{x}) - \mathrm{id}\|_{s,s+\Delta} \le C'|\tau - t| e^{|\tau - t| C_*}, \tag{1.19}$$

where $s = d$. Applying Proposition 1.4 we find that this estimate holds for any $s \in [d - \Delta - d_J, d + \Delta]$.

We have proved the following result:

**Proposition 1.5.** *Let us assume that the assumptions of Theorem 1.3 hold with $C_* < \infty$, that $\|V_{f*}(t, \mathfrak{x})\|_{d,d+\Delta} \le C'$ for all $\mathfrak{x} \in O$ with some $\Delta \ge 0$, and that the form $\alpha_2$ is a constant coefficient, namely $\alpha_2 = \bar{J} d\mathfrak{x} \wedge d\mathfrak{x}$, where $\bar{J}$ defines an isomorphism of the scale of order $-d_J$. Then the flow maps $S^{\tau}_t : O_1 \to O$ satisfy estimates (1.19) for all $s \in [d - \Delta - d_J, d + \Delta]$, provided that they map $O_1$ to $O$.*

Now we consider Hamiltonian equations corresponding to vector fields which define non-linear morphisms of the scale of a positive order:

**Theorem 1.3′** *Let us assume that the Hamiltonian vector field $V_f$ defines a $C^1$-smooth map $\mathbb{R} \times O \to X_{d-d_1}$, where $O \subset X_d$ and $d_1 \le 2d + d_J$. Let a point $\mathfrak{x}_0 \in O$ be such that the solution $\mathfrak{x}(t) = S^t_{t_0}(\mathfrak{x}_0)$ of (1.18) exists for $t_0 \le t \le T$ and for these $t$'s the flow maps $S^t_{t_0**}(\mathfrak{x}_0)$ for the linearized equation are well defined in $X_d$. Then these maps are symplectic.*

*Proof.* We have to check that the maps $S^{\tau}_{t_0**} = S^{\tau}_{t_0**}(\mathfrak{x}(t_0))$, $t_0 \le \tau \le T$, are such that

$$\alpha_2(\mathfrak{x}(\tau))\big[S^{\tau}_{t_0**}\xi, S^{\tau}_{t_0**}\eta\big] = \alpha_2(\mathfrak{x}(t_0))[\xi, \eta]$$

for any $\xi, \eta \in X_d$. Since the map $S^{t_0}_{t_0} = \mathrm{id}$, then in order to prove the theorem we have to check that the function

$$l(\tau) := \alpha_2(\mathfrak{x}(\tau))[S^{\tau}_{t_0*}\xi, S^{\tau}_{t_0*}\eta]$$

is a $\tau$-independent constant.

Since the curve $S^t_{t_0**}\xi =: \xi(t)$ satisfies the linearized equation $\dot{\xi} = V_f(t, \mathfrak{x}(t))_* \xi$, and similarly with $S^t_{t_0**}\eta = \eta(t)$, then $l(t) = \alpha_2(\mathfrak{x}(t))[\xi(t), \eta(t)]$. Therefore we should check that $(d/dt)\langle \bar{J}(\mathfrak{x}(t))\xi(t), \eta(t)\rangle = 0$, or

$$\langle \bar{J}'_{V_f}(\mathfrak{x})\xi, \eta\rangle + \langle \bar{J}(\mathfrak{x})\dot{\xi}, \eta\rangle + \langle \bar{J}(\mathfrak{x})\xi, \dot{\eta}\rangle = 0,$$

where $\bar{J}'_{V_f}$ stands for a derivative of the operator-valued map $\bar{J}(\mathfrak{x})$ in the direction $V_f$. The three terms in the l.h.s. are well defined. Indeed, $\dot{\eta}$ is a continuous curve in the space $X_{d-d_1}$ and $\bar{J}(\mathfrak{x})\xi$ is a continuous curve in the space $X_{d+d_J}$; since $d_1 \le 2d + d_J$, the third term is a well-defined continuous function of $t$, etc. Since $V_{f*}(t, \mathfrak{x})\xi = J(\mathfrak{x})(\nabla f)_*\xi + J'_\xi \nabla f$, then

$$\langle \bar{J}\dot{\xi}, \eta\rangle = \langle \bar{J} V_{f*}(t, \mathfrak{x})\xi, \eta\rangle = \langle \bar{J} J (\nabla f)_*\xi, \eta\rangle + \langle \bar{J} J'_\xi \nabla f, \eta\rangle$$
$$= -\langle (\nabla f)_*\xi, \eta\rangle + \langle \bar{J} J'_\xi \nabla f, \eta\rangle.$$

Transforming the third term in a similar way we find that we have to check the following relation:

$$\langle \bar{J}'_{V_f}\xi, \eta\rangle - \langle (\nabla f)_*\xi, \eta\rangle + \langle \bar{J} J'_\xi \nabla f, \eta\rangle + \langle \bar{J}\xi, J(\nabla f)_*\eta\rangle + \langle \bar{J}\xi, J'_\eta \nabla f\rangle = 0.$$

The fourth term equals $\langle \xi, (\nabla f)_*\eta\rangle$ and cancels the second since they equal $d^2 f(\xi, \eta)$ and $-d^2 f(\xi, \eta)$ correspondingly. Since $\nabla f = -\bar{J} V_f$, it remains to prove that

$$\langle \bar{J}'_{V_f}\xi, \eta\rangle - \langle \bar{J} J'_\xi \bar{J} V_f, \eta\rangle + \langle \bar{J} J'_\eta \bar{J} V_f, \xi\rangle = 0. \tag{1.20}$$

Differentiating the equality $J\bar{J} = -\text{id}$ in the direction $\xi$ we find that $J'_\xi \bar{J} + J\bar{J}'_\xi = 0$, or $\bar{J} J'_\xi \bar{J} = \bar{J}'_\xi$. Similarly $\bar{J} J'_\eta \bar{J} = \bar{J}'_\eta$. Now (1.20) follows since using the Cartan formula (1.2) in the relation $d\alpha_2[V, \xi, \eta] = 0$ we get that $\langle \bar{J}'_V \xi, \eta\rangle - \langle \bar{J}'_\xi V, \eta\rangle + \langle \bar{J}'_\eta V, \xi\rangle = 0$ for any $V, \xi, \eta \in X_d$. □

**Corollary.** *If (1.18) is a semilinear equation as in Proposition 1.2 and a non-linear part $V^0$ of the vector field $V_f = B + V^0$ defines a $C^1$-smooth map $\mathbb{R} \times O \to X_d$, then the flow maps $S^t_{t_0}$ are $C^1$-smooth symplectomorphisms.*

*Proof.* The flow maps $S^t_{t_0}$ are $C^1$-smooth by Proposition 1.2. So $S^t_{t_0**} = S^t_{t_0*}$ are bounded linear maps and the theorem applies. □

Let $O$ be a domain in a symplectic space $(X_d, \alpha_2 = \bar{J}(\mathfrak{x})\, d\mathfrak{x} \wedge d\mathfrak{x})$.

**Definition 1.3.** Let $C^1$-smooth functions $H_1$ and $H_2$ on $O$ define continuous gradient maps of orders $d_1$ and $d_2 \le 2d$ such that

$$d_1 + d_2 + d_J \le 2d. \tag{1.21}$$

Then the Poisson bracket $\{H_1, H_2\}$ of $H_1, H_2$ is the continuous on $O$ function $\{H_1, H_2\}(\mathfrak{x}) = \langle J(\mathfrak{x})\nabla H_1(\mathfrak{x}), \nabla H_2(\mathfrak{x})\rangle$.

The scalar product $\langle J\nabla H_1, \nabla H_2\rangle(\mathfrak{x})$ is well defined and is continuous in $\mathfrak{x}$ due to (1.5). The Poisson bracket is skew-symmetric,

$$\{H_1, H_2\} = -\{H_2, H_1\},$$

since the operator $J$ defines an anti-self-adjoint morphism which satisfies (1.5). In particular, $\{H, H\} \equiv 0$ (if $\text{ord}\, \nabla H \le d - d_J/2$ for the Poisson bracket to be well defined).

Let the functions $H_1$ and $H_2$ and the operator $\bar{J}$ be $\gamma$-analytic on a domain $O \subset X_d$ and $\text{ord}\, \nabla H_1 \le -d_J$. Let $Q$ be a sub-domain of $O$ such that $\text{dist}_{X_d}(Q, X_d \setminus O) \ge \delta$. We consider the Hamiltonian equation in $O$ with the Hamiltonian $H_1$:

$$\dot{\mathfrak{x}} = J(\mathfrak{x})\nabla H_1(\mathfrak{x}) =: V_1(\mathfrak{x}), \tag{1.22}$$

and denote by $S^\tau$ its flow maps.

**Theorem 1.4.** *Let us assume that the vector field $V_1$ analytically extends to a complex neighbourhood $O+\gamma \subset X_d^c$, where its norm $\| V_1 \|_d$ is bounded by some constant $K$. Then the maps $S^\tau : Q \to O$, $0 \le \tau < \delta/K$, are well-defined analytic symplectomorphisms and*

$$H_2(S^\tau(\mathfrak{x})) = H_2(\mathfrak{x}) + \tau\{H_1, H_2\} + O(\tau K)^2$$

*for $\mathfrak{x} \in Q$. In particular,*

$$(d/dt)H_2(S^t(\mathfrak{x})) \mid_{t=0} = \{H_1, H_2\}(\mathfrak{x}).$$

*Proof.* The flow maps $S^\tau$ are well defined for sufficiently small $\tau$ since the vector field $V_1$ is Lipschitz by the Cauchy estimate. If $\mathfrak{x} \in Q$, then $\|S^\tau(\mathfrak{x}) - \mathfrak{x}\|_d \le \tau K$ and $S^\tau(\mathfrak{x})$ stays in $O$ for $0 \le \tau < \delta/K$. So the first assertion follows from Theorem 1.3.

Since $V_1(S^\tau(\mathfrak{x})) = V_1(\mathfrak{x}) + O(\tau K^2)$ due to the Cauchy estimate, then $S^\tau(\mathfrak{x}) = \mathfrak{x} + \tau V_1(\mathfrak{x}) + O(\tau K)^2$ in $X_d$. Hence, $H_2(S^\tau(\mathfrak{x})) - H_2(\mathfrak{x})$ equals

$$\langle \nabla H_2(\mathfrak{x}), S^\tau(\mathfrak{x}) - \mathfrak{x}\rangle + O(\|S^\tau(\mathfrak{x}) - \mathfrak{x}\|_d^2) = \tau\langle \nabla H_2, J\nabla H_1\rangle + O(\tau K)^2$$

and the theorem is proved. □

If $\operatorname{ord} \nabla H_1 = d_1 > -d_J$, then the vector field $V_1 = J\nabla H_1$ is unbounded and to state a version of the theorem we have to assume that the domain $O = O_d$ belongs to a system of compatible domains $\{O_s \subset X_s \mid d_0 \le s \le d\}$, where $d_0 = d - d_1 - d_J$:

**Theorem 1.4′.** *Let us assume that $C^1$-smooth functions $H_1$ and $H_2$ on the domain $O \subset X_d$ as above define the continuous gradient maps $\nabla H_1 : O_s \to X_{s-d_1}$ and $\nabla H_2 : O_s \to X_{s-d_2}$ for $s \in [d_0, d]$. Let $\operatorname{ord} V_1 = d_1 + d_J \ge 0$, the numbers $d_1$ and $d_2$ satisfy (1.21), and $d_0 \ge d_2/2$. Then for any solution $\mathfrak{x}(t)$ of (1.22) we have $(d/dt)H_2(\mathfrak{x}(t)) = \{H_1, H_2\}(\mathfrak{x}(t))$.*

*Proof.* Since $d_0 - d_2 \ge -d_0$, where $d_2 = \operatorname{ord} \nabla H_2$, then $H_2$ is a $C^1$-smooth function on $O_{d_0}$. Since the curve $\mathfrak{x}(t)$ is $C^1$-smooth in $O_{d_0}$ by the definition of a solution, then

$$\frac{d}{dt}H_2(\mathfrak{x}(t)) = \langle \nabla H_2(\mathfrak{x}), \dot{\mathfrak{x}}\rangle = \langle \nabla H_2(\mathfrak{x}), J(\mathfrak{x})\nabla H_1(\mathfrak{x})\rangle = \{H_1, H_2\}(\mathfrak{x}).$$ □

An immediate consequence is that if $d \ge d_1 + \frac{1}{2}d_J$ and $\nabla H_1$ defines a $C^1$-smooth morphism of order $d_1 \ge 0$ for $d_0 \le s \le d$, then $H_1$ is an integral of motion for eqn (1.22). That is, $H_1(\mathfrak{x}(t))$ is a time-independent quantity for any solution $\mathfrak{x}(t)$. If $d' \le d$ is such that the functional $H_1$ is continuous in $O_{d'}$ as well as the flow maps $S^t$, then by continuity $H(S^t(u))$ is $t$-independent for any $u \in O_{d'}$.

**Example 1.7 (NLS equation).** A non-linear Schrödinger equation

$$\dot{u}(t,x) = i(-u_{xx} + P(|u|^2)u), \quad x \in S^1, \qquad \text{(NLS)}$$

where $P$ is a real analytic function, can be written in the Hamiltonian form (1.7) in the symplectic scale of Sobolev spaces $(\{Z_s = H^s(S^1;\mathbb{C})\}, \omega_2)$. We view these spaces as real and provide them with real inner products. In particular, the scalar product in $Z_0$ is

$\langle u, v\rangle = \mathrm{Re}\int u\bar{v}\,dx$. The symplectic structure is defined by the form $\omega_2 = J\,du \wedge du$, where $Ju(x) = iu(x)$. For the Hamiltonian $h$ one should take $h(u) = \frac{1}{2}\int\left(|u_x|^2 + Q(|u|^2)\right)dx$, where $Q'(t) = P(t)$. The gradient map $\nabla h : Z_d \to Z_{d-2}$ is an analytic morphism of the scale of order two and its non-linear part $u \mapsto P(|u|^2)u$ defines an analytic morphism of zero order if $d > \frac{1}{2}$. So (NLS) is a semilinear equation and its flow maps $S^t$ are well defined in $Z_d$, $d > \frac{1}{2}$, locally in time.[8]

Now $d_J = 0$, $\mathrm{ord}\nabla h = 2$, and the hamiltonian is continuous in $Z_1$. So $h(S^t u) = \text{const}$ for $u \in Z_1$.

For $d \in (\frac{1}{2}, 1)$ the flow maps are continuous in $Z_d$ but the Hamiltonian is not. However, the assertion $h(S^t u) = \text{const}$ remains true if for $u \in Z_d \setminus Z_1$ we set $h(u) = \infty$. □

Theorems 1.3 and 1.4 admit obvious reformulations for Hamiltonian equations in sub-domains of the symplectic manifold $(\mathcal{Y}_d, \beta_2)$, $\beta_2 = dp \wedge dq + \bar{\Upsilon}\,dy \wedge dy$ (see (1.3) and the end of Section 1.4). In this case

$$\{H_1(p,q,y), H_2(p,q,y)\} = \nabla_p H_1 \cdot \nabla_q H_2 - \nabla_q H_1 \cdot \nabla_p H_2 + \langle \Upsilon \nabla_y H_1, \nabla_y H_2 \rangle. \tag{1.23}$$

Corresponding versions of the theorems are used below.

## 1.6 A Darboux lemma

The classical Darboux lemma states that locally near a point any closed non-degenerate 2-form in $\mathbb{R}^{2n}$ can be written as $dp \wedge dq$. This result has several versions which put a closed non-degenerate 2-form on a manifold to different normal forms in the vicinity of a closed set (for the classical lemma the set is a point); see Arnold and Givental (1989), Hofer and Zehnder (1994). Some of these results admit direct infinite-dimensional reformulations which can be proven by the same arguments due to Moser and Weinstein. In this section we present an analytic version of the Darboux lemma which will be used later.

Let $\mathcal{Y}_d = \mathbb{R}^n \times \mathbb{T}^n \times Y_d$ and $W$ be its subset of the form $W = P \times \mathbb{T}^n \times \{0\}$, where $P$ is a bounded domain in $\mathbb{R}^n$. By $O, O_1, \ldots$ we denote $\delta$-neighbourhoods of $W$ in $\mathcal{Y}_d$ with different $\delta > 0$ and suppose that in a neighbourhood $O$ we are given two closed analytic 2-forms $\omega_0$ and $\omega_1$. We write these forms as $\omega_j = \bar{J}^j(\mathfrak{y})\,d\mathfrak{y} \wedge d\mathfrak{y}$, where $\mathfrak{y} = (p, q, y)$, and assume that:

(1) $\omega_0 = \omega_1$ in $TO\,|_W$,

(2) for all $t \in [0,1]$ and all $\mathfrak{y} \in O$ the map $\bar{J}^t(\mathfrak{y}) = (1-t)\bar{J}^0 + t\bar{J}^1$ defines an isomorphism $\bar{J}^t : Z_d \underset{\sim}{\longrightarrow} Z_{d+d_J}$, where $Z_d = \mathbb{R}^n \times \mathbb{R}^n \times Y_d$ and $d_J \geq 0$.

By (2), the map $J^t(\mathfrak{y}) = (-\bar{J}^t)^{-1} : Z_{d+d_J} \underset{\sim}{\longrightarrow} Z_d$ is well defined and analytically depends on $\mathfrak{y}$. By Poincaré's lemma (see Lemma 1.3 above), the form $\omega_1 - \omega_0$ is a

---

[8] That is, for any $u_0 \in Z_d$ the flow maps are defined and analytic in a neighbourhood of $u_0$ for $|t| \leq T(\|u_0\|_d)$, $T > 0$.

differential $d\alpha$ of some analytic 1-form $\alpha = a(\mathfrak{y})\, d\mathfrak{y}$ such that $a(p, q, y) = O(\|y\|_d^2)$. We specify smoothness of the map $a$ assuming that

(3) the map $O_1 \to Z_{d+d_J}$, $\mathfrak{y} \mapsto a$, is Lipschitz analytic in $O_1$.

**Lemma 1.4 (Moser–Weinstein).** *Under assumptions (1)–(3) there exists a neighbourhood $O_2$ and an analytic diffeomorphism $\varphi : O_2 \to O$ such that $\varphi\mid_W = id$, $\varphi_*\mid_W = id$, and $\varphi^*\omega_1 = \omega_0$. Moreover, $\varphi$ equals the time-1 flow map $S_0^1$, corresponding to the non-autonomous equation $\dot{\mathfrak{y}} = J^t(\mathfrak{h})a(\mathfrak{y}) =: V(t, \mathfrak{y})$.*

*Proof.* For $0 \le t \le 1$ let us consider the 2-forms $\omega_t = (1-t)\omega_0 + t\omega_1 = \bar{J}^t\, d\mathfrak{y} \wedge d\mathfrak{y}$. These forms are closed as well as the forms $\omega_0$ and $\omega_1$ and are non-degenerate in a neighbourhood $O_3$ since $\omega_t = \omega_0 = \omega_1$ on $W$ by (1). Now we denote by $\varphi^t$ the flow maps $S_0^t$ of equation $\dot{\mathfrak{y}} = V(t, \mathfrak{y})$; so $\varphi^0 = id, \varphi^1 = \varphi$ and $(\varphi^1 - id)(p, q, y) = O(\|y\|_d^2)$. The lemma will be proved if we check that $(\varphi^t)^*\omega_t =$ const. Because of Cartan's identity (Lemma 1.2), we have to verify that

$$\frac{\partial \omega_t}{\partial t} + d(V \rfloor \omega_t) = 0.$$

Since $\partial\omega_t/\partial t = \omega_1 - \omega_0 = d\alpha$, it remains to check that $\alpha + V\rfloor\omega_t \equiv 0$. But $V\rfloor\omega_t = V\rfloor \bar{J}^t\, d\mathfrak{y} \wedge d\mathfrak{y} = (\bar{J}^t V) d\mathfrak{y}$. So $\alpha + V\rfloor\omega_t = (a + \bar{J}^t V)\, d\mathfrak{y} = (a + \bar{J}^t J^t a)\, d\mathfrak{y} = 0$ and the lemma is proved. □

## Appendix 1. Time-quasiperiodic solutions

The main goal of this book is to study time-quasiperiodic solutions $\mathfrak{x}(t)$ of some Hamiltonian equations (1.7). Here we recall the corresponding basic definitions.

**Definition.** A $C^1$-curve $\gamma : \mathbb{R} \to X$ in a Banach space or a manifold $X$ is called *quasiperiodic (QP) with $\le n$ frequencies* if there exists a $C^1$-smooth map $\Gamma : \mathbb{T}^n \to X$, a vector $\omega \in \mathbb{R}^n$, and a point $q_0 \in \mathbb{T}^n$ such that

$$\gamma(t) \equiv \Gamma(q_0 + \omega t)\,. \tag{A1.1}$$

The vector $\omega$ is called the *frequency vector* and $q_0$ is called the *phase*. The minimal $n$ such that $\gamma(t)$ admits a representation (A1.1) is called the number of independent frequencies; corresponding numbers $\omega_1, \dots, \omega_n$ are called the *basic frequencies*.

*Remark.* We note that the vector $\omega$ formed by the basic frequencies is defined only up to a unimodular transformation $L$[9] since the curve $\gamma(t)$ can also be written as $\gamma(t) = \Gamma_L(Lq_0 + L\omega t)$, where $\Gamma_L(q) = \Gamma(L^{-1}q)$. What is uniquely defined, is the $\mathbb{Z}$-module $\mathbb{Z}\,\omega_1 + \mathbb{Z}\,\omega_2 + \dots + \mathbb{Z}\,\omega_n \subset \mathbb{R}$. We shall usually ignore this subtlety. □

---

[9] $L$ is a volume-preserving linear operator in $\mathbb{R}^n$ such that its matrix has integer entries. It defines an automorphism of the torus $\mathbb{T}^n = \mathbb{R}^n/2\pi\mathbb{Z}^n$.

Let $\gamma(t)$ be a QP curve (A1.1) with a $C^1$-smooth map $\Gamma$ of maximal rank.[10] If components of the vector $\omega$ are rationally independent (i.e. $\omega \cdot s \not\equiv 0$ for each non-zero integer $n$-vector $s$), then the solenoid $q_0 + t\omega$ is dense in $\mathbb{T}^n$ (see Arnold 1974, Section 51) and the closure $\overline{\gamma(\mathbb{R})}$ of the curve $\gamma$ equals $\Gamma(\mathbb{T}^n)$. So $n$ equals the Hausdorff dimension $\dim_{\mathcal{H}} \overline{\gamma(\mathbb{R})}$[11] and equals to the number of frequencies (if $\gamma$ admitted a representation (A1.1) with a smaller $n'$, then $\dim_{\mathcal{H}} \overline{\gamma(\mathbb{R})}$ would be $\leq n-1$). If components of $\omega$ are rationally dependent, then the solenoid $q_0 + t\omega$ lies in a sub-torus $\mathbb{T}^m \subset \mathbb{T}^n$ with $m < n$ and the number of frequencies is less than $n$. Finally, a curve (A1.1) with a $C^1$-smooth map $\Gamma$ of maximal rank has $n$ frequencies if and only if components of the frequency vector $\omega$ are rationally independent.

The closure $\overline{\gamma(\mathbb{R})}$ is called the *hull* of $\gamma$. If components of $\omega$ are rationally independent, then the hull equals $\Gamma(\mathbb{T}^n)$.

**Example.** Let $f(t)$ be a periodic real-valued function with a period $2\pi/\omega$ and a mean value $f_0$. Then its integral modulo $2\pi$, $x(t) = \int_0^t f(\tau)\,d\tau \in S^1 := \mathbb{R}/2\pi$, is a QP function with frequencies $f_0$ and $\omega$. Indeed, $x(t) = f_0 t + F(t)$, where $F(t) = \int_0^t (f(\tau) - f_0)\,d\tau$ is a $2\pi/\omega$-periodic function. So we can write $x(t)$ as

$$x(t) = \Gamma(f_0 t, \omega t), \quad \Gamma : \mathbb{T}^2 \to S^1, \quad \Gamma(q_1, q_2) = q_1 + F(q_2/\omega) \bmod 2\pi. \qquad \square$$

We call a solution $\mathfrak{x}$ of (1.7) a (time-) quasiperiodic solution, if the curve $\mathfrak{x} : \mathbb{R} \to X_d$ is QP, and call it analytic quasiperiodic if the corresponding map $\Gamma$ is analytic of maximal rank. The hull $\Gamma(\mathbb{T}^n)$ of an analytic QP solution of the form (A1.1) with $n$ basic frequencies is an invariant analytic $n$-torus of the equation. This torus is an analytic submanifold of $X$ if the map $\Gamma$ is an immersion.

## Appendix 2. Hilbert matrices and the Schur criterion

Let $X$ and $Y$ be two Hilbert spaces with the bases $\{\varphi_j \mid j \in \mathcal{J}\}$ and $\{\psi_l \mid l \in \mathcal{L}\}$, respectively ($\mathcal{J}$ and $\mathcal{L}$ are some countable sets). A bounded linear operator $A : X \to Y$ defines an infinite matrix $\boldsymbol{a} = \{a_{jl} \mid j \in \mathcal{J},\ l \in \mathcal{L}\}$, where

$$A\Big(\sum_{j\in\mathcal{J}} x_j \varphi_j\Big) = \sum_{l\in\mathcal{L}} \Big(\sum_{j\in\mathcal{J}} a_{lj} x_j\Big)\psi_l.$$

Clearly,

$$a_{lj} = \langle A\varphi_j, \psi_l\rangle_Y. \tag{A2.1}$$

The matrix $\boldsymbol{a}$ is called a *Hilbert matrix* of the operator $A$.

Applying the operator $A$ to vectors $\varphi_j$ we readily get that

$$\sum_l a_{lj}^2 \leq \|A\|_{X,Y}^2, \quad \forall j.$$

[10] I.e. rank $\Gamma_*(q') = n$ for some $q' \in \mathbb{T}^n$.

[11] I.e. $\text{mes}_n^{\mathcal{H}} \overline{\gamma(\mathbb{R})}$ is finite and positive, see Appendix 8 in Chapter 8, Part II.

The following result, which estimates the operator norm of $A$ in terms of the matrix $\boldsymbol{a}$, is known as the *Schur criterion*:

**Theorem.** *Let us define the numbers* $K_1$ *and* $K_2$ *as* $K_1 = \sup_l \sum_j |a_{lj}|$ *and* $K_2 = \sup_j \sum_l |a_{lj}|$. *Then* $\|A\|^2_{X,Y} \le K_1 K_2$.

*Proof.* For any $x = \sum x_j \varphi_j$, we use the Schwartz inequality to get that

$$\|Ax\|_Y^2 = \sum_l \Big(\sum_j a_{lj} x_j\Big)^2 \le \sum_l \sum_j |a_{lj}| \sum_j |a_{lj}||x_j|^2$$
$$\le K_1 \sum_j |x_j|^2 \sum_l |a_{lj}| \le K_1 K_2 \|x\|_X^2.$$

Now the assertion follows. □

Let $\{X_s\}$ and $\{Y_s\}$ be two Hilbert scales with bases $\{\varphi_j \mid j \in \mathbb{Z}_0\}$ and $\{\tilde\varphi_j \mid j \in \mathbb{Z}_0\}$, corresponding to the sequences $\{\vartheta_j\}$ and $\{\tilde\vartheta_j\}$ as in Section 2.3. For any $s$ and $r$, $\{\vartheta_j^{-s}\varphi_j\}$ and $\{\tilde\vartheta_j^{-r}\tilde\varphi_j\}$ are Hilbert bases of the spaces $X_s$ and $Y_r$. According to (A2.1), for an operator $A : X_s \to Y_r$ its Hilbert matrix is the matrix $\{a_{ij} \mid i, j \in \mathbb{Z}_0\}$, where

$$a_{ij} = \vartheta_j^{-s}\vartheta_i^r \langle A\varphi_j, \varphi_i\rangle. \tag{A2.2}$$

For the Hilbert scales as above let $\{X_s^c\}$ and $\{Y_s^c\}$ be the corresponding complexified scales. For the bases of these scales we shall often choose the complex bases $\{\psi_j \mid j \in \mathbb{Z}_0\}$ and $\{\tilde\psi_j \mid j \in \mathbb{Z}_0\}$, where

$$\psi_j = \tfrac{1}{\sqrt{2}}(\varphi_j - i\varphi_{-j}), \qquad \psi_{-j} = \tfrac{1}{\sqrt{2}}(\varphi_j + i\varphi_{-j}), \quad \forall j \in \mathbb{N},$$

and similarly with $\{\tilde\psi_j\}$. Since $\langle\psi_j, \bar\psi_k\rangle = \langle\psi_j, \psi_{-k}\rangle = \delta_{j,k}$ for any $j, k$ (we recall that $\langle\cdot,\cdot\rangle$ is the complex-linear paring), then the Hilbert matrix for a complex-linear operator $A : X_s^c \to Y_r^c$ has the entries

$$a_{ij} = \vartheta_j^{-s}\vartheta_i^r \langle A\psi_j, \psi_{-i}\rangle. \tag{A2.3}$$

# 2

# Integrable subsystems of Hamiltonian equations and Lax-integrable equations

We consider a real Hilbert scale $\{Z_s\}$ as in Section 1.2, defined by means of a sequence $\{\theta_k \mid k \in \tilde{\mathbb{Z}} \subset \mathbb{Z}\}$ of algebraic growth: $0 < \theta_k = C|k|^m + o(|k|^m)$ (if originally the parameter set $\tilde{\mathbb{Z}}$ was an even subset of $\mathbb{Z}^n$, we re-parameterize it by points of $\mathbb{Z}$ or $\mathbb{Z} \setminus \{0\}$). Stretching the index $s$ linearly we achieve $m = 1$, see Proposition 1.1. Accordingly, below we assume that

$$C^{-1}k \leq \vartheta_k \leq Ck, \quad k \in \tilde{\mathbb{Z}} \subset \mathbb{Z}.$$

We provide the scale with a symplectic structure by means of a constant coefficient 2-form $\alpha_2 = \bar{J}\, dz \wedge dz$, where $\bar{J}$ defines an anti-self-adjoint automorphism of the scale of a non-positive order $-d_J \leq 0$. To a Hamiltonian $\mathcal{H}$,

$$\mathcal{H} = \tfrac{1}{2}\langle Az, z\rangle + H(z),$$

where $A$ is a self-adjoint morphism of the scale of order $d_A$, the symplectic structure corresponds the Hamiltonian equation

$$\dot{u} = J\nabla\mathcal{H}(u) = J(Au + \nabla H(u)) =: V_{\mathcal{H}}(u), \quad J = (-\bar{J})^{-1}. \tag{2.1}$$

We assume that the Hamiltonian $\mathcal{H}$ is analytic quasilinear; that is, the functional $H$ is analytic on a domain $O_d \subset Z_d$, $d \geq d_A/2$, and defines an analytic gradient map of order $d_H < d_A$:

$$\nabla H : O_d \to Z_{d-d_H}.$$

By interpolation, for any $u \in O_d$ the map $\nabla H(u)_*$ defines a self-adjoint morphism of the scale $\{Z_s\}$ of order $d_H$ for $s \in [-d + d_H, d]$ (see the Corollary in Section 1.2).

Denoting by $d_1$ an order of the vector field $V_{\mathcal{H}}$ we have

$$d_1 = d_A + d_J \leq 2d + d_J.$$

We *do not* assume that the flow maps of the equation are defined on the whole domain $O_d$ (i.e. we do not assume that the equation can be solved for any initial condition $u(0) \in O_d$).

Quasilinear Hamiltonian PDEs with analytic coefficients have the form (2.1), where $O_d$ usually equals the whole space $Z_d$ and the gradient map $\nabla H$ is analytic of some order $d_H$ for all sufficiently smooth spaces $Z_d$. The following three examples and their perturbations will be referred to constantly throughout the book.

## 2.1 Three examples

**Example 2.1 (KdV equation;** cf. Example 1.4**).** Let us take for a scale $\{Z_s \mid s \in \mathbb{Z}\}$ the scale $\{H_0^s(S^1;\mathbb{R})\}$ of $2\pi$-periodic Sobolev functions with zero mean value, defined in Example 1.1. We choose $J = \partial/\partial x$, $A = \frac{1}{4}\partial^2/\partial x^2$, and $H(u) = \frac{1}{4}\int u^3\,dx$, so $\mathcal{H}(u) = \int\big(-\frac{1}{8}u'^2 + \frac{1}{4}u^3\big)dx$. Then eqn (2.1) becomes the Korteweg–de Vries (KdV) equation:

$$\dot{u} = \frac{1}{4}\frac{\partial}{\partial x}(u_{xx} + 3u^2). \tag{KdV}$$

It is considered under the zero mean-value periodic boundary conditions:

$$u(t,x) \equiv u(t,x+2\pi), \qquad \int_0^{2\pi} u(t,x)\,dx \equiv 0,$$

which are satisfied automatically since we are looking for solutions in a space $H_0^s$. The gradient map $Z_d \to Z_d$, $u \mapsto \nabla H = \frac{3}{4}u^2$, is analytic of zero order for $d \geq 1$ (see Example 1.6).

Now we have ord $A = d_A = 2$ and ord $J = d_J = 1$. □

**Example 2.2 (higher KdV equations).** The KdV equation is an equation from an infinite hierarchy of Hamiltonian PDEs, called the KdV hierarchy (Dubrovin *et al.* 1976; McKean and Trubowitz 1976; Zakharov *et al.* 1980). The $l$th equation from the hierarchy can be written in the form (2.1) in the same symplectic Hilbert scale $(\{H_0^s\}, \langle J\,du, du\rangle)$ as in Example 2.1. It has a Hamiltonian $\mathcal{H}_l$ of the form

$$\mathcal{H}_l(u) = K_l \int_0^{2\pi} \Big((-1)^l u^{(l)2} + \langle\text{higher order terms with } \leq l-1 \text{ derivatives}\rangle\Big)\,dx,$$

where $K_l$ is a non-zero constant ($\mathcal{H}_1$ is just the KdV Hamiltonian). In particular, the Hamiltonian $\mathcal{H}_2$ has the form

$$\mathcal{H}_2 = \frac{1}{8}\int \big(u_{xx}^2 - 5u^2u_{xx} - 5u^4\big)\,dx$$

and the corresponding Hamiltonian equation is the fifth-order partial differential equation

$$\dot{u} = \frac{1}{4}u^{(5)} - \frac{1}{4}\frac{\partial}{\partial x}(5u_x^2 + 10uu_{xx} + 10u^3).$$

The gradient map of the non-quadratic part of the Hamiltonian $\mathcal{H}_2$,

$$u(x) \mapsto -\frac{1}{4}(5u_x^2 + 10uu_{xx} + 10u^3),$$

defines an analytic morphism of the Sobolev scale $\{H_0^s\}$ of order $d_H = 2$ for $s \geq 2$. The order $d_A$ of the linear part equals 4 and $d_J = 1$. □

**Example 2.3 (Sine–Gordon equation).** The Sine–Gordon (SG) equation on the circle

$$\ddot{u} = u_{xx}(t, x) - \sin u(t, x), \quad x \in S = \mathbb{R}/2\pi\mathbb{Z}, \tag{SG}$$

is a semilinear equation with a bounded non-linearity. Multiplying the equation by $\dot{u}(t, \cdot)$ in $L_2(S)$, we get the a priori estimate

$$\frac{1}{2}\frac{d}{dt}\left(|\dot{u}|_{L_2}^2 + |u_x'|_{L_2}^2\right) \leq C|\dot{u}|_{L_2},$$

which implies for $|t| \leq T$ with any $T$ a bound for the norm $r(t) = |\dot{u}(t)|_{L_2} + |u_x'(t)|_{L_2}$ in terms of $r(0)$ and $T$. Accordingly, for any $u_0 \in H^1(S)$ and $u_1 \in L_2(S)$ the equation has a unique solution $u(t, x)$,

$$u \in C(\mathbb{R}, H^1) \cap C^1(\mathbb{R}, L_2) \cap C^2(\mathbb{R}, H^{-1}),$$

such that $u(0, x) = u_0$ and $\dot{u}(0, x) = u_1$. Moreover, if $u_0 \in H^s(S)$ and $u_1 \in H^{s-1}(S)$, then $u \in C(\mathbb{R}, H^s) \cap C^1(\mathbb{R}, H^{s-1})$. This is almost obvious; see Pazy (1983).

Equation (SG) can be written in a Hamiltonian form in different ways.

1. The most straightforward way is to write (SG) as

$$\dot{u} = -v, \qquad \dot{v} = -u_{xx} + \sin u(t, x). \tag{2.2}$$

To see that these equations are Hamiltonian, we take the symplectic scale $\left(\{Z_s = H^s(S) \times H^s(S)\}, \alpha_2 = \langle \bar{J}d\xi, d\xi\rangle\right)$, where $\xi = (u, v) \in Z_s$ and $J(u, v) = (-v, u)$ (so $\bar{J} = J$). For a Hamiltonian $\mathcal{H}$ we choose $\mathcal{H} = \frac{1}{2}\langle A\xi, \xi\rangle + H(\xi)$, where $A(u, v) = (-u_{xx}, v)$ and $H(u, v) = -\int \cos u(x)\, dx$. Then $\nabla H(u, v) = (\sin u, 0)$ and the Hamiltonian equation $\dot{\xi} = J\nabla\mathcal{H} = J(A\xi + \nabla H(\xi))$ coincides with (2.2).

The Hamiltonian form (2.2) is traditional (cf. McKean 1981; Faddeev and Takhtajan 1986) and is convenient to study explicit ('finite-gap') solutions of (SG), but not to carry out detailed analysis of the equation since the linear operator $A$ as above defines a self-adjoint morphism of the scale $\{Z_s\}$ of order 2, which is not an order-2 automorphism (the inverse map $A^{-1}$ defines a morphism of order 0, not $-2$).

2. To derive a Hamiltonian form of the SG equation, convenient for its analysis, we take the shifted Sobolev scale $\{Z_s = H^{s+1}(S) \times H^{s+1}(S)\}$, where the space $Z_0$ is given the $H^1$-scalar product

$$\langle \xi_1, \xi_2\rangle = \int_S \left(\xi_{1x}' \cdot \xi_{2x}' + \xi_1 \cdot \xi_2\right) dx\,, \quad \xi_j = (u_j(x), w_j(x)),$$

and any space $Z_s$ is given the product $\langle \xi_1, \xi_2\rangle_s = \langle A^s\xi_1, \xi_2\rangle$. Here $A^s$ stands for the $s$th degree of the differential operator $A = -\partial^2/\partial x^2 + 1$. Obviously, the operator $A^{s/2}$ defines an order-$s$ self-adjoint automorphism of the scale.

The operator

$$J(u,w) = \big(-\sqrt{A}\,w, \sqrt{A}\,u\big)$$

defines an order-1 anti-self-adjoint automorphism. For a symplectic 2-form in the scale $\{Z_s\}$ we take the form $\beta_2 = \langle \bar{J}\,d\xi, d\xi\rangle$.

By $\operatorname{Cos} u$ we denote the function $\operatorname{Cos} u = -\cos u + 1 - \frac{1}{2}u^2$, and consider the functional

$$H(u,w) = \int_S \operatorname{Cos} u(x)\,dx.$$

It is analytic in any space $Z_s$ with $s \geq 0$ and its gradient (with respect to the $H^1$-scalar product $\langle\cdot,\cdot\rangle$) is[13]

$$\nabla H(u,w) = (A^{-1}\operatorname{Cos}' u(x), 0) = (A^{-1}(\sin u - u), 0).$$

Since $\operatorname{ord} A^{-1} = -1$, then $\nabla H$ is a 1-smoothing analytic map $\nabla H : Z_s \to Z_{s+1}$, if $s \geq 0$.

The functional $\mathcal{H}(\xi) = \frac{1}{2}\langle\xi,\xi\rangle + H(\xi)$ is a Hamiltonian of the equation

$$\dot{\xi} = J\nabla\mathcal{H}(\xi), \tag{2.3}$$

which can be written as the system

$$\dot{u} = -\sqrt{A}\,w, \qquad \dot{w} = \sqrt{A}\,(u + A^{-1}(\operatorname{Cos}' u(x)). \tag{2.4}$$

The $u$-component of a solution for (2.4) satisfies the equation

$$\ddot{u} = -A(u + A^{-1}(\operatorname{Cos}' u(x))) = -Au - \sin u + u = u_{xx} - \sin u,$$

i.e. eqn (SG).

In accordance with discussions in item 1, the flow maps of eqn (2.3), $S^t : Z_s \to Z_s$, are well defined for any $t$ if $s \geq 0$. These maps are $C^1$-smooth. This is obvious for integer $s$ and remain true for real $s$ (Pazy 1983). In particular, flow maps of the linearized equation are well defined in $Z_s$ and equal linearizations of the flow maps $S^t$; so by Theorem 1.3′ they are symplectomorphisms.

We note that the $(u,v)$ variables as in eqn (2.2) and the $(u,w)$ variables as in (2.4) are related by the linear isomorphism $(u,v) \mapsto (u, A^{-1/2}v) = (u,w)$. This map *is not* symplectic with respect to the symplectic forms $\alpha_2$ and $\beta_2$.

3. (Even periodic boundary conditions.) If $u(t,x)$ is any solution of (SG) such that the initial conditions $(u_0(x), u_1(x)) = (u(0,x), \dot{u}(0,x))$ are even periodic functions, i.e.

$$u_0(x) \equiv u_0(x+2\pi) \equiv u_0(-x) \tag{EP}$$

and similarly for $u_1$, then $u^-(t,x) = u(t,-x)$ is another $2\pi$-periodic solution for (SG) with the same initial conditions. Since a solution of the initial-value problem for (SG) is

13 Indeed, $\langle\nabla H(\xi), \xi_1\rangle = dH(\xi)(\xi_1) = \int \operatorname{Cos}' u(x) u_1(x)\,dx = \langle A^{-1}(\operatorname{Cos}' u, 0), \xi_1\rangle$.

unique, then $u^-(t,x) \equiv u(t,x)$. That is, the space of even periodic functions is invariant under the SG flow and we can study the equation under the boundary conditions (EP). These conditions clearly imply Neumann boundary conditions on the half-period:

$$(u'_{0x}, u'_{1x})(0) = (u'_{0x}, u'_{1x})(\pi) = (0,0). \tag{N}$$

The former can be viewed as a 'smoother version' of the latter since for any smooth even periodic function all its odd-order derivatives (not only the first one) coincide at $x = 0$ and $x = \pi$.

Denoting for any real $s$ by $Z_s^e$ a subspace of $Z_s$, formed by even functions, we observe that the equation (SG) + (EP) can be written in the Hamiltonian form (2.3) = (2.4) in the symplectic scale $(\{Z_s^e\},\ \beta_2 = \langle \bar{J}d\xi, d\xi\rangle)$.

As before, the flow maps of eqn (2.3), (EP), define $C^1$-smooth symplectomorphisms of the symplectic spaces $(Z_s^e, \beta_2)$, $s \geq 0$.

We note that for $s = 1$ the space $Z_1^e$ is formed by the vector functions from $H^2[0,\pi]\times H^2[0,\pi]$ which satisfy (N) (the functions are assumed to be extended to the segment $[0, 2\pi]$ in the even way). That is, for solutions of eqn (SG) in the Sobolev space $H^2$, the boundary conditions (OP) and (N) are equivalent.

4. (Odd periodic boundary conditions.) Similarly, eqn (SG) under the odd periodic boundary conditions

$$u(t,x) \equiv u(t, x+2\pi) \equiv -u(t,-x) \tag{OP}$$

can be written in the Hamiltonian form (2.3) = (2.4) in the symplectic scale $(\{Z_s^o\},\ \beta_2 = \langle \bar{J}d\xi, d\xi\rangle)$, where

$$Z_s^o = \{\xi(x) \in Z_s \mid \xi \text{ satisfies (OP)}\}.$$

These boundary conditions imply the Dirichlet:

$$(u_0, u'_1)(0) = (u_0, u'_1)(\pi) = (0,0). \tag{D}$$

For $s = 0$ or 1, the space $Z_s^o$ is formed by even extensions to the segment $[0, 2\pi]$ of vector functions from $H^{s+1}([0,\pi];\mathbb{R}^2)$ which satisfy (D). So for solutions of (2.4) in the Sobolev spaces $H^1$ and $H^2$ the boundary conditions (OP) and (D) are equivalent. In this case (i.e. for $s = 0$ and $s = 1$) it is convenient to replace the odd periodic boundary conditions by Dirichlet (cf. Part II, Chapter 9, Section 9.4). In particular, for $s = 0$ the phase space is $\overset{\circ}{H}{}^1([0,\pi];\mathbb{R}^2)$, where the space $\overset{\circ}{H}{}^1$ is formed by vector functions from $H^1$ which vanish for $x = 0$ and $x = \pi$ (while in terms of the $(u, v)$ variables the phase-space is $\overset{\circ}{H}{}^1[0,\pi] \times L_2[0,\pi]$). □

## 2.2 Integrable subsystems

We assume that eqn (2.1) possesses an invariant submanifold $\mathcal{T}^{2n} \subset O_d \cap Z_\infty$, such that restriction of the equation to $\mathcal{T}^{2n}$ is integrable. For some important examples the manifold $\mathcal{T}^{2n}$ may have singularities and the restricted symplectic form $\alpha_2\mid_{\mathcal{T}^{2n}}$ may degenerate at some points. Since our objects are analytic, these degenerations can only happen on singular subsets of positive codimension and do not affect the final KAM results which neglect subsets of small measure: at some point we shall cut out the singular subsets with their small neighbourhoods. But our preliminary arguments are *global*. To carry them out we have to develop global notations. We shall do it to an extent which is sufficient to cover the main examples of integrable Hamiltonian PDEs.

We assume that $\mathcal{T}^{2n} = \Phi_0(R \times \mathbb{T}^n)$ where $\mathbb{T}^n = \{\mathfrak{z}\}$ is the standard $n$-torus and $R$ is a connected $n$-dimensional real analytic set which is the real part of a connected complex analytic subset $R^c$ of a domain $\Pi^c \subset \mathbb{C}^N$.[14] By $R_s^c$ we denote any proper analytic subset of $R^c$ which contains its singular part and denote by $R_s$ the real part of $R_s^c$, i.e. $R_s = R_s^c \cap R$.

We assume that the map $\Phi_0$ is analytic and the form $\alpha_2 \mid_{\mathcal{T}^{2n}}$ does not degenerate identically:

1. The map $\Phi_0 : R \times \mathbb{T}^n \to Z_l$ is analytic for each $l$. That is, for some $\delta > 0$ it extends to an analytic map $\Pi^c \times \{|\mathrm{Im}\, \mathfrak{z}| < \delta\} \to Z_l^c$.
2. $R^c$ contains a proper analytic subset $R_{s1}^c$ such that the analytic 2-form $\Phi_0^* \alpha_2$ is non-degenerate in $(R \setminus (R_s \cup R_{s1})) \times \mathbb{T}^n$, where $R_{s1} = R_{s1}^c \cap R$.

We call $R_{s1}^c$ and its real part the *sets of degeneracy of the form* $\Phi_0^* \alpha_2$. For brevity we re-denote $R_s := R_s \cup R_{s1}$ and similarly re-denote $R_s^c$. We set $R_0^c = R^c \setminus R_s^c$ and $R_0 = R \setminus R_s$. Since $R_s$ and $R_s^c$ comprise singularities of the analytic sets $R$ and $R^c$ as well as of the map $\Phi_0$ (i.e. they contain the points where the linearization is not well defined or its rank drops), then the sets $R_0$ and $R_0^c$ are smooth analytic manifolds and the map

$$\Phi_0 : R_0 \times \mathbb{T}^n \to Z_l\,, \qquad \Phi_0(R_0 \times \mathbb{T}^n) =: \mathcal{T}_0^{2n},$$

is an analytic immersion.

Now we specify the integrability of eqn (2.1), restricted to $\mathcal{T}^{2n}$:

3. The set $\mathcal{T}_0^{2n}$ is a smooth analytic submanifold of each space $Z_l$, invariant for eqn (2.1), as well as the tori $T^n(r) = \Phi_0(\{r\} \times \mathbb{T}^n)$, $r \in R_0$. The equation restricted to $T^n(r)$ takes the form $\dot{\mathfrak{z}} = \omega(r)$, where $\omega$ extends to an analytic map $\omega : \Pi^c \to \mathbb{C}^n$.

Owing to 2 and 3, the manifold $\mathcal{T}_0^{2n}$ is filled with smooth time-quasiperiodic solutions of eqn (2.1).

The *frequency map* $r \mapsto \omega(r)$ is assumed to be non-degenerate:

4. For almost all $r \in R_0$,

$$\text{the tangent map} \quad \omega_*(r) : T_r R_0 \to \mathbb{R}^n \quad \text{is an isomorphism.} \qquad (2.5)$$

---

[14] That is, $R^c$ is formed by the zeros of an analytic map $\Pi^c \to \mathbb{C}^{N-n}$ such that at some points of $\Pi^c$ its linearization has full rank. For elementary facts concerning analytic sets, real and complex, see Milnor (1968) and Gunning and Rossi (1965, Sections II and III).

The non-degeneracy property (2.5) can be viewed as an *amplitude–frequency modulation*: by changing the amplitude vector $r$ one can change the frequency vector $\omega$ in a prescribed direction.

By Theorem 1.2 the equation restricted to the symplectic manifold $\mathcal{T}_0^{2n}$ is Hamiltonian. Because of conditions 3 and 4, this equation is integrable:

**Lemma 2.1.** *A Hamiltonian equation (2.1) which satisfies 1–4 is Liouville–Arnold integrable in* $\mathcal{T}_0^{2n}$.

*Proof.* Since the map $r \mapsto \omega(r)$ is analytic non-degenerate, then for almost all $r$ the components of the vector $\omega(r)$ are rationally independent and the flow $\dot{\mathfrak{z}} = \omega(r)$ on $T^n(r)$ is ergodic (see Arnold 1974). A torus $T^n(r)$ with ergodic flow is Lagrangian in $\mathcal{T}_0^{2n}$. Indeed,[15] since the flow maps of eqn (2.1) are symplectic, then their restrictions to the torus preserve the form $\Omega_2 = \alpha_2 \mid_{T^n(r)}$. Since the flow on the torus is ergodic, then $\Omega_2 = \sum a_{ij}\, d\mathfrak{z}_i \wedge d\mathfrak{z}_j$ with some constant coefficients $a_{ij}$. A coefficient $a_{ij}$ equals the averaging of $\Omega_2$ along the 2-torus $\{\mathfrak{z} \mid \mathfrak{z}_l = 0 \text{ if } l \neq i, j\}$. So it vanishes because the form $\Omega_2$ is exact as well as the form $\alpha_2$. By continuity, all the tori $T^n(r)$ are Lagrangian.

For any $r \in R_0$ we choose coordinates $r_1, \dots, r_n$ in the vicinity of the torus $T^n(r)$ in $\mathcal{T}_0^{2n}$ and consider the functions

$$f_j : \mathcal{T}_0^{2n} \ni \Phi_0(r, \mathfrak{z}) \mapsto r_j, \qquad j = 1, \dots, n.$$

Since the $f_j$'s are constant on each torus $T^n(r)$, then for any $\mathfrak{z} \in T^n(r)$ and every tangent vector $\xi \in \Pi := T_{\mathfrak{z}} T^n(r)$ we have

$$0 = \langle df_j(\mathfrak{z}), \xi \rangle = -\omega_2(V_{f_j}(\mathfrak{z}), \xi).$$

Thus, the vectors $V_{f_j}(\mathfrak{z})$ lie in the skew-orthogonal complement to $\Pi$, which equals $\Pi$ because the torus $T^n(r)$ is Lagrangian. Hence, the $f_j$'s are in involution: $\{f_j, f_k\} = V_{f_j}(f_k) = 0$ for all $j, k$. Similarly, each $f_j$ commutes with the Hamiltonian of the equation and the lemma follows since the equation has $n$ commuting integrals of motion. □

By the previous lemma and the Liouville–Arnold theorem, $R_0$ can be covered by a countable system of domains $R_{0j}$, $R_0 = R_{01} \cup R_{02} \cup \dots$, such that eqn (2.1) restricted to each manifold $\mathcal{T}_j^{2n} = \Phi_0(R_{0j} \times \mathbb{T}^n)$ admits action–angle variables $(p, q)$ with actions $p \in P_j \Subset \mathbb{R}^n$ and angles $q \in \mathbb{T}^n$. That is, $\omega_2 = dp \wedge dq$ and the equation restricted to $\mathcal{T}_j^{2n}$ takes the form

$$\dot{p} = 0, \qquad \dot{q} = \nabla h(p), \qquad h = \mathcal{H} \circ \Phi_0. \tag{2.6}$$

Moreover, the actions $p$ depend only on $r$.

[15] We repeat arguments from Herman (1989).

Constructing the action–angles $(p, q)$ we can choose the cycles $Q_1, \ldots, Q_n$,

$$Q_l = \{(q_1, \ldots, q_n) \mid q_l \in S^1 \text{ and } q_j = 0 \text{ for } j \neq l\},$$

to be homotopic to any $n$ cycles forming a basis of $H_1(\mathbb{T}^n; \mathbb{Z})$. Our choice is

$$Q_l \sim \mathfrak{z}_l := \{pt\} \times \cdots \times S^1 \times \cdots \times \{pt\} \subset \mathbb{T}^n \tag{2.7}$$

(the circle stands on the $l$th place).

**Lemma 2.2.** *Under assumptions 1–3 and the choice (2.7) the gradient $\nabla h(p(r))$ equals $\omega(r)$. If, in addition, (2.5) holds, then $q = \mathfrak{z} + q^0(r)$.*

*Proof.* Since $\partial h / \partial p_j$ equals the large time limit of the number of intersections of any trajectory on $T^n(r)$ with the cycle $Q_j$, divided by the time, and $\omega_j$ equals the similar limit with $Q_j$ replaced by the homotopic cycle $\mathfrak{z}_j$, then the first assertion follows.

To prove the second we note that $(d/dt)(q - \mathfrak{z}) = \nabla h - \omega = 0$, so $q - \mathfrak{z} = \text{const}$ along each trajectory. By (2.5), the trajectories are dense in a torus $T^n(r)$ with typical $r$ and the second assertion follows by continuity. □

## 2.3 Lax-integrable equations

Let us consider a Hamiltonian PDE, supplemented by appropriate boundary conditions, and write it in the Hamiltonian form

$$\dot{u} = J \nabla H(u) \tag{2.8}$$

in some symplectic Hilbert scale $(\{Z_s\}, \alpha_2 = \langle \bar{J} dz, dz \rangle)$. This equation is called *Lax-integrable* (or an equation of Lax type)[16] if there exist linear operators $\mathcal{L}_u$ and $\mathcal{A}_u$ which depend on $u \in Z_\infty$ and define linear morphisms of finite orders of some additional real or complex Hilbert scale $\{\mathfrak{Z}_s\}$, such that a curve $u(t)$ is a smooth solution of (2.8) if and only if

$$\frac{d}{dt} \mathcal{L}_{u(t)} = [\mathcal{A}_{u(t)}, \mathcal{L}_{u(t)}]. \tag{2.9}$$

The operators $\mathcal{L}_u$ and $\mathcal{A}_u$ are said to form an $\mathcal{L}$–$\mathcal{A}$ pair (or a *Lax pair* of eqn (2.8)).

We specify dependence of the $\mathcal{L}$ and $\mathcal{A}$ operators on $u$ and assume from now on the existence of integers $s'$ and $d'$ such that for all $s \geq s'$ the maps

$$Z_s \to L(\mathfrak{Z}_d, \mathfrak{Z}_{d-d'}), \quad u \mapsto \mathcal{L}_u, \quad \text{and} \quad u \mapsto \mathcal{A}_u \tag{2.10}$$

are analytic, provided that $d \leq s$. (This is a non-restrictive assumption which holds for all 'classical' Lax-integrable PDEs.) Because of this assumption, the l.h.s. of (2.9) is well defined for any $C^1$-smooth curve $u(t) \in Z_s$ if $s \geq s'$.

---

[16] It would be more systematic to introduce the notion of a Lax-integrable boundary value problem, but we do not wish to change the received terminology.

We abbreviate $\mathcal{L}_t = \mathcal{L}_{u(t)}$ and $\mathcal{A}_t = \mathcal{A}_{u(t)}$, where $u(t)$ is a smooth solution for (3.4). A crucial property of the $\mathcal{L}$ and $\mathcal{A}$ operators is that the spectrum of the operator $\mathcal{L}_t$ is time-independent and its eigenvectors are preserved by the flow, defined by the operators $\mathcal{A}_t$:

**Lemma 2.3.** *Let $\chi_0 \in \mathfrak{Z}_\infty$ be a smooth eigenvector of $\mathcal{L}_0$, i.e. $\mathcal{L}_0\chi_0 = \lambda\chi_0$. Let us also assume that the initial-value problem*

$$\dot{\chi} = \mathcal{A}_t\chi, \qquad \chi(0) = \chi_0, \tag{2.11}$$

*has a unique smooth solution $\chi(t) \in \mathfrak{Z}_\infty$. Then*

$$\mathcal{L}_t\chi(t) = \lambda\chi(t) \tag{2.12}$$

*for every $t$.*

*Proof.* Let us denote the l.h.s. of (2.12) by $\xi(t)$, the r.h.s. by $\eta(t)$, and calculate their derivatives. We have

$$\frac{d}{dt}\xi = \frac{d}{dt}\mathcal{L}\chi = [\mathcal{A}, \mathcal{L}]\chi + \mathcal{L}\mathcal{A}\chi = \mathcal{A}\mathcal{L}\chi = \mathcal{A}\xi$$

and

$$\frac{d}{dt}\eta = \frac{d}{dt}\lambda\chi = \lambda\mathcal{A}\chi = \mathcal{A}\eta\,.$$

Thus, both $\xi(t)$ and $\eta(t)$ solve the problem (2.11), with $\chi_0$ replaced by $\lambda\chi_0$, and coincide for all $t$ by the uniqueness assumption. □

In many important examples of Lax-integrable equations, $\{Z_s\}$ is the Sobolev scale of $L$-periodic in $x$ (vector-) functions and $\mathcal{L}$ and $\mathcal{A}$ are $u$-dependent differential operators, acting on complex vector-functions, which can be specified to satisfy various boundary conditions. In particular, one can take for the scale $\{\mathfrak{Z}_s\}$ the Sobolev scale of $L$-periodic complex vector-functions. So the $L$-periodic (discrete smooth) spectrum of the operator $\mathcal{L}_u$ is an integral of motion for eqn (2.8) if the linear equation (2.11) defines a flow in the space of smooth $L$-periodic vector-functions. The set of integrals which we obtain in this way usually is incomplete. To obtain the missing integrals we note that an $L$-periodic in $x$ solution $u(t, x)$ can also be treated as an $Lm$-periodic solution for any $m \in \mathbb{N}$. Accordingly, we can consider the same $\mathcal{L}$ and $\mathcal{A}$ operators under $mL$-periodic boundary conditions and take for $\{\mathfrak{Z}_s\}$ the Sobolev scale of $mL$-periodic vector-functions. Due to the lemma, the $mL$-periodic spectrum of $\mathcal{L}$ is an integral of motion if eqn (2.11) defines a flow in the corresponding space $\mathfrak{Z}_\infty$. This set of integrals contains the initial one since any $L$-periodic in $x$ eigenfunction of $\mathcal{L}$ is an $mL$-periodic eigenfunction as well.

Similarly, the $L$-antiperiodic smooth spectrum of the operator $\mathcal{L}_t$ is an integral of motion provided that the operators $\mathcal{L}_t$ and $\mathcal{A}_t$ define linear morphisms of the corresponding scale and eqn (2.11) defines a flow in the space of smooth $L$-antiperiodic functions.

In many cases the set of integrals of motion of a Lax-integrable equation, formed by the $L$-periodic and $L$-antiperiodic spectra, is complete and can be used to construct invariant manifolds $\mathcal{T}^{2n}$ as above.

Both the KdV and SG equations are of Lax type. Below we show how to use the periodic and antiperiodic spectra of their $\mathcal{L}$-operators to obtain the manifolds $\mathcal{T}^{2n}$ for these equations.

# 3
# Finite-gap manifolds for the KdV equation and theta formulas

In this chapter we study the famous finite-gap solutions of the KdV equation under zero-mean-value periodic boundary conditions:

$$\dot{u} = \frac{1}{4}\frac{\partial}{\partial x}(u_{xx} + 3u^2), \qquad u(t,x) \equiv u(t, x+2\pi), \qquad \int_0^{2\pi} u\,dx \equiv 0. \qquad \text{(KdV)}$$

The finite-gap solutions fill invariant submanifolds $\mathcal{T}^{2n} \subset H_0^s(S^1)$ with integrable dynamics on them, as in Section 2.2. To study the manifolds $\mathcal{T}^{2n}$ we use the Its–Matveev formula which represents the finite-gap solutions in terms of theta functions. This formula does not apply well to small-amplitude solutions and to study the manifolds near the origin we use normal form techniques. The two approaches jointly provide us with the information we need to study embeddings of the manifolds $\mathcal{T}^{2n}$ to function spaces and examine their persistence under perturbations of the Hamiltonian.

The approach to studying finite-gap manifolds we develop in this section is rather general. In the next section we apply it to the Sine–Gordon equation.

## 3.1 Finite-gap manifolds

The $\mathcal{L}$ and $\mathcal{A}$ operators for the KdV equation are

$$\mathcal{L}_u = -\frac{\partial^2}{\partial x^2} - u, \qquad \mathcal{A}_u = \frac{\partial^3}{\partial x^3} + \frac{3}{2}u\frac{\partial}{\partial x} + \frac{3}{4}u_x.$$

Indeed, calculating the commutator $[\mathcal{A}, \mathcal{L}]v$ one sees that most of the terms cancel and there is nothing left except $(\frac{1}{4}u_{xxx} + \frac{3}{2}uu_x)v$. Thus, $[\mathcal{A}, \mathcal{L}]$ is an operator of multiplication by the r.h.s. of KdV and the equation can be written in the form (2.9). For the scale $\{\mathfrak{Z}_s\}$ we take one of the following scales of complex Sobolev functions: the scale of $2\pi$-periodic functions, or the scale of $2\pi$-antiperiodic functions, or the scale of $4\pi$-periodic functions.

It is well known (Marčenko 1977; McKean and Trubowitz 1976) that for any continuous $2\pi$-periodic potential $u$ the spectrum of the Sturm–Liouville operator $\mathcal{L}_u$ acting on twice differentiable functions of period $4\pi$ is a sequence of simple or double eigenvalues $\{\lambda_j \mid j \geq 0\}$, tending to infinity:

$$\lambda_0 < \lambda_1 \leq \lambda_2 < \lambda_3 \leq \lambda_4 < \cdots \nearrow \infty.$$

Corresponding eigenfunctions are smooth if the potential $u(x)$ is. The spectrum $\{\lambda_j\}$ can also be described without doubling the period: it equals the union of the periodic and antiperiodic spectra of the operator $\mathcal{L}_u$, considered on the segment $[0, 2\pi]$. Below we denote $\boldsymbol{\lambda} = \{\lambda_0, \lambda_1, \dots\}$ and refer to the sequence $\boldsymbol{\lambda} = \boldsymbol{\lambda}(u)$ as the periodic/antiperiodic spectrum of the operator $\mathcal{L}_u$.

**Example 3.1.** For $u = 0$ we have $\lambda_{2k} = k^2/4$, $k \geq 0$, and $\lambda_{2l-1} = l^2/4$, $l \geq 1$. Corresponding eigenfunctions are $(2\pi)^{-1/2} \cos kx/2$ and $(2\pi)^{-1/2} \sin lx/2$. □

If $u(t, x)$ is a smooth $x$-periodic function, then the linear equation

$$\dot{v} = \mathcal{A}_{u(t,x)} v, \qquad v(0, x) = v_0(x),$$

has a unique smooth $x$-periodic solution $v(t, x)$ for any given smooth periodic initial data $v_0(x)$ (this follows from an abstract theorem in Pazy 1983, Section 5.2). Hence, Lemma 2.3 with $\{\mathfrak{Z}_s = H^s(\mathbb{R}/4\pi\mathbb{Z})\}$ implies that the sequence $\boldsymbol{\lambda}$ is an integral of motion:

$$\boldsymbol{\lambda}(u(t, \cdot)) \text{ is time-independent if } u(t, x) \text{ is a solution of the KdV.} \tag{3.1}$$

The segment $\Delta_j = [\lambda_{2j-1}, \lambda_{2j}]$, $j = 1, 2, \dots$, is called the $j$th spectral gap. The gap $\Delta_j$ is *open* if $\lambda_{2j} > \lambda_{2j-1}$ and is *closed* if $\lambda_{2j} = \lambda_{2j-1}$. See Fig. 3.1.

Let us fix any integer $n$-vector $\mathcal{V}$,

$$\mathcal{V} = (\mathcal{V}_1, \dots, \mathcal{V}_n) \in \mathbb{N}^n, \quad \mathcal{V}_1 < \cdots < \mathcal{V}_n,$$

and consider a set $\mathcal{T}_{\mathcal{V}}^{2n}$,

$$\mathcal{T}_{\mathcal{V}}^{2n} = \{u(x) \mid \text{the gap } \Delta_j(u) \text{ is open iff } j \in \{\mathcal{V}_1, \dots, \mathcal{V}_n\}\}.$$

This set equals the union of isospectral subsets $T^n(r) = T_{\mathcal{V}}^n(r)$ with prescribed lengths of the open gaps:

$$\mathcal{T}_{\mathcal{V}}^{2n} = \bigcup_{r \in \mathbb{R}_+^n} T_{\mathcal{V}}^n(r), \quad \text{where } T_{\mathcal{V}}^n(r) = \{u(x) \in \mathcal{T}_{\mathcal{V}}^{2n} \mid |\Delta_{\mathcal{V}_j}| = r_j \ \forall j\}.$$

By (3.1) each set $T_{\mathcal{V}}^n(r)$ is invariant for the KdV flow.

Remarkably, the whole spectrum $\boldsymbol{\lambda}$ of an $n$-gap potential is defined by the $n$-vector $r$ and analytically depends on it (Garnet and Trubowitz 1987). Each set $T_{\mathcal{V}}^n(r)$ is not empty and is an analytic $n$-torus in any space $H_0^s = H_0^s(S^1)$. The tori $T_{\mathcal{V}}^n(r)$ are analytically

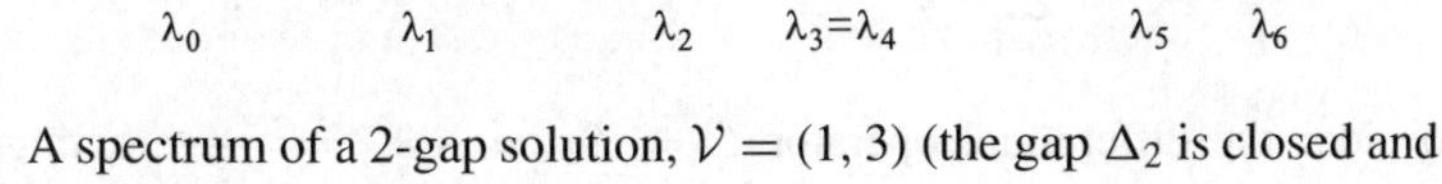


**Fig. 3.1:** A spectrum of a 2-gap solution, $\mathcal{V} = (1, 3)$ (the gap $\Delta_2$ is closed and the gaps $\Delta_1$ and $\Delta_3$ are open).

glued together, so $\mathcal{T}^{2n}_{\mathcal{V}}$ is an analytic submanifold of each space $H^s_0$ (moreover, each finite-gap potential $u(x) \in T^n_{\mathcal{V}}(r)$ is an analytic function!). These are well-known results from the inverse spectral theory of the Sturm–Liouville operator $\mathcal{L}_u$; see Marčenko(1977), Moser (1980), Garnet and Trubowitz (1987), and McKean and Trubowitz (1976).

So the inverse spectral theory provides us with KdV-invariant $2n$-manifolds foliated to invariant $n$-tori. In Section 3.2 we shall construct analytic maps $\Phi_0$ which represent these manifolds in the form $\Phi_0(R \times \mathbb{T}^n)$, as in Section 2.

When any gap—say, $\Delta_{\mathcal{V}_n}$—shrinks to a point, the $n$-gap potential $u(x) \in T^n_{\mathcal{V}}(r)$ degenerates to an $(n-1)$-gap potential from $\mathcal{T}^{2n}_{(\mathcal{V}_1,\dots,\mathcal{V}_{n-1})}$. This degeneration occurs in an analytic way:

**Theorem 3.1.** *The closure $\overline{\mathcal{T}^{2n}_{\mathcal{V}}}$ of $\mathcal{T}^{2n}_{\mathcal{V}}$ in any space $H^s_0$, $s \geq 1$, is a $2n$-dimensional analytic submanifold of $H^s_0$, diffeomorphic to $\mathbb{R}^{2n} = \{z\}$. This manifold contains all finite-gap manifolds $\mathcal{T}^{2m}_{\mathcal{V}^m}$, where $\mathcal{V}^m \subset \mathcal{V}$ $(m < n)$. It passes through the origin and its tangent space there is spanned by the vectors $e^{\pm}_l \in H^s_0$, $l = 1, \dots, n$, where*

$$e^+_l = \frac{1}{\sqrt{\pi}} \cos \mathcal{V}_l x = \frac{\partial}{\partial z_{2l-1}}, \qquad e^-_l = -\frac{1}{\sqrt{\pi}} \sin \mathcal{V}_l x = \frac{\partial}{\partial z_{2l}}. \tag{3.2}$$

*For any function $u = \pi^{1/2} \sum_{j=1}^{\infty} (u^+_j \cos jx - u^-_j \sin jx)$ from $\mathcal{T}^{\leq 2n}_{\delta}$ we have*

$$z_{2k-1} = u^+_{\mathcal{V}_k} + O(\|u\|^2_s), \qquad z_{2k} = u^-_{\mathcal{V}_k} + O(\|u\|^2_s), \quad k = 1, \dots, n. \tag{3.3}$$

*The $z$-coordinates are such that*

$$z^2_{2j-1} + z^2_{2j} = r^2_j, \quad \forall\, j.$$

The second assertion of the theorem justifies the notation

$$\overline{\mathcal{T}^{2n}_{\mathcal{V}}} = \mathcal{T}^{\leq 2n}_{\mathcal{V}} = \mathcal{T}^{\leq 2n},$$

which we use from now on. We call both manifolds $\mathcal{T}^{\leq 2n}$ and $\mathcal{T}^{2n}$ the *n-gap manifolds*.

For the proof of the theorem see Garnet and Trubowitz (1987), McKean and Trubowitz (1976), Kappeler (1991), and Bättig *et al.* (1996). For our purposes we need only a local version of this result, related to the set $\mathcal{T}^{\leq 2n}_{\delta} = \overline{\mathcal{T}^{2n}} \cap \mathcal{O}_\delta(H^s_0)$. Below we state it and give an elementary proof.

**Theorem 3.1′.** *The set $\mathcal{T}^{\leq 2n}_{\delta}$ with sufficiently small positive $\delta$ satisfies obvious local versions of all assertions of Theorem 3.1.*

*Proof.* To simplify notation we suppose that $\mathcal{V} = (1, \dots, n)$ and abbreviate $\mathcal{O}_\delta(H^s_0)$ to $\mathcal{O}_\delta$. Let us take any function $u(x) \in \mathcal{O}_\delta$ and write it using the trigonometric basis (1.1):

$$u(x) = \pi^{-1/2} \sum_{k=1}^{\infty} (u^+_k \cos kx - u^-_k \sin kx), \qquad \|u\|_s = \gamma < \delta.$$

Let us consider the differential operator $\mathcal{L} = \mathcal{L}_u = -\partial^2/\partial x^2 - u$, acting on $4\pi$-periodic functions. It is a $\gamma$-small perturbation of the operator $\mathcal{L}_0 = -\partial^2/\partial x^2$. Its eigenvalues $\lambda_{2j-1}(u), \lambda_{2j}(u)$ are $C\gamma$-close to the double eigenvalue $j^2/4$ of the operator $\mathcal{L}_0$ since by Rellich's theorem (Kato 1966) they analytically depend on $u$. An invariant plane $\Pi_j = \Pi_j(u)$ of the operator $\mathcal{L}_u$, corresponding to the eigenvalues $\lambda_{2j-1}(u)$ and $\lambda_{2j}(u)$, is $C\gamma^2$-close to the eigenplane $\Pi_j^0$ of the operator $\mathcal{L}_0$, spanned by the vectors $\phi_{j0} = (2\pi)^{-1/2}\cos jx/2$ and $\phi_{-j0} = -(2\pi)^{-1/2}\sin jx/2$ (see Example 3.1).[17] Since the plane $\Pi_j$ analytically depends on $u$, then it has a uniquely defined analytic in $u$ basis $\{\phi_j(u), \phi_{-j}(u)\}$ such that:

(1) the basis is orthonormal with respect to the scalar product in $L_2(\mathbb{R}/4\pi\mathbb{Z})$,
(2) for $u = 0$ it equals $\{\phi_{j0}, \phi_{-j0}\}$, and
(3) $\phi_j(u)$ is a unit vector in $\Pi_j$ which is the closest to the subspace formed by even functions.

This basis is well defined if $\delta$ is not too large. Since the plane $\Pi_j(u)$ is $O(\gamma^2)$-close to the plane $\Pi_j^0$, then the vectors $\phi_{\pm j}(u)$ are $O(\gamma^2)$-close to $\phi_{\pm j0}$.

Let us take $u$ be equal to $\varepsilon v$, where $v(x) = \pi^{-1/2}\sum_{k=1}^{\infty}(v_k^+\cos kx - v_k^-\sin kx) \in H_0^s$ and $\varepsilon \ll 1$. For $j \geq 1$ let $M_j(\varepsilon v)$ be a matrix of the self-adjoint operator $-\mathcal{L}_{\varepsilon v}\mid_{\Pi_j}$ in the basis, constructed above. It analytically depends on $\varepsilon$. Since $\phi_{\pm j}(\varepsilon v)$ is $\varepsilon^2$-close to $\phi_{\pm j0}$, then $\frac{\partial}{\partial\varepsilon}M_j(\varepsilon v)\mid_{\varepsilon=0}$ equals the derivative in $\varepsilon$ at $\varepsilon = 0$ of a matrix of the quadratic form of the operator $-\mathcal{L}_{\varepsilon v}$, restricted to the plane $\Pi_j^0$ and calculated in the basis $\{\phi_{\pm j0}\}$. Therefore,

$$\frac{\partial}{\partial\varepsilon}M_j(\varepsilon v)\mid_{\varepsilon=0} = \begin{pmatrix} a_1^j & a_{12}^j \\ a_{12}^j & a_2^j \end{pmatrix},$$

where

$$a_1^j = \int_0^{4\pi} v(x)\varphi_{j0}(x)^2\,dx = \frac{1}{\pi}\int_0^{2\pi} v(x)\cos^2\frac{1}{2}jx\,dx = \frac{1}{2}v_j^+,$$

$$a_2^j = \int_0^{4\pi} v(x)\varphi_{j0}(x)\varphi_{-j0}(x)\,dx = \frac{1}{\pi}\int_0^{2\pi} v(x)\sin^2\frac{1}{2}jx\,dx = -\frac{1}{2}v_j^+,$$

$$a_{12}^j = \int_0^{4\pi} v(x)\varphi_{-j0}(x)^2\,dx = -\frac{1}{\pi}\int_0^{2\pi} v(x)\sin\frac{1}{2}jx\cos\frac{1}{2}jx\,dx = \frac{1}{2}v_j^-.$$

For a $2\times 2$-matrix $M$ its *deviator* $M^D$ equals the traceless matrix $M - \left(\frac{1}{2}\mathrm{tr}\,M\right)E$, where $E$ is the identity $2\times 2$-matrix. Following Kappeler (1991) we consider the map

$$\mathbf{M}^D : u(x) \mapsto (M_1^D(u), M_2^D(u), \dots), \quad u \in \mathcal{O}_\delta.$$

---

[17] To prove this assertion one can write the spectral projection to $\Pi_j$ as a contour integral (see Kato 1966), decompose it in series in $\gamma$, and observe that the term corresponding to $\gamma$ vanishes.

Let $\mathfrak{H}^s$ be the space of all sequences $\mathbf{L} = (L_1, L_2, \dots)$ of traceless symmetric $2 \times 2$-matrices with the finite norm $\left(\sum_{j=1}^{\infty} j^{2s}|L_j|^2\right)^{1/2}$, and let $\mathfrak{H}_n^s$ be a subspace formed by the sequences $(L_1, \dots, L_n, 0, \dots)$. Then $\mathcal{T}_\delta^{\le 2n} = \left(\mathbf{M}^D\right)^{-1}(\mathfrak{H}_n^s)$. Straightforward calculations show that the map $\mathbf{M}^D : \mathcal{O}_\delta \longrightarrow \mathfrak{H}^s$ is analytic if $s \ge 1$.[18] Owing to our preceding arguments linearization of this map at zero sends a function $v = \pi^{-1/2} \sum_{j=1}^{\infty}(v_j^+ \cos jx - v_j^- \sin jx)$ to the sequence $(M_1^D, M_2^D, \dots)$, where

$$M_j^D(v) = \frac{1}{2}\begin{pmatrix} v_j^+ & v_j^- \\ v_j^- & -v_j^+ \end{pmatrix},$$

so it defines an isomorphism of the two spaces. Now by the implicit function theorem (see Lang 1972), the set $\mathcal{T}_\delta^{\le 2n} = (\mathbf{M}^D)^{-1}(\mathfrak{H}_n^s)$ is an analytic submanifold of $\mathcal{O}_\delta$ such that

(1) the map $\mathbf{M}^D$ composed with the natural projection $\mathfrak{H}^s \to \mathfrak{H}_n^s$ defines its analytic isomorphism with a neighbourhood of the origin in $\mathfrak{H}_n^s$, and
(2) the tangent space $T_0\mathcal{T}_\delta^{\le 2n}$ equals $(\mathbf{M}^D(0)_*)^{-1}\mathfrak{H}_n^s$.

By (2), the tangent space $T_0\mathcal{T}_\delta^{\le 2n}$ is spanned by the vectors $e_1^\pm, \dots e_n^\pm$ defined in (3.2), as stated in Theorem 3.1′.

For $j = 1, \dots, n$ let us write $M_j^D(u)$ as

$$M_j^D = \frac{1}{2}\begin{pmatrix} z_{2j-1} & z_{2j} \\ z_{2j} & -z_{2j-1} \end{pmatrix}.$$

Then $z = (z_1, \dots, z_{2n})$ is a coordinate system in $\mathfrak{H}_n^s$, so by (1) the functions $z_j \circ \mathbf{M}^D$ form a coordinate system on $\mathcal{T}_\delta^{\le 2n}$. For any function $u \in \mathcal{T}_\delta^{\le 2n}$ the relations (3.3) clearly hold. So the tangent vector $\partial/\partial z_{2l-1} \in T_0\mathcal{T}_V^{\le 2n}$ equals $e_l^+$ and $\partial/\partial z_{2l}$ equals $e_l^-$.

By construction of the matrix $M_j^D$, the size of the $j$th open gap $r_j = |\Delta_j|$ equals the difference of its eigenvalues and equals $z_{2j-1}^2 + z_{2j}^2$. The theorem is proved. □

For further use we note that our calculations prove the following small-gap spectral asymptotic for a small-amplitude potential $u = \pi^{-1/2} \sum_{k=1}^{\infty} (u_k^+ \cos kx - u_k^- \sin kx)$:

$$\begin{aligned} \lambda_{2j-1} &= \frac{j^2}{4} - \frac{1}{2}\left(|u_j^+|^2 + |u_j^-|^2\right)^{1/2} + O(\|u\|_s^2), \\ \lambda_{2j} &= \frac{j^2}{4} + \frac{1}{2}\left(|u_j^+|^2 + |u_j^-|^2\right)^{1/2} + O(\|u\|_s^2), \\ \lambda_0 &= O(\|u\|_s^2), \end{aligned} \tag{3.4}$$

for any $s \ge 1$. Indeed, $\lambda_{2j-1}$ and $\lambda_{2j}$ are eigenvalues of the matrix

$$M_j(u) = \frac{1}{2}\begin{pmatrix} u_j^+ & u_j^- \\ u_j^- & -u_j^+ \end{pmatrix} + O(\|u\|_s^2),$$

[18] Even if $s = 0$—see Kappeler (1991).

so the first two relations in (3.4) follows since the eigenvalues analytically depend on $u$. The classical perturbation theory, applied to the single eigenvalue $\lambda_0$, implies the last relation.

The way to study the local (near the origin) structure of finite-gap manifolds we have described, is rather general and applies to other Lax-integrable equations: locally they are quite similar. On the contrary, the *global* structure of finite-gap manifolds can be rather different. Cf. Chapter 4 and see Kappeler (1991) and Bättig *et al.* (1996) for global coordinates on the finite-gap manifolds in the KdV case.

Since restriction of the symplectic form $\alpha_2$ to the tangent space $T_0\mathcal{T}_{\mathcal{V}}^{\leq 2n}$ is non-degenerate by (3.2), then it also is non-degenerate in the manifold $\mathcal{T}_\delta^{\leq 2n}$, provided that $\delta > 0$ is sufficiently small. It has been known since the first works on space-periodic solutions of KdV (Novikov 1974; Lax 1975) that each torus $T^n(r)$ (and the whole manifold $\mathcal{T}_\delta^{\leq 2n}$) are invariant for the vector fields of all equations from the KdV hierarchy; see Example 2.2 and Dubrovin *et al.* (1976), McKean and Trubowitz (1976), and Zakharov *et al.* (1980). So KdV restricted to $\mathcal{T}_\delta^{\leq 2n}$ has $n$ commuting integrals of motion $\mathcal{H}_0, \dots, \mathcal{H}_{n-1}$ (where $\mathcal{H}_1$ is the KdV Hamiltonian). Since $\mathcal{H}_j = \text{const} \int {u^{(j)}}^2 + \cdots dx$ (the dots stand for higher-order terms) and $u(x) = \pi^{-1/2} \sum \left(u_k^+ \cos kx - u_k^- \sin kx\right)$, where $u_{\mathcal{V}_k}^+ = z_{2k-1} + O(|z|^2)$, $u_{\mathcal{V}_k}^- = z_{2k} + O(|z|^2)$, and $u_l^{\pm} = O(|z|^2)$ if $l \neq \mathcal{V}_k$ for all $k$, then near the origin a Hamiltonian $\mathcal{H}_m \mid_{\mathcal{T}_\delta^{\leq 2n}}$ has the following form:

$$\mathcal{H}_m(z) = C_m \sum_{j=1}^{n} \mathcal{V}_j^{2m} \left(z_{2j-1}^2 + z_{2j}^2\right) + O(|z|^3).$$

The system of quadratic forms $\sum \mathcal{V}_j^{2m} (z_{2j-1}^2 + z_{2j}^2)$, $m = 0, \dots, n-1$, is non-degenerate in the sense that the determinant of the matrix $\{\mathcal{V}_j^{2m} \mid 1 \leq j \leq n,\ 0 \leq m \leq n-1\}$ is non-zero (due to Vandermonde). Therefore Vey's version of the Liouville–Arnold theorem near a singularity provides us with analytic Birkhoff coordinates; see Vey (1978) and Ito (1989). Also see Appendix 1 in Bobenko and Kuksin (1995*a*), where this result is obtained without Vey's theorem and without the extra integrals of motion, using instead Lemma 3.3 given below in Section 3.2 (the lemma's proof, presented in Appendix 6, is independent of Theorem 3.2). We arrive at a result which specifies Theorem 3.1′:

**Theorem 3.2.** *If $\delta$ is sufficiently small and $s \geq 1$, then there exists $\delta_1 > 0$ and an analytic map*

$$U : \mathcal{O}_{\delta_1}(\mathbb{R}_y^{2n}) \to \mathcal{T}_{\mathcal{V}}^{\leq 2n} \subset H_0^s, \quad y \mapsto U(\cdot; y),$$

*such that its image is contained in $\mathcal{T}_\delta^{\leq 2n}$. The transformation $y \mapsto z = z(U(\cdot, y))$ is a diffeomorphism of the form $z = y + O(|y|^2)$ (so $y(0) = 0$). Besides,*

*(1)* $U^*\alpha_2 = \sum_{l=1}^n \mathcal{V}_l^{-1}\, dy_{2l-1} \wedge dy_{2l}$;

*(2) pull-back under this map of the Hamiltonian of the KdV equation is an analytic function $h^n$ of the arguments $y_1^2 + y_2^2, \dots, y_{2n-1}^2 + y_{2n}^2$;*

*(3) for any $l \leq n$, the submanifold formed by potentials $u(x)$ such that $|\Delta_{\mathcal{V}_l}| = 0$ corresponds to the subspace $\{y \mid y_{2l-1} = y_{2l} = 0\}$;*

*(4) the finite-gap tori $T^n(r)$ in the $y$-coordinates take the form $\{y_{2l-1}^2 + y_{2l}^2 = C_l(r)\}$.*

The last assertion holds since by the Vey theorem the Hamiltonians $\mathcal{H}_0, \dots, \mathcal{H}_{n-1}$ all are functions of $y_{2l-1}^2 + y_{2l}^2$ and since they are constant on each finite-gap torus.

The coordinate $y$ provides us with analytic action-angle variables $(I, q)$ on the manifold $\mathcal{T}_{\mathcal{V}}^{2n}$, where

$$I_j = \frac{1}{2\mathcal{V}_j}(y_{2j-1}^2 + y_{2j}^2), \qquad q_j = \arg(y_{2j-1} + iy_{2j}). \tag{3.5}$$

These coordinates are symplectic since $U^*\alpha_2 = dI \wedge d\varphi$ by the first assertion of Theorem 3.2. The KdV Hamiltonian is an analytic function $h^n(I)$ of the actions $I$ and the KdV equation restricted to $\mathcal{T}_{\overline{\mathcal{V}}}^{\leq 2n}$ takes the form

$$\dot{I} = 0, \qquad \dot{q} = \nabla h^n(I).$$

Abusing notation, we denote the map $U$, written in the $(I, q)$ variables, also as $U$. Then the finite-gap solutions that fill the $n$-gap manifold $\mathcal{T}_\delta^{2n}$ can be written as

$$u(t, x) = U(x; I, q + t\nabla h^n(I)). \tag{3.6}$$

For further usage we note that since a point $U(y)$ has a $z$-coordinate $z = y + O|y|^2$ and since by (3.2) a point in $\mathcal{T}_{\overline{\mathcal{V}}}^{\leq 2n}$ with a coordinate $z$ is the function $\pi^{-1/2}\sum(z_{2l-1}\cos\mathcal{V}_l x - z_{2l}\sin\mathcal{V}_l x) + O|z|^2$, then

$$\begin{aligned} U(x; I, q) &= \pi^{-1/2}\sum\sqrt{2\mathcal{V}_l I_l}\,(\cos q_l \cos\mathcal{V}_l x - \sin q_l \sin\mathcal{V}_l x) + O(I) \\ &= \pi^{-1/2}\sum\sqrt{2\mathcal{V}_l I_l}\,\cos(q_l + \mathcal{V}_l x) + O(I). \end{aligned}$$

By the last assertion of Theorem 3.2 the actions $I_j$ are functions of the radii $r_1 > 0, \dots, r_n > 0$. These functions analytically extend to the origin:

**Lemma 3.1.** *Each action $I_j$ is an analytic at zero function of $r_1^2, \dots, r_n^2$ of the form $I_j = (r_j^2/2\mathcal{V}_j)(1 + O(|r|^2))$.*

*Proof.* We recall that $r_j^2 = z_{2j-1}^2 + z_{2j}^2$ and denote by $w_{\pm j}$ the complex numbers

$$w_j = z_{2j-1} + iz_{2j} = r_j e^{i\varphi_j}, \qquad w_{-j} = \overline{w}_j, \quad j = 1, \dots, n.$$

Since $I_j$ is an analytic at zero function of $z$, then it can be written as a convergent series $I_j = \sum_{s\in\mathbb{Z}_{\geq 0}^{2n}} a_s w^s$, where $\mathbb{Z}_{\geq 0} = \mathbb{N} \cup \{0\}$ and $w^s = w_{-n}^{s_{-n}} \dots w_n^{s_n}$. Or

$$I_j = \sum_{s\in\mathbb{Z}_{\geq 0}^{2n}} a_s^j \prod_{p=1}^{n} r_p^{s_p + s_{-p}} e^{i\varphi_p(s_p - s_{-p})}.$$

Since each $I_j$ does not depend on the angles $\varphi$ but only on the radii $r_1, \dots, r_n$, then $a_s \neq 0$ only if $s_p = s_{-p}$ for each $p$, i.e. $s = (l, l)$ for some $n$-vector $l \in \mathbb{Z}_{\geq 0}^n$. Then $I_j = \sum_{l\in\mathbb{Z}_{\geq 0}^n} b_l^j r^{2l}$, $b_l^j = a_{l,l}^j$. By the third assertion of Theorem 3.2, $I_j$ vanishes with $r_j$. This means that $b_l^j = 0$ if $l_j = 0$; so $I_j$ equals $r_j^2/2\mathcal{V}_j$ times an analytic function of $r_1^2, \dots, r_n^2$. Since $y = z + O(|z|^2)$, then $I_j^2 - r_j^2/(2\mathcal{V}_j) = O(|z|^3)$ and the analytic function as above is $(1 + O(|r|^2))$. □

## 3.2 The Its–Matveev theta formulas

To check that the $n$-gap manifolds $\mathcal{T}_{\mathcal{V}}^{\leq 2n}$ of the KdV equation possess the properties 1–4 from Section 2.2, we have to present an analytic map $\Phi_0$ as in Section 2.2 and to study its properties. We shall write the map $\Phi_0$ in terms of theta functions, following Dubrovin (1981) and Belokolas *et al.* (1994). An alternative presentation of the small-amplitude part $\mathcal{T}_{\delta}^{\leq 2n}$ of the $n$-gap manifold $\mathcal{T}_{\mathcal{V}}^{\leq 2n}$ in the desired form is given by Theorem 3.2, and formula (3.6) can be used to construct the map $\Phi_0(r, \mathfrak{z})$ for $|r| \ll 1$. The reader can skip this section and just take for granted that each $n$-gap torus $T_{\mathcal{V}}^n$ is filled with solutions, given by the formula (3.17) below, where the function $\Phi_0(r, \mathfrak{z})(x)$ and the vector $W(r)$ are analytic in $\mathfrak{z} \in \mathbb{T}^n$, $x \in S$ and $r \in \mathbb{R}_+^n$.

Our notations 'almost' agree with Belokolos *et al.* (1994) and mostly agree with Dubrovin (1981). All results on Riemann surfaces, given without a reference, can be found in Springer (1981).

Let us take any $n$-gap potential $u(x) \in T_{\mathcal{V}}^n(r)$ and denote by $E_1(r) < E_2(r) < \cdots < E_{2n+1}$ end points of the open gaps plus $\lambda_0$ (so $E_1 = \lambda_0$ and $\Delta_{\mathcal{V}_1} = [E_2, E_3], \ldots, \Delta_{\mathcal{V}_n} = [E_{2n}, E_{2n+1}]$); see Figs 3.1 and 3.2. The Riemann surface $\Gamma = \Gamma(r)$ of genus $n$,

$$\Gamma = \left\{ P = (\lambda, \mu) \mid \mu^2 = R(\lambda; r) := \prod_{j=1}^{2n+1} (\lambda - E_j(r)) \right\},$$

has branching points at $E_1, \ldots, E_{2n+1}$ and $\infty$.

After the curve $\Gamma$ has been cut along the ovals that lie above the segments $[E_1, E_2], \ldots, [E_{2n-1}, E_{2n}], [E_{2n+1}, \infty]$, it falls into two sheets $\Gamma_+$ and $\Gamma_-$, chosen in such a way that $\mu$ is positive on the upper edge of the cut $[E_{2n+1}, E_\infty]$ in $\Gamma_+$. We denote by $\pi$ the projection

$$\pi : \Gamma \to \mathbb{C} \cup \{\infty\}, \qquad \pi(P) = \lambda,$$

and by $\tau$ the anti-holomorphic involution of $\Gamma$,

$$\tau : \Gamma \to \Gamma, \quad (\lambda, \mu) \mapsto (\bar{\lambda}, -\bar{\mu})$$

(its linearizations define half-linear complex maps). The cuts as above are invariant for $\tau$, as well as the sheets $\Gamma_+$ and $\Gamma_-$.

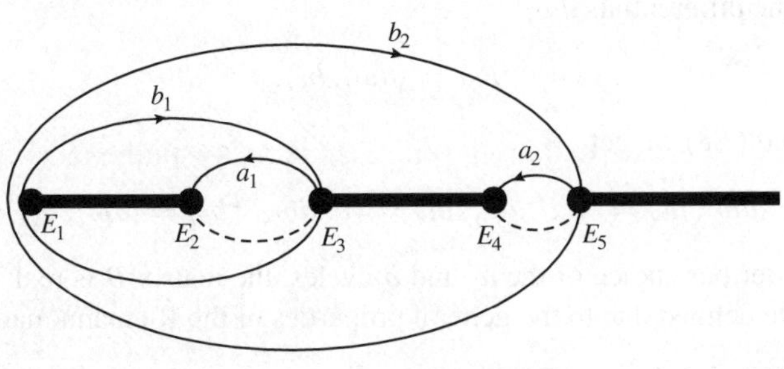


**Fig. 3.2:** Circles on $\Gamma$ ($n = 2$).

Let $a_1, \ldots, a_n$ be the ovals in $\Gamma$ lying above the open gaps $\Delta_{\nu_1}, \ldots, \Delta_{\nu_n}$ (i.e. $a_j = \pi^{-1}\Delta_{\nu_j}$). We supplement them by $n$ $b$-circles $b_1, \ldots, b_n$ as in Fig. 3.2.

The $b$-circles lie in $\Gamma_+$ and we choose them in such a way that for each $j$ the circle $\tau(b_j)$ equals $b_j$ as a set.[19] Since $\tau$ inverts orientations of the circles, then

$$\tau(b_j) = -b_j, \quad j = 1, \ldots, n. \tag{3.7}$$

Because $R(\lambda)$ is negative on the gaps $(E_{2j}, E_{2j+1})$, the $\mu$-components of the points from $a$-ovals are purely imaginary and the ovals are fixed for $\tau$:

$$\tau(a_j) = a_j, \quad j = 1, \ldots, n. \tag{3.7'}$$

Moreover, there are no fixed points of $\tau$ outside these ovals. The $a$- and $b$-circles are chosen in such a way that they have the canonical intersection matrix:

$$a_i \circ a_j = b_i \circ b_j = 0, \quad a_i \circ b_j = \delta_{ij}.$$

Next we take a basis $d\omega_1, \ldots, d\omega_n$ of holomorphic differentials on $\Gamma$, normalized by the conditions

$$\langle d\omega_j, a_k\rangle := \oint_{a_k} d\omega_j = 2\pi i \delta_{jk}.$$

These differentials exist and are uniquely defined by the normalization. Since $\langle d\omega_j, a_k\rangle = \langle \tau^* d\omega_j, \tau a_k\rangle = \overline{\langle \tau^* d\omega_j, a_k\rangle}$, then $\langle -\overline{\tau^* d\omega_j}, a_k\rangle = -\overline{\langle d\omega_j, a_k\rangle} = 2\pi i \delta_{jk}$. Each differential $-\overline{\tau^* d\omega_j}$ is holomorphic and meets the normalization. So it equals $d\omega_j$:

$$-\overline{\tau^* d\omega_j} = d\omega_j. \tag{3.8}$$

Since the differentials $(\lambda^l/\mu)\, d\lambda$, $l = 0, \ldots, n-1$, are holomorphic in $\Gamma$ and the space of holomorphic differentials is $n$-dimensional (see Springer 1981 and Zakharov *et al.* 1980), then each $d\omega_j$ can be written as

$$d\omega_j = \frac{\text{polynomial of } \lambda \text{ degree } \leq n-1}{\mu}\, d\lambda. \tag{3.9}$$

By (3.8) the polynomial in the numerator has real coefficients.

The Riemann matrix $B = B(r) = (B_{jk})$ of the curve $\Gamma$ is defined as the matrix of $b$-periods of the differentials $d\omega_j$:

$$B_{jk} = \langle d\omega_j, b_k\rangle.$$

Using (3.7) and (3.8) we get

$$\overline{B_{jk}} = \langle \overline{d\omega_j}, b_k\rangle = -\langle \tau^* d\omega_j, b_k\rangle = \langle \tau^* d\omega_j, \tau b_k\rangle = \langle d\omega_j, b_k\rangle = B_{jk}.$$

Therefore, under our choice of the $a$- and $b$-cycles, the matrix $B$ is real. Its symmetric part is negative defined due to the general properties of the Riemann matrices.

---

19 To this end the loops $\pi(b_j)$ should be invariant for the complex conjugation $\lambda \mapsto \bar{\lambda}$.

Now we define the theta function $\theta$ of the curve $\Gamma = \Gamma(r)$:

$$\theta = \theta(z;r) = \sum_{s\in\mathbb{Z}^n} \exp\left(\frac{1}{2}(B(r)s,s) + (z,s)\right), \qquad z \in \mathbb{C}^n$$

(the sum converges due to the properties of the Riemann matrix $B$). Clearly, the function is $2\pi$-periodic in imaginary directions:

$$\theta(z + 2\pi i e_k) = \theta(z),$$

where $e_k$ is the $k$th basis vector of $\mathbb{C}^n$.

The differentials $d\omega_j$ analytically depend on the parameter $r \in \mathbb{R}^n_+$ as well as the matrix $B(r)$, formed by their $b$-periods.[20] Therefore the function $\theta(z;r)$ is analytic in $r \in \mathbb{R}^n_+$.

Since the matrix $B$ is real, $\theta$ is real and even:

$$\overline{\theta(z)} = \theta(\bar{z}), \qquad \theta(z) = \theta(-z).$$

In particular, this function is real in both the real and purely imaginary directions:

$$\theta(z),\ \theta(iz) \in \mathbb{R}, \quad \text{if } z \in \mathbb{R}^n.$$

Next, on the surface $\Gamma(r)$ we consider Abelian differentials of the second kind $d\Omega_1$ and $d\Omega_3$ with vanishing $a$-periods, and with the only poles at infinity of the form

$$\begin{aligned} d\Omega_1 &= dk + (c + O(k^{-2}))\,dk^{-1}, \quad k = i\sqrt{\lambda} \to \infty, \\ d\Omega_3 &= dk^3 + O(1)\,dk^{-1}, \end{aligned} \tag{3.10}$$

where $c$ is an unknown constant. The normalization (3.10) defines the differentials uniquely; see Springer (1981), Zakharov *et al.* (1980) and Belokolos *et al.* (1994).

The following lemma, proved in Appendix 4, comprises some useful properties of these differentials :

**Lemma 3.2.** *The differentials $d\Omega_1$ and $d\Omega_3$ can be written in the form*

$$d\Omega_1 = \frac{i}{2}\frac{\lambda^n + \cdots}{\mu}\,d\lambda, \qquad d\Omega_3 = -\frac{3}{2}i\,\frac{\lambda^{n+1} + \cdots}{\mu}\,d\lambda, \tag{3.11}$$

*where the dots stand for real polynomials of degree $n-1$. Each open interval $(E_{2j}, E_{2j-1})$, $j = 1, \ldots, n$, contains exactly one zero of $d\Omega_1(\lambda)$ and a zero of $d\Omega_3(\lambda)$.*

Let us define the complex $n$-vectors $i\boldsymbol{V}(r)$ and $i\boldsymbol{W}(r)$ as the vectors of $b$-periods of these differentials:

$$iV_j = \langle d\Omega_1, b_j\rangle, \qquad iW_j = \langle d\Omega_3, b_j\rangle.$$

The vector $\boldsymbol{V}$ is called the *wave-number vector* and $\boldsymbol{W}$ the *frequency vector*.

[20] In Appendix 4 we prove similar statement for the differentials $d\Omega_1$ and $d\Omega_3$ (defined below) and for their $b$-periods.

Since the circle $b_j$ can be deformed to $[E_1, E_{2j}] \cup [E_{2j}, E_1]$ (the first segment stands for a path through the upper edge of the cut and the second through the lower edge), since by (3.11) $d\Omega_{1,2}$ changes its sign when we cross a cut, and since integrals of $d\Omega_{1,2}$ along the open gaps vanish due to normalization (cf. Appendix 4), then

$$iV_j = 2 \int_{[E_1, E_{2j}]} d\Omega_1, \qquad iW_j = 2 \int_{[E_1, E_{2j}]} d\Omega_3.$$

Since the dots in (3.11) stand for real polynomials, then

$$\tau^* \, d\Omega_1 = \frac{i}{2} \frac{\bar\lambda^n + \cdots}{-\bar\mu} \, d\bar\lambda = \overline{d\Omega_1}, \qquad \tau^* \, d\Omega_3 = \overline{d\Omega_3}.$$

That is, the differentials $d\Omega_{1,2}$ are *real* (with respect to the anti-holomorphic involution $\tau$). Accordingly,

$$\overline{iV_j} = \langle \overline{d\Omega_1}, b_j \rangle = \langle \tau^* \, d\Omega_1, b_j \rangle = -\langle \tau^* \, d\Omega_1, \tau b_j \rangle = -\langle d\Omega_1, b_j \rangle = -iV_j$$

(we use (3.7)). Thus, the vector $\boldsymbol{V}$ is real. Similarly with $\boldsymbol{W}$:

$$\boldsymbol{V}, \ \boldsymbol{W} \in \mathbb{R}^n.$$

One of the top achievements of the finite-gap theory is the Its–Matveev formula, which represents any $n$-gap potential $u(x) \in T^n(r)$ in the form

$$u(x) = u(x; r, \mathfrak{z}) = 2 \frac{\partial^2}{\partial x^2} \ln \theta(i\mathbf{V}x + i\mathfrak{z}; r) + 2c. \tag{3.12}$$

Here the constant $c$ is the same as in (3.10) and the phase $i\mathfrak{z}$ is

$$i\mathfrak{z} = -A(\mathcal{D}) - \boldsymbol{K},$$

where $\boldsymbol{K}$ is the vector of Riemann constants (see Dubrovin 1981 and Belokolos *et al.* 1994 or Appendix 3 below) and $A(\mathcal{D})$ is the Abel transformation of a positive divisor $\mathcal{D} = \mathcal{D}(u)$, $\mathcal{D} = D_1 \cdots D_n$, $D_j \in a_j$. That is, $A(\mathcal{D})$ is a complex $n$-vector such that its $j$th component $A(\mathcal{D})_j$ equals

$$A(\mathcal{D})_j = \sum_{r=1}^{n} \int_{\infty}^{D_r} d\omega_j,$$

where $\{d\omega_j\}$ are the holomorphic differentials on $\Gamma$ as above. The divisor $\mathcal{D}$ is a divisor of Dirichlet eigenvalues, i.e. $D_j = (\lambda_j, \mu_j)$, where $\lambda_j$ is an eigenvalue of the operator $\mathcal{L}_u$ subject to the Dirichlet boundary conditions $\varphi(0) = \varphi(2\pi) = 0$ (each gap $\Delta_j$ contains exactly one point from the Dirichlet spectrum; see Marčenko 1977 and McKean and Trubowitz 1976).[21] In particular, every point $D_j$ analytically depends on the potential $u$.

[21] This divisor can also be described as a divisor of poles of the Baker–Akhiezer eigenfunction $\varphi(x; P)$ of the operator $\mathcal{L}_u$, $\mathcal{L}_u\varphi = \pi(P)\varphi$, normalized at infinity as $\varphi \sim e^{i\sqrt{\lambda}x}$. See Dubrovin (1981) and Belokolos *et al.* 1994 or Section 6.2 below, where this function is denoted as $\chi$ (the notation $\varphi$ agrees with Dubrovin 1981 and Belokolos *et al.* 1994).

The phase vector $\mathfrak{z}$ turns out to be real (see Appendix 3(i)), so $\theta(i\mathbf{V}x + i\mathfrak{z})$ is a real valued function of $x$. The theta function is non-zero at any imaginary point $i\xi \in i\mathbb{R}^n$ (see in Belokolos *et al.* 1994, Lemma 3.7 on p.68, and its proof). Since this function is periodic, then

$$|\theta(i\xi)| \geq C(r) > 0, \quad \forall\, \xi \in \mathbb{R}^n. \tag{3.13}$$

Hence, the r.h.s. of (3.12) is analytic in $\mathfrak{z} \in \mathbb{T}^n$.

Owing to periodicity, we can treat $\mathfrak{z}$ as a point in the torus $\mathbb{T}^n$. Thus we get an analytic map:

$$T^n(r) \to \mathbb{T}^n, \qquad u(\cdot) \mapsto \mathfrak{z}.$$

This map has the analytic inverse given by formula (3.12).[22] The coordinates $\mathfrak{z}$ on $T^n(r)$ are called the *theta angles.*

The r.h.s. of (3.12) defines a quasiperiodic function with the frequencies $V_1, \ldots, V_n$ (see Appendix 1). Since $u(x)$ is $2\pi$-periodic, then the wave-number vector is integer:

$$\mathbf{V} \in \mathbb{Z}^n. \tag{3.14}$$

Condition (3.14) is clearly sufficient for the periodicity. Its necessity is 'obvious' but still has to be proven. We prove it in Appendix 3(iii).

Since the mean value of the r.h.s. in (3.12) equals $2c$, then we must have

$$c = 0. \tag{3.15}$$

In Appendix 4 we show that

$$V_j = -i\langle d\Omega_1, b_j\rangle \to \mathcal{V}_j \quad \text{as } \mathcal{T}^{2n}_{\mathcal{V}} \ni u \to 0.$$

Comparing this relation with (3.14) we get that

$$\mathbf{V} \equiv \mathcal{V}.$$

*Everywhere below we write* **V** *instead of* $\mathcal{V}$. In particular, we denote the $n$-gap manifolds as $\mathcal{T}^{\leq 2n}_{\mathbf{V}}$ and $\mathcal{T}^{2n}_{\mathbf{V}}$.

The time evolution $u(t, x)$ of the $n$-gap potential $u(x) \in T^n(r)$ as in (3.12) along the KdV flow is given by the following formula, also due to Its–Matveev:

$$u(t, x; r, \mathfrak{z}) = 2\frac{\partial^2}{\partial x^2} \ln \theta(i(\mathbf{V}x + \mathbf{W}(r)t + \mathfrak{z}); r) \tag{3.16}$$

(we use that $c = 0$ by (3.15)).

[22] Strictly speaking, we have to check that for each $\mathfrak{z} \in \mathbb{T}^n$ the vector $i\mathfrak{z}$ can be represented in the form $i\mathfrak{z} = -A(\mathcal{D}) - K$. We prove this in Appendix 3 (see (A3.3)).

Let us denote by $\Phi_0(r, \mathfrak{z})(x)$ the function of $x$, defined by the r.h.s. of (3.16) with $t = 0$. The map $(r, \mathfrak{z}) \mapsto \Phi_0(r, \mathfrak{z})(\cdot)$ represents the $n$-gap torus in the form

$$T^n(r) = \Phi_0(r, \mathbb{T}^n) \subset H_0^d .$$

In terms of the function $\Phi_0(r, \mathfrak{z})(\cdot)$ the $n$-gap solution (3.16) can be written as

$$u(t, x; r, \mathfrak{z}) = \Phi_0(r, \mathfrak{z} + \boldsymbol{W}(r)t)(x). \tag{3.17}$$

This shows that in the $(r, \mathfrak{z})$ variables the KdV-flow on $\mathcal{T}^{2n}$ takes the form

$$\dot{\mathfrak{z}} = \boldsymbol{W}(r).$$

That is, the theta angles $\mathfrak{z}$ integrate the KdV equation on any torus $T^n(r)$.

Let $R$ be a sub-cube of the octant $\mathbb{R}^n_+$ of the form

$$R = \{r \in \mathbb{R}^n_+ \mid 0 < r_j < K\}$$

with some $K > 0$, and

$$\mathcal{T}^{2n} = \Phi_0(R \times \mathbb{T}^n) \subset \mathcal{T}_V^{2n}$$

for any fixed wave-number vector $\boldsymbol{V}$. The set $\mathcal{T}^{2n} \subset H_0^d$, $d \geq 1$, is an invariant manifold of the KdV equation. It meets assumptions 1–3 from Section 2.2 since the map $\Phi_0$ is an analytic embedding and $\mathcal{T}^{2n}$ is an analytic submanifold of $H_0^d$. The form $\Phi_0^* \alpha_2$ is analytic and is non-degenerate for small $r$ by (3.2), so the set of its degeneracies is a proper analytic subset of the cube $R$ (in fact, it is empty—see the Amplification to Theorem 6.2 and its proof in Chapter 6).

The non-degeneracy assumption 4 also holds for KdV, as stated in the following non-degeneracy lemma, proved in Appendix 6:

**Lemma 3.3.** *The determinant* $\det\{\partial W_j / \partial r_k\}$ *is non-zero almost everywhere.*

## 3.3 Small-gap solutions

In this section we fix any finite-gap manifold $\mathcal{T}_V^{\leq 2n}$ and prove that the corresponding frequency vector $\boldsymbol{W}$ depends on the small radii-vector $r$ in the following way:

$$W_j(r) = -\frac{1}{4} V_j^3 + \frac{3}{8\pi V_j} r_j^2 + \cdots, \quad j = 1, \ldots, n. \tag{$*$}$$

This asymptotic is important for forthcoming constructions since it implies the non-resonance relations we have to check in order to apply our abstract theorems to the KdV equation. To prove $(*)$ we have to consider a moduli manifold $\mathfrak{G}$, formed by all surfaces $\Gamma(r)$ such that $0 < r_j \leq \delta$ for each $j$, and to study its closure $\overline{\mathfrak{G}}$. It turns out that $\overline{\mathfrak{G}}$ is an analytic manifold and the frequency map $\mathfrak{G} \to \boldsymbol{W}$ analytically extends to $\overline{\mathfrak{G}}$. It remains to expand $\boldsymbol{W}$ to series of $\mu$, where $\mu$ is a coordinate in the vicinity of the point $r = 0$ in $\overline{\mathfrak{G}}$, and to check that this expansion coincides with $(*)$.

There are classical ways to construct the analytic coordinate $\mu$ (i.e. to 'normalize $\bar{\mathfrak{G}}$'), see Fay (1973) and Belokolos *et al.* (1994, Section 5). These coordinates can be used to prove (∗) (see Bobenko and Kuksin 1991). Since (∗) implies that $\det \partial \boldsymbol{W}/\partial r \not\equiv 0$, then in this way one also gets an alternative proof of Lemma 3.3.

Unfortunately, the classical ways to normalize $\overline{\mathfrak{G}}$ and to decompose specific functions on $\Gamma = \Gamma(\mu)$ (like components of the frequency vector $\boldsymbol{W}$) to series in $\mu$ are very technical, and this book hardly is a proper place to present them. Below we choose another (a 'non-classical') way to normalize $\overline{\mathfrak{G}}$, using the $y$-coordinates provided by the Vey theorem (Theorem 3.2). To calculate the first two terms of a decomposition of $\boldsymbol{W}$ to series of $y$, needed to check (∗), we examine more closely small-amplitude two-gap solutions. This way of expanding $\boldsymbol{W}$ to series of $r$ is general and straightforwardly applies to other Lax-integrable equations.

In Appendices 2 and 3 we present elementary calculations that specify the small-gap behaviour of the frequencies $W_j$:

$$W_j = -i\langle d\Omega_3, b_j\rangle \longrightarrow -\frac{1}{4}V_j^3 \quad \text{as } \mathcal{T}_{\mathbf{V}}^{2n} \ni u \to 0. \tag{3.18}$$

To study small-gap solutions from $\mathcal{T}_{\mathbf{V}}^{\le 2n}$ further, we shall use the Birkhoff coordinates $y = (y_1, \dots, y_{2n})$. Since in the action-angle variables $(I, q)$ (see (3.5)) the KdV Hamiltonian is an analytic function $h^n(I)$, then by (3.16) and Lemmas 2.2 and 3.3 we have that

$$\nabla h^n = \mathbf{W} \tag{3.19}$$

and

$$q - \mathfrak{z} = q^0(r). \tag{3.20}$$

Let us denote

$$\mathcal{R}_j = \sqrt{y_{2j-1}^2 + y_{2j}^2} = \sqrt{2V_j I_j}.$$

Then the symplectic form $U^*\alpha_2$ equals $\frac{1}{2}\sum(1/V_j)d\mathcal{R}_j^2 \wedge dq_j$ and

$$\boldsymbol{W} \quad \text{is an analytic function of } \mathcal{R}_1^2, \dots, \mathcal{R}_n^2$$

because of (3.19) and item (2) of Theorem 3.2. By Lemma 3.1,

$$\mathcal{R}_j = r_j\big(1 + O(|r|^2)\big), \quad j = 1, \dots, n. \tag{3.21}$$

Below, in order to study small-gap solutions we use the $\mathcal{R}$-variables rather than $r$.

Let us take any $n$-gap solution $u(t, \cdot) \in T^n(r)$ such that $|r| \ll 1$. Using (3.6) and (3.19) we write it as $u = U(x; \mathcal{R}, t\boldsymbol{W}(\mathcal{R}) + q)$. Since this solution can also be written in the form (3.16), then $U(x; \mathcal{R}, t\boldsymbol{W}(\mathcal{R}) + q) = U(0; \mathcal{R}, t\boldsymbol{W}(\mathcal{R}) + xV + q)$. Therefore, denoting

$$G(q, \mathcal{R}) = U(0; \mathcal{R}, q)$$

we write the solution $u$ as

$$u(t,x;\mathcal{R},q) = G(W(\mathcal{R})t + Vx + q, \mathcal{R}). \tag{3.22}$$

The function $G$ is analytic in $q \in \mathbb{T}^n$ and in $\mathcal{R}$, $|\mathcal{R}| \ll 1$. Using the small-gap limit for the map $U$, given after Theorem 3.2, we find that

$$G(q,\mathcal{R}) = \frac{1}{\sqrt{\pi}} \sum \mathcal{R}_j \cos q_j + O(|\mathcal{R}|^2). \tag{3.23}$$

Since the map $U$ is analytic in the $y$-variables, then the function $G$ is analytic in the $y$-variables as well as in the complex variables $w_{\pm j}$, $j = 1, \ldots, n$, where $w_j = y_{2j-1} + i y_{2j} = \mathcal{R}_j e^{iq_j}$ and $w_{-j} = \bar{w}_j$. Hence,

$$G(q,\mathcal{R}) = \sum_{s \in \mathbb{Z}^{2n}_{\geq 0}} C_s w^s = \sum_{s \in \mathbb{Z}^{2n}_{\geq 0}} C_s \prod_{p=1}^{n} \mathcal{R}_p^{s_p + s_{-p}} e^{iq_p(s_p - s_{-p})}, \tag{3.24}$$

where $w^s = w_{-n}^{s_{-n}} \cdots w_n^{s_n}$ and $\mathbb{Z}_{\geq 0} = \mathbb{N} \cup \{0\}$.

**Example 3.2 (one-gap potentials).** For $n = 1$ and for $\mathbf{V} = V_1 = k$ the one-gap manifold $\mathcal{T}_k^2$ is a union of time-periodic solutions $w(t,x)$ for the (KdV) of the form $w = G(kx + Wt + q; \mathcal{R})$. Here $G(Y;\mathcal{R})$ is an analytic function, $2\pi$-periodic in $Y$, and $W$ is analytic in $\mathcal{R}^2$. Since $\int w\,dx = 0$, then $\int G\,dY = 0$. Using (3.18) and (3.23) we write the functions $G$ and $W$ as follows:

$$G(Y,\mathcal{R}) = \mathcal{R}\frac{1}{\sqrt{\pi}} \cos Y + \mathcal{R}^2 g_2(Y) + \mathcal{R}^3 g_3(Y) + \cdots,$$

$$W = -\frac{1}{4}k^3 + \mathcal{R}^2 W_2 + \cdots.$$

Substituting $w$ into the KdV equation we get that $WG' = \frac{1}{4}k^3 G''' + \frac{3}{4}k(G^2)'$, where the prime stands for $\partial/\partial Y$. Or

$$k^3 G'' - 4WG + 3kG^2 = \text{const}.$$

Terms of first order in $\mathcal{R}$ in the l.h.s. cancel. Equating to zero terms of second and the third order, we obtain the following two equations:

$$k^3 g_2'' + k^3 g_2 + \frac{3k}{\pi}\cos^2 Y = \text{const},$$

$$k^3 g_3'' + k^3 g_3 - 4W_2 \cos Y + 6kg_2 \cos Y = \text{const}.$$

Since $\int g_2\,dY = 0$, then from the first equation we find that $g_2 = (1/2k^2\pi)\cos 2Y$. So the second one takes the form:

$$-k^3(g_3'' + g_3) = \left(\frac{3}{2k\pi} - 4W_2\right)\cos Y + \frac{3}{2k}\cos 3Y.$$

For this equation to be solvable we must have $W_2 = 3/(8k\pi)$.

Thus, one-gap solutions from a torus $T_k^1(\mathcal{R})$, $\mathcal{R} \ll 1$, have the form

$$w(t, x; \mathcal{R}, q) = \mathcal{R} w_1(Y) + \mathcal{R}^2 w_2(Y) + \cdots , \tag{3.25}$$

where

$$w_1 = \frac{1}{\sqrt{\pi}} \cos Y, \qquad w_2 = \frac{1}{2k^2\pi} \cos 2Y$$

and $Y = kx + Wt + q$, with

$$W(\mathcal{R}) = -\frac{k^3}{4} + \frac{3\mathcal{R}^2}{8k\pi} + O(\mathcal{R}^2). \tag{3.26}$$

□

For any $n$-vector $U$ and any $m \le n$ we denote by $U^{\hat{m}}$ the $(n-1)$-vector obtained by dropping the $m$th component, i.e. $U^{\hat{m}} = (U_1, \ldots, \hat{U}_m, \ldots, U_n)$.

For $m \le n$ let us consider the $(n-1)$-gap submanifold $\mathcal{T}^{2n-2}_{V^{\hat{m}}}$ of $\mathcal{T}^{\le 2n}_V$, obtained by closing the $m$th open gap. Since $\boldsymbol{W} = \nabla h^n$ and $h^n \mid_{\mathcal{R}_m = 0} = h^{n-1}$ by Theorem 3.2, then

$$\boldsymbol{W}^{\hat{m}}(\mathcal{R}) \mid_{\mathcal{R}_m=0} = \boldsymbol{W}(\mathcal{R}^{\hat{m}}), \tag{3.27}$$

where the $(n-1)$-vector in the r.h.s. is a frequency vector corresponding to the manifold $\mathcal{T}^{2n-2}_{V^{\hat{m}}}$.

**Proposition 3.1.** *(1) For any $m \le n$ and for a sufficiently small vector $\mathcal{R} \in \mathbb{R}^n$ such that $\mathcal{R}_m = 0$ and $\mathcal{R}_l > 0$ for $l \ne m$, the function*

$$u_{n-1}(t, x; \mathcal{R}^{\hat{m}}, q) = G(\boldsymbol{V}x + \boldsymbol{W}t + q; \mathcal{R})$$

*is an $(n-1)$-gap solution from $T^{n-1}_{V^{\hat{m}}}(\mathcal{R}^{\hat{m}})$ with the frequency vector $\boldsymbol{W}^{\hat{m}}$. This solution is independent of $q_m$.*

*(2) Let $\mathcal{R}^\varepsilon$ be the vector $(\mathcal{R}_1, \ldots, \varepsilon, \ldots, \mathcal{R}_n)$ ($\varepsilon$ stands on the m*th *place). Then for any $q_m \in S^1$ the function $v = (\partial/\partial\varepsilon) G(\boldsymbol{V}x + \boldsymbol{W}t + q, \mathcal{R}^\varepsilon) \mid_{\varepsilon=0}$ solves the KdV equation, linearized about $u_{n-1}$:*

$$\dot{v} - \frac{1}{4} v_{xxx} = \frac{3}{2} \frac{\partial}{\partial x}(u_{n-1} v). \tag{3.28}$$

*Proof.* The first part of the first statement follows from item (3) of Theorem 3.2 and from (3.27). By formula (3.24) the function $G \mid_{\mathcal{R}_m=0}$ is $q_m$-independent; therefore $u_{n-1}$ is $q_m$-independent as well.

The second statement is obvious: since the solution $G(\boldsymbol{V}x + \boldsymbol{W}t + q, \mathcal{R}^\varepsilon)$ smoothly depends on $\varepsilon$, then its $\varepsilon$-derivative at zero satisfies (3.28). □

The example to this result given below is straightforward and technical. It is important since it implies the asymptotic (*) which is the main goal of this section.

**Example 3.3 (two-gap potentials).** Let us choose any $m \neq k$ and consider a two-gap solution $u \in T^2(\mathcal{R}_k, \mathcal{R}_m) \subset T^{\le 2}_{k,m}$, where $0 \le \mathcal{R}_m \ll \mathcal{R}_k \ll 1$:

$$u(t,x;\mathcal{R}_k,\mathcal{R}_m,q) = G(\boldsymbol{V}x + \boldsymbol{W}t + q;\mathcal{R}_k,\mathcal{R}_m), \tag{3.29}$$

where $\boldsymbol{V} = (k,m)$, $\boldsymbol{W} = (W_k, W_m)$, and $q = (q_k, q_m)$ (we abuse notation and write $(W_k, W_m)$ and $(q_k, q_m)$ instead of $(W_1, W_2)$ and $(q_1, q_2)$; besides possibly $k > m$). The function $w(t,x;\mathcal{R}_k) = u(t,x;\mathcal{R}_k,0)$ is the one-gap potential from Example 3.2 with $\mathcal{R} = \mathcal{R}_k$ (see (3.25)). By Proposition 3.1, the function $v = u'_{\mathcal{R}_m}(t,x;\mathcal{R}_k,0)$ solves the linearized equation (3.28) with $u_{n-1} = w$. Owing to (3.27), $W_k(\mathcal{R}_k,0)$ equals to the frequency $W(\mathcal{R}_k)$, so $W_k$ satisfies asymptotic (3.26) with $\mathcal{R} = \mathcal{R}_k$. This function is analytic in $\mathcal{R}_k$ and in $q_k$ and $q_m$.

Below we abbreviate $\mathcal{R}_k$ to $\mathcal{R}$.

Since the frequency vector $\boldsymbol{W}(\mathcal{R}_k,\mathcal{R}_m)$ is an analytic function of $\mathcal{R}_k^2$ and $\mathcal{R}_m^2$, then $\boldsymbol{W}'_{\mathcal{R}_m}(\mathcal{R},0) = 0$. So differentiating (3.29) we get that $u'_{\mathcal{R}_m}\,|_{\mathcal{R}_m=0} = G'_{\mathcal{R}_m}(V + Wt + q;\mathcal{R},0)$. Analysing (3.24) we see that non-zero contributions to $G'_{\mathcal{R}_m}\,|_{\mathcal{R}_m=0}$ come from terms with $s_m = 1, s_{-m} = 0$ and $s_m = 0, s_{-m} = 1$. Hence, denoting

$$Z = mx + W_m t + q_m, \qquad Y = kx + W_k t + q_k,$$

we can write $v$ in the form

$$v = C_1(1 + f(Y,\mathcal{R}))e^{iZ} + C_2(1 + g(Y,\mathcal{R}))e^{-iZ},$$

where $f(Y,0) = g(Y,0) = 0$ and $|C_1| + |C_2| \neq 0$ (the latter holds since by Theorem 3.2 linearization at zero of the map $y \mapsto U(\cdot;y) \in H_0^s$ is non-degenerate). Constructing an appropriate linear combination of solutions $v$ with shifted phase $q_m$ (or taking $\bar{v}$ instead of $v$ if $C_1 = 0$) we obtain a solution for (3.28) of the form

$$v = e^{iZ}H(Y,\mathcal{R}), \qquad H = 1 + \mathcal{R}h_1(Y) + \mathcal{R}^2 h_2(Y) + \cdots.$$

This function satisfies eqn (3.28) with $u_{n-1} = w$. Substituting there $v = e^{iZ}H$ and multiplying the equation by $e^{-iZ}$ we get that

$$e^{-iZ}\left(\frac{\partial}{\partial t} - \frac{1}{4}\frac{\partial^3}{\partial x^3}\right)e^{iZ}H = \frac{3}{2}e^{-iZ}\frac{\partial}{\partial x}(we^{iZ}H). \tag{3.30}$$

Owing to (3.18), the function $W_m(\mathcal{R},0)$ has the form $W_m(\mathcal{R},0) = -m^3/4 + \omega_2\mathcal{R}^2 + O(\mathcal{R}^4)$ with some unknown $\omega_2$. Hence,

$$e^{-iZ}\left(\frac{\partial}{\partial t} - \frac{1}{4}\frac{\partial^3}{\partial x^3}\right)e^{iZ} = i\omega_2\mathcal{R}^2 + O(\mathcal{R}^4).$$

Noting that

$$\frac{\partial H}{\partial t} = W_k H'_Y(Y) = -\frac{k^3}{4}H'_Y(Y) + O(\mathcal{R}^3)$$

(since $W_k = -k^3/4 + O(\mathcal{R}^2)$ and $H'_Y = O(\mathcal{R})$) and that

$$\frac{\partial^p H}{\partial x^p} = k^p H_Y^{(p)}(Y)$$

for any $p$, we get

$$\mathrm{e}^{-iZ}\left(\frac{\partial}{\partial t}-\frac{1}{4}\frac{\partial^3}{\partial x^3}\right)\mathrm{e}^{iZ}H = i\omega_2\mathcal{R}^2 H + \left(\frac{\partial}{\partial t}-\frac{1}{4}\frac{\partial^3}{\partial x^3}\right)H$$
$$-\frac{3}{4}\mathrm{e}^{-iZ}\left(\frac{\partial}{\partial x}\mathrm{e}^{iZ}\frac{\partial^2}{\partial x^2}H+\frac{\partial^2}{\partial x^2}\mathrm{e}^{iZ}\frac{\partial}{\partial x}H\right)+O(\mathcal{R}^4)$$
$$= i\omega_2\mathcal{R}^2 H - \frac{k}{4}\frac{\partial}{\partial Y}M\left(\frac{\partial}{\partial Y}\right)H + O(\mathcal{R}^4),$$

where $M(\partial/\partial Y) = M$ is the following differential operator: $M(f(Y)) = k^2 f'' + 3imkf' + (k^2-3m^2)f$. Hence, the l.h.s. of (3.30) is

$$\mathrm{e}^{-iZ}\left(\frac{\partial}{\partial t}-\frac{1}{4}\frac{\partial^3}{\partial x^3}\right)\mathrm{e}^{iZ}H = -\mathcal{R}\frac{k}{4}\frac{\partial}{\partial Y}Mh_1 + \mathcal{R}^2\left(i\omega_2 - \frac{k}{4}\frac{\partial}{\partial Y}Mh_2\right)+\cdots.$$

Using (3.25) we find that the r.h.s. of (3.30) equals

$$\frac{3}{2}\mathrm{e}^{-iZ}\frac{\partial}{\partial x}(w\mathrm{e}^{iZ}H) = \frac{3}{2}imwH + \frac{3}{2}k\frac{\partial}{\partial Y}(wH)$$
$$= \frac{3}{2}\mathcal{R}\left(imw_1 + k\frac{\partial}{\partial Y}w_1\right)$$
$$+\frac{3}{2}\mathcal{R}^2\left(imw_2 + imw_1h_1 + k\frac{\partial}{\partial Y}(w_1h_1+w_2)\right)+\cdots.$$

Now we equate the terms of first and the second order in $\mathcal{R}$ in (3.30) to obtain two equations:

$$-\frac{k}{4}Mh_1 = \frac{3}{2}im\left(\frac{\partial}{\partial Y}\right)^{-1}w_1 + \frac{3}{2}kw_1 = \frac{3}{2\sqrt{\pi}}(im\sin Y + k\cos Y),$$
$$-\frac{k}{4}Mh_2 = i\left(\frac{\partial}{\partial Y}\right)^{-1}\left[\frac{3}{2}m(w_2+w_1h_1) - \omega_2\right] + \frac{3}{2}k(w_1h_1+w_2).$$

From the first equation we find that $h_1 = -(i/\sqrt{\pi}m)\sin Y$. For the r.h.s. of the second one to be well defined, the mean value of the function in the square brackets must vanish:

$$0 = \left\langle \frac{3}{2}m(w_2+w_1h_1) - \omega_2\right\rangle = \left\langle \frac{3}{2}m\left(\frac{\cos 2Y}{2k^2\pi} - \frac{i}{m\pi}\sin Y\cos Y\right) - \omega_2\right\rangle = -\omega_2\,,$$

where the angle brackets stand for averaging in $Y$. So $\omega_2 = 0$ and the solution $v$ we are discussing has the form

$$v = \mathrm{e}^{i(mx+W_mt+q_m)}\left(1 - \frac{i\mathcal{R}}{m\sqrt{\pi}}\sin(kx+W_kt+q_k) + O(\mathcal{R}^2)\right),$$

where $W_m = -m^3/4 + O(\mathcal{R}^4)$. Since $W_k(\mathcal{R})$ satisfies (3.26), then the frequency vector $\boldsymbol{W} = \boldsymbol{W}(\mathcal{R}_k,\mathcal{R}_m) = (W_k, W_m)$ obeys the following asymptotics as $\mathcal{R}_k = \mathcal{R} \to 0$

and $\mathcal{R}_m = 0$:

$$W_k = -\frac{k^3}{4} + \frac{3\mathcal{R}^2}{8k\pi} + O(\mathcal{R}^4), \qquad W_m = -\frac{m^3}{4} + O(\mathcal{R}^4). \tag{3.31}$$

□

**Lemma 3.4.** *For any finite-gap manifold $\mathcal{T}_V^{\leq 2n}$ the corresponding frequency vector $W(\mathcal{R})$ has the following asymptotic as $\mathcal{R} = (\mathcal{R}_1, \ldots, \mathcal{R}_n) \to 0$:*

$$W_j(\mathcal{R}) = -\frac{1}{4}V_j^3 + \frac{3}{8\pi V_j}\mathcal{R}_j^2 + O(|\mathcal{R}|^4), \quad j = 1, \ldots, n.$$

*This result remains true with $\mathcal{R}$-variables replaced by $r$-variables.*

*Proof.* The zero-order term of this asymptotic follows from (3.18). For small $\mathcal{R}$ each $W_j$ is an analytic function of the arguments $\mu_l = \mathcal{R}_l^2$, $l = 1, \ldots, 0$. Applying (3.27) iteratively we get that $W_j(0, \ldots, \mathcal{R}_j, \ldots, 0)$ is the frequency of the one-gap solution from Example 3.2, so (3.26) implies that $\partial W_j / \partial \mu_j(0) = 3/(8\pi V_j)$. Using (3.27) once again we find that the function $W_j(0, \ldots, \mathcal{R}_j, 0, \ldots, \mathcal{R}_l, \ldots, 0)$ with $l \neq j$ is a first component of the frequency vector of a two-gap solution. Applying (3.31) with $m = j$ and $k = l$ to $W_j \mid_{\mathcal{R}_j = 0}$ we get that $\partial W_j / \partial \mu_l(0) = 0$ and the asymptotic follows.

The last assertion results from (3.21). □

## 3.4 Higher equations from the KdV hierarchy

Let us take any $n$-gap manifold $\mathcal{T}_V^{2n}$. The manifold itself and each torus $T^n(r) \subset \mathcal{T}_V^{2n}$ are invariant for all Hamiltonian equations with the Hamiltonians $\mathcal{H}_0, \mathcal{H}_1, \ldots$ from the KdV hierarchy (see Example 2.2). The flow of any $l$th KdV equation on $\mathcal{T}_V^{2n}$ is very similar to the KdV flow: it is given by the theta formula (3.16) where the frequency vector $\boldsymbol{W}$ should be replaced by an $n$-vector $\boldsymbol{W}^{(l)}$ with $i W_j^{(l)}$ equal to the $b_j$-period of an Abelian differential $d\Omega_{2l+1}$, normalized by the conditions that its $a$-periods vanish and near infinity it has the form:

$$d\nu^{-2l-1} + \text{ regular part}, \quad \nu = \frac{1}{i\sqrt{\lambda}} \tag{3.32}$$

(see Dubrovin *et al.* 1976 and Zakharov *et al.* 1980; cf.(3.10) where $l = 1$).

All results of Sections 3.1–3.3 up to Proposition 3.1 have obvious reformulations for the higher KdV equations, valid for the same arguments as in the KdV case. Our proof of Lemma 3.4 is rather concrete. Instead of trying to repeat its calculations for a general $l$th equation from the KdV hierarchy, it is easier to expand the vector $\boldsymbol{W}^{(l)}$ to series of $r$ using the classical coordinates on the moduli manifold $\mathfrak{G}$ mentioned in Section 3.3. We state the corresponding result without a proof: The vector $\boldsymbol{W}^{(l)}$ is analytic in $r_1^2, \ldots, r_n^2$ and

$$W_j^{(l)}(r) = W_{j0}^{(l)} + W_{j1}^{(l)} r^2 + O(|r|^4) \tag{3.33}$$

for any $j = 1, \ldots, n$, with some non-zero constants $W_{j1}^{(l)}$.

Any manifold $\mathcal{T}_V^{2n}$ treated as an invariant manifold of an $l$th KdV equation satisfies assumptions 1–4 for the same reason as for $l = 1$ (i.e. as in the KdV case).

## Appendix 3. On the Its–Matveev formulas

Here we prove that the vector $\mathfrak{z}(\mathcal{D})$, defined by the relation

$$i\mathfrak{z}(\mathcal{D}) = -A(\mathcal{D}) - \boldsymbol{K}, \quad \mathcal{D} = D_1 \cdots D_n, \quad D_j \in a_j,$$

is real, that for each $\mathfrak{z} \in \mathbb{T}^n$ the formula (3.12) defines a finite-gap solution, and prove that the vector $\boldsymbol{V}$ in (3.12) has to be integer for a function $u(x)$ to be $2\pi$-periodic.

(i) The vector $\boldsymbol{K}$ is equal to the minus one-half of the Abel transformation of the *canonical class* $C$ of $\Gamma$, where $C$ is an equivalence class of the divisor of zeros and poles of any Abel differential $d\Omega$ (see Dubrovin 1981, Section 2.7). Let us choose for $d\Omega$ the differential

$$d\Omega = (\lambda - E_2) \cdots (\lambda - E_{2n})\mu^{-1}\, d\lambda.$$

It has a double zero in each $E_{2j}$ and a double pole at infinity. Therefore,

$$K_j = -\sum_{r=1}^{n} \int_{\infty}^{E_{2r}} d\omega_j.$$

Since $D_r \in a_r$, then

$$i\mathfrak{z}_j(\mathcal{D}) = \sum_{r=1}^{n} \left( \int_{\infty}^{E_{2r}} d\omega_j - \int_{\infty}^{D_r} d\omega_j \right) = \sum_{r=1}^{n} \int_{D_r}^{E_{2r}} d\omega_j, \quad \forall\, j. \tag{A3.1}$$

Since $\overline{d\omega_j} = -\tau^*\, d\omega_j$ (see (3.8)) and each $a_r$ is a fixed oval for the anti-holomorphic involution $\tau$ (see (3.7$'$)), then

$$\overline{i\mathfrak{z}_j(\mathcal{D})} = \sum_r \int_{D_r}^{E_{2r}} \overline{d\omega_j} = -\sum_r \int_{D_r}^{E_{2r}} \tau^*\, d\omega_j = -\sum_r \int_{D_r}^{E_{2r}} d\omega_j = -i\mathfrak{z}_j(D).$$

Thus the vector $\mathfrak{z}$ is real as stated.

(ii) Now let us take any point $\mathfrak{z}^1$ from the real $n$-torus $\mathbb{T}^n$, and consider the following equation for a divisor $\mathcal{D} = D_1 \cdots D_n$ in $\Gamma$:

$$A(\mathcal{D}) = i\mathfrak{z}^1 - \mathbf{K} =: \eta^1 \tag{A3.2}$$

(the equality holds in the Jacobian of $\Gamma$, i.e. modulo periods of the theta function). By the Riemann theorem (see Dubrovin 1981 and Belokolos *et al.* 1994) this equation has a unique solution $\mathcal{D}$ if the function on $\Gamma$ which sends $P$ to $\theta(A(P) - \eta^1 - \mathbf{K}) = \theta(A(P) - i\mathfrak{z}^1)$ does not vanish identically. At infinity the function equals $\theta(i\mathfrak{z}^1)$ which is not zero (see (3.13)), so (A3.2) has a unique solution $\mathcal{D} = D_1 \cdots D_n$. The divisor $\mathcal{D}$ satisfies (A3.1) with $\mathfrak{z}$ replaced by $\mathfrak{z}^1$.

Now we show that the points $D_j$, forming $\mathcal{D}$, are $\tau$-invariant. Conjugating relation (A3.1) with $\mathfrak{z} = \mathfrak{z}^1$ and making use of (3.8) we get that

$$i\mathfrak{z}_j^1 = -\overline{i\mathfrak{z}_j^1} = \sum_{r=1}^{n} \int_{D_r}^{E_{2r}} \overline{d\omega_j} = -\sum_{r=1}^{n} \int_{D_r}^{E_{2r}} \tau^* \, d\omega_j = \sum_{r=1}^{n} \int_{\tau D_r}^{E_{2r}} d\omega_j, \quad \forall j.$$

Thus, the divisor $\tau\mathcal{D}$ also solves (A3.2), so it must equal $\mathcal{D}$.

To show that the points $D_j$ are $\tau$-invariant, we take a point $\eta^0 = i\mathfrak{z}^0 - K$ with any $i\mathfrak{z}^0$ of the form $i\mathfrak{z}^0 = A(\mathcal{D}^0) + \boldsymbol{K}$, where the divisor $\mathcal{D}^0$ is as in item (i) (i.e. $\mathcal{D}_j^0 \in a_j$) and denote by $\mathfrak{z}^t$, $0 \le t \le 1$, any curve in $\mathbb{T}^n$ which connects $\mathfrak{z}^0$ with $\mathfrak{z}^1$. For $t \in [0, 1]$ eqn (A3.2) with $\mathfrak{z}^1$ replaced by $\mathfrak{z}^t$ has a unique solution $\mathcal{D}^t$. This solution continuously depends on $\mathfrak{z}^t$ and is $\tau$-invariant. Since for $t = 0$ we have $\tau D_j^t = D_j^t$, $j = 1, \dots, n$, since $|\mathcal{D}^t| \equiv n$, and since the $\tau$-invariant circles $a_j$ do not intersect, then during the deformation each $a_j$ still contains exactly one point of $\mathcal{D}^t$. So $\tau D_j^t = D_j^t$ for all $t$ and $j$. We have proved that

$$\begin{aligned}&\text{for each } \mathfrak{z}^1 \in \mathbb{T}^n \text{ there exists a unique divisor } \mathcal{D},\\ &\mathcal{D} = D_1 \cdots D_n,\ D_j \in a_j, \quad \text{which satisfies} \quad (\text{A3.2}).\end{aligned} \tag{A3.3}$$

(iii) Now we show that the vector $\boldsymbol{V}$ corresponding to any (periodic) $n$-gap potential $u(x) \in T^n(r)$ is integer. Since $\boldsymbol{V}$ is analytic in $r \in \mathbb{R}_+^n$ (see Appendix 4), then it suffices to prove that it is integer for small $r$ or, equivalently, for small $\mathcal{R}$.

Let us consider an $n$-gap potential (3.22) with $t = 0$, with zero phase $q$ and small $\mathcal{R}$. Since the function $G$ is analytic in $q$ and $\mathcal{R}$, then using (3.23) we write it as

$$G(q, \mathcal{R}) = \frac{1}{\sqrt{\pi}} \sum_{j=1}^{n} \mathcal{R}_j \cos q_j + \sum_{s \in \mathbb{Z}^n} g_s(\mathcal{R}) \cos s \cdot q + \langle\text{sine-series}\rangle,$$

where the Fourier coefficients $g_s = O(|\mathcal{R}|^2)$ are analytic in $\mathcal{R}$. We fix any $j$, extract from the second sum all terms corresponding to $s$ such that $s \cdot \boldsymbol{V}(r) \equiv V_j(r)$, and write the $n$-gap potential as

$$\begin{aligned}u(x; \mathcal{R}) = &-\frac{1}{\sqrt{\pi}} \big(\mathcal{R}_j + O(|\mathcal{R}|^2)\big) \cos V_j(\mathcal{R}) x \\ &+ \sum f_s(r) \cos(s \cdot \boldsymbol{V}(\mathcal{R})) x + \langle\text{sine-series}\rangle,\end{aligned}$$

where the sum is taken over all $s$ such that $s \cdot \boldsymbol{V} \ne V_j$ for almost all $\mathcal{R}$. Since $u$ is $2\pi$-periodic in $x$, then all Fourier coefficients corresponding to $\cos \lambda x$ with a non-integer $\lambda$ must vanish. Since $\mathcal{R}_j + O(|\mathcal{R}|^2)$ is non-zero for small $\mathcal{R}$, then $V_j$ must be integer for almost all small $\mathcal{R}$, and therefore it must be integer for all $\mathcal{R}$ and $r$. □

## Appendix 4. On the vectors $V$ and $W$

Here we study the differentials $d\Omega_1$ and $d\Omega_3$ on a surface $\Gamma(r)$, and the vectors $\boldsymbol{V}(r)$ and $\boldsymbol{W}(r)$, corresponding to an $m$-gap potential $u(x)$ with open gaps $\Delta_{\nu_1}, \ldots, \Delta_{\nu_m}$, where $|\Delta_{\nu_j}| = r_j$.

The differential $(i/2)(\lambda^m\, d\lambda/\mu)$ is holomorphic outside infinity, where it has the form $dk + f(k^{-1})\, dk^{-1}$ with $k = i\sqrt{\lambda}$ and with some analytic at zero function $f$. Since $d\Omega_1$ has the same asymptotics at infinity (see (3.10)), then the former differential differs from the latter by a holomorphic differential. Hence,

$$d\Omega_1 = \frac{i}{2}\frac{\lambda^m + f_{m-1}\lambda^{m-1} + \cdots + f_0}{\mu}\, d\lambda = \frac{i}{2}\frac{P_m(\lambda)}{\mu}\, d\lambda \tag{A4.1}$$

(cf. the arguments used to prove (3.9)). Since $d\Omega_1$ has zero $a$-periods, then the coefficients $f_{m-1}, \ldots, f_0$ satisfy the following system of linear equations:

$$\sum_{j=1}^{m} f_{m-j}\left\langle \frac{i}{2}\frac{\lambda^{m-j}\, d\lambda}{\mu}, a_l \right\rangle = -\left\langle \frac{i}{2}\frac{\lambda^m\, d\lambda}{\mu}, a_l \right\rangle, \quad l = 1, \ldots, m.$$

This system has real coefficients and its solution is unique since the differential $d\Omega_1$ is uniquely defined. Hence, $f_{m-1}, \ldots, f_0$ are real numbers, analytic in $r \in \mathbb{R}^m_+$: *the differential $d\Omega_1$ analytically depends on $r \in \mathbb{R}^m_+$.*

For $j = 1, \ldots, m$ we have

$$0 = \langle \Omega_1, a_j \rangle = -2 \int_{E_{2j-1}}^{E_{2j}} \frac{P_m(\lambda)}{-2i\sqrt{R(\lambda)}}\, d\lambda.$$

In the interval $(E_{2j-1}, E_{2j})$ the denominator $-2i\sqrt{R(\lambda)}$ is a non-vanishing real function of a constant sign. Hence, the polynomial $P_m(\lambda)$ has a root in $(E_{2j-1}, E_{2j})$. Thus, all $m$ roots of $P_m$ are localized and the differential $d\Omega_1$ has the form stated in Lemma 3.2

Quite similarly, the *differential $d\Omega_3$ has the form (3.11) and analytically depends on $r$.*

Now we examine the limiting behaviour of the differentials $d\Omega_1$ and $d\Omega_3$ when $r = (r_1, \ldots, r_m) \to 0$. Denoting $Q_j(\lambda) = (\lambda - x_j)/\sqrt{(\lambda - E_{2j})(\lambda - E_{2j+1})}$, where $x_j = x_j(r)$ is a root of the polynomial $P_m$ (see (A4.1)) in the interval $(E_{2j-1}, E_{2j})$, we write $d\Omega_1$ as

$$d\Omega_1 = \frac{i}{2\sqrt{\lambda - E_1}} Q_1(\lambda) \cdots Q_m(\lambda)\, d\lambda.$$

Elementary calculations show that for any $l$ the integral of the function $|Q_l(\lambda)|$ over the interval $\Delta_{\nu_l} + |r|^{1/2}$ converges to zero when $r \to 0$; at the same time $Q_l$ converges to one uniformly outside this interval.

Since $E_1 \to 0$ and $E_{2j}, E_{2j+1} \to \mathcal{V}_j^2/4$ (see (3.4)), then we get the small-gap limits of the wave numbers:

$$iV_j(r) = 2\int_{E_1}^{E_{2j}} d\Omega_1 \to i\int_0^{\mathcal{V}_j^2/4} \lambda^{-1/2}\, d\lambda = i\mathcal{V}_j \quad \text{as } r \to 0.$$

That is,

$$V_j(r) \to \mathcal{V}_j \quad \text{as } r \to 0. \tag{A4.2}$$

Since $\lambda_{2p} \to p^2/4$ as $r \to 0$ for any $p$, then also

$$\int_{E_1}^{\lambda_{2p}} d\Omega_1 \to \frac{i}{2}\int_0^{p^2/4} \lambda^{-1/2}\, d\lambda = \frac{ip}{2}. \tag{A4.3}$$

Now let us take a vector $r \in \mathbb{R}_+^m$ and fix any $j \le m$. We denote by $r^{\hat{j}}$ the $(m-1)$-vector $(r_1, \ldots, \hat{r}_j, \ldots, r_m)$ and denote $r_\varepsilon = (r_1, \ldots, \varepsilon, \ldots, r_m)$ ($\varepsilon$ stands instead of $r_j$). Repeating the arguments above we get that for a suitable sequence $\varepsilon_M \to 0$ the differential $d\Omega_1^{(m)}$ with $r = r_{\varepsilon_M}$ converges to a limit

$$i\,\frac{Q_1(\lambda)\cdots\hat{Q}_j(\lambda)\cdots Q_m(\lambda)\, d\lambda}{2\sqrt{\lambda - E_1}}, \tag{A4.4}$$

where $Q_l(\lambda)$ as above depends on a limiting point $x_l$. The limiting vector $(x_1, \ldots, \hat{x}_j, \ldots, x_m)$ a priori depends on the sequence $\{\varepsilon_M\}$. Any limiting differential (A4.4) is a holomorphic differential on the Riemann surface $\Gamma(r^{\hat{j}})$ of genius $m-1$. It inherits from $d\Omega_1$ the following normalizations:

$$0 = \int_{E_{2l}}^{E_{2l+1}} \frac{i}{2\sqrt{\lambda - E_1}}\, Q_1\cdots\hat{Q}_j\cdots Q_m\, d\lambda, \quad l = 1, \ldots, \hat{j}, \ldots, m.$$

Hence, this differential equals $d\Omega_1^{(m-1)}(r^{\hat{j}})$. Since the limit does not depend on the sequence $\{\varepsilon_M\}$, the convergence to $d\Omega_1^{(m-1)}$ holds as $r_j = \varepsilon \to 0$. Passing to the limit in the formula for $iV_j^{(m)}(r)$, we get that

$$V_j^{(m)}(r) = -2i\int_{E_1}^{E_{2j}(r)} d\Omega_1^{(m)}(r) \longrightarrow -2i\int_{E_1}^{E_{2j}=E_{2j+1}(r^{\hat{j}})} d\Omega_1^{(m-1)}(r^{\hat{j}}) \quad \text{as } r_j \to 0.$$

Applying the same arguments to the differential $d\Omega_3$ we get that

$$\int_{E_1}^{\lambda_{2p}} d\Omega_3 \longrightarrow -\frac{3i}{2}\int_0^{p^2/4} \sqrt{\lambda}\, d\lambda = -i\left(\frac{p}{2}\right)^3 \quad \text{as } r \to 0$$

for any $p$. In particular, we have recovered small-gap limits of the frequencies

$$W_j \longrightarrow -\frac{1}{4}\mathcal{V}_j^3 \quad \text{as } r \to 0. \tag{A4.5}$$

A similar asymptotic holds when we shrink only one open gap:

$$W_j^{(m)}(r) \to -2i \int_{E_1}^{E_{2j}=E_{2j+1}(r^{\hat{j}})} d\Omega_3^{(m-1)}(r^{\hat{j}}) \quad \text{as } r_j \to 0. \tag{A4.6}$$

Moreover, since the forms $d\Omega_1$ and $d\Omega_3$ and the eigenvalues $E_1, \ldots, E_{2m+1}$ are analytic in $r$, then *the vectors* $\mathbf{V}$ *and* $\mathbf{W}$ *are analytic in* $r \in \mathbb{R}_+^m$.

## Appendix 5. A small-gap limit for the theta functions

Here we discuss some elementary properties of the theta functions, corresponding to the potentials $u(x)$ from $T^m(r)$, where $|r| \ll 1$. These potentials are small: $\|u\|_s \le C_s|r|$ for each $s$ (see Marčenko 1986 or see the proof of Theorem 3.1′ in Section 3.1).

Let us consider any holomorphic differential $d\omega_j$ as in Section 3.2, written in the form (3.9). Since each $a_p$-period of the differential with any $p \neq j$ vanishes, then the numerator in its polynomial presentation (3.9) has a root $y_p^j$ in each open gap $\Delta_{V_p}$ except the gap $\Delta_{V_j}$. So we can write $d\omega_j$ as

$$d\omega_j = C_j \frac{Q_{j,1}(\lambda)\cdots\widehat{Q_{j,j}(\lambda)}\cdots Q_{j,m}(\lambda)}{\sqrt{(\lambda-E_1)(\lambda-E_{2j})(\lambda-E_{2j+1})}}\,d\lambda,$$

where $Q_{j,p} = (\lambda - y_p^j)/\sqrt{(\lambda-E_{2p})(\lambda-E_{2p+1})}$. Using again the vanishing of the $a_p$-periods of $d\omega_j$ with $p \neq j$ we find that the point $y_p^j$ is close to the middle of the $p$th gap: $y_p^j = \frac{1}{2}(E_{2p}+E_{2p+1}) + O(|r|^2)$.[23] Hence,

$$Q_{j,p}(\lambda) = 1 + O(|r|^2) \quad \text{if dist}\,(\lambda, \Delta_p) \ge C > 0. \tag{A5.1}$$

[23] Proof: On the $(m-1)$-cube $K = \{-1 \le x_p \le 1 \mid 1 \le p \le m,\ p \neq j\}$ we consider the vector field $F(x) = (F_1, \ldots, \widehat{F_j}, \ldots, F_m)$, where $F_p(x)$ equals the $a_p$-period of the form as above with $y_p^j = y_p^j(x) = (E_{2p}+E_{2p+1})/2 + Cr_p^2 x_p$. Straightforward estimates show that $F_p > 0$ if $x_p = 1$ and $F_p < 0$ if $x_p = -1$, provided that $r$ is sufficiently small and $C$ was chosen sufficiently large. Now degree arguments (see Nirenberg 1974) show that $F$ vanishes at some point $x \in K$. Corresponding points $y_p^j(x)$ $(p \le m,\ p \neq j)$ define the form $d\omega_j$.

Since $E_1 = O(|r|^2)$, $E_{2j-1} = V_j^2/4 - r_j/2 + O(|r|^2)$, and $E_{2j} = V_j^2/4 + r_j/2 + O(|r|^2)$ (see (3.4)), then we can use (A5.1) to write the normalization $\langle d\omega_j, a_j\rangle = 2\pi i$ as

$$\pi = \int_{E_{2j}}^{E_{2j+1}} \frac{C_j(1+O(|r|^2))\,d\lambda}{\sqrt{(\lambda - E_1)(\lambda - E_{2j})(E_{2j+1} - \lambda)}}$$

$$= \frac{C_j(1+O(|r|^2))}{V_j/2} \int_0^1 \frac{dx}{\sqrt{x(1-x)}}.$$

Since $\int_0^1 (x(1-x))^{-1/2}\,dx = \pi$, then $C_j = V_j/2 + O(|r|^2)$.

So far we have examined integrals of $d\omega_j$ over open gaps. Now we use (A5.1) to estimate integrals over the intervals between them:

$$\int_{E_{2p-1}}^{E_{2p}} d\omega_j = \begin{cases} C_p + O(|r|), & p \neq j, j+1, \\ \ln r_p + C_p + O(|r|), & p = j, \\ -\ln r_p + C_p + O(|r|), & p = j+1. \end{cases}$$

Here the first asymptotic follows from (3.4) and (A5.1), while the last two result from calculations similar to those used to estimate the integral over $\Delta_{V_j}$.[24] Thus,

$$B_{jp} = 2\int_{E_1}^{E_{2p}} d\omega_j = \begin{cases} O(1), & j \neq p, \\ 2\ln r_p + O(1), & j = p, \end{cases}$$

and for any integer vector $s$ we have $e^{\frac{1}{2}(Bs,s)} = C_s(r)\prod r_j^{s_j^2}$, where $C_s, C_s^{-1} = O(1)$ as $r \longrightarrow 0$. We arrive at the following small-gap asymptotic for the theta function:

$$\theta(iz) = 1 + \sum_{j=1}^m C_j r_j \frac{e^{iz_j} + e^{-iz_j}}{\sqrt{\pi}} + O(|r|^2)$$

$$= 1 + 2\sum_{j=1}^m C_j r_j \frac{\cos z_j}{\sqrt{\pi}} + O(|r|^2), \tag{A5.2}$$

where $\{C_j\}$ are some new constants and $O(|r|^2)$ stands for a function on $\mathbb{T}^m$ such that each its $C^k$-norm is $O(|r|^2)$.

Because of (A5.2), for small $r$ the $n$-gap potential (3.16) $|_{t=0}$ is equal to

$$2\frac{\partial^2 \ln\theta(i\mathbf{V}x + i\mathfrak{z}, r)}{\partial x^2} = -4\sum_{j=1}^m C_j V_j^2 r_j \frac{\cos(V_j x_j + \mathfrak{z}_j)}{\sqrt{\pi}} + O(|r|^2).$$

[24] Different signs for the integrals along the upper edges of the cuts $[E_{2p-1}, E_{2p}]$ and $[E_{2p+1}, E_{2p+2}]$ in $\Gamma^+$ are due to the fact that the function $\sqrt{(\lambda - E_{2p})(\lambda - E_{2p+1})}$ is negative on the former and positive on the latter: for small $r_p$ it behaves there like $\lambda - (E_{2p} + E_{2p+1})/2$.

By (3.4), its $V_j$th gap (i.e. the $j$th open gap) has the size $4C_j V_j^2 r_j + O(|r|^2)$. Hence, $4C_j V_j^2 = 1 + O(|r|)$ and we arrive at a small-gap limit for the theta function:

$$\theta(iz; r) = 1 + \frac{1}{2\sqrt{\pi}} \sum V_j^{-2} r_j \cos z_j + O(|r|^2). \tag{A5.3}$$

For the same reason as in Appendix 4, the differentials $d\omega_j$ analytically depend on $r \in \mathbb{R}_+^n$ (i.e. the multi-valued functions $(d\omega_j/d\lambda)(\lambda; r)$ are analytic in $r \in \mathbb{R}_+^n$ and in $\lambda$ outside the singularities). Hence, *the Riemann matrix and the theta function are both analytic in* $r \in \mathbb{R}_+^n$.

## Appendix 6. A non-degeneracy lemma

In this appendix we prove a stronger statement that implies Lemma 3.3.

Let $E_1 < \cdots < E_{2n+1}$ be any real numbers and $\Gamma = \Gamma(E)$ be a Riemann surface of the equation $\mu^2 = R(\lambda) := \prod(\lambda - E_j)$. We define the differentials $d\Omega_1$ and $d\Omega_3$ and the vectors $\boldsymbol{V}$ and $\boldsymbol{W}$ as in Section 3.2. Now the vector $\mathbf{V}$ may be non-integer and formula (3.12) defines a function $u(x)$ which is a quasiperiodic $n$-gap potential, see Dubrovin (1981), Dubrovin*et al.* (1976) and Belokolos *et al.* (1994) (the properties of $u(x)$ are irrelevant for results of this appendix).

**Theorem.** The analytic map

$$\mathbf{E} = (E_1 < E_2 < \cdots < E_{2n+1}) \mapsto (\boldsymbol{V}, \boldsymbol{W}, c) \in \mathbb{R}^{2n+1} \tag{A6.1}$$

is non-degenerate everywhere ($c$ stands for the same constant as in (3.10) and (3.12)).

The theorem implies the assertion of Lemma 3.3 in a stronger form. Indeed, since $\boldsymbol{W}(r)$ is a restriction of the map (A6.1) to the $n$-manifold which is a pre-image of the $n$-dimensional affine space $\{\boldsymbol{V} = \text{const}, c = 0\}$, then the map $r \mapsto \boldsymbol{W}(r)$ is non-degenerate *everywhere* in $\mathbb{R}_+^n$.

The proof we present below is based on a scheme, proposed by Krichever (1991), which was completed in full details in Bikbaev and Kuksin (1993*b*).

*Proof of the theorem.* We shall need the following properties of the zeros of the differentials $d\Omega_1$ and $d\Omega_3$:

**Proposition.** *(1) All zeros of the differential* $d\Omega_1$ *lie outside the branching points of* $\Gamma$*; (2) at least* $2n$ *zeros of* $d\Omega_3$ *lie outside the branching points; (3) zeros of the differential* $d\Omega_1$ *lie outside the zeros of* $d\Omega_3$.

*Proof.* The first two assertions follow from Lemma 3.2. Moreover, due to the lemma, $d\Omega_1$ has $2n$ roots of the form $P_j^\pm = (\lambda_j, \pm\mu_j)$, $j = 1, \dots, n$, where each interval $\Delta_j^0 = (E_{2j}, E_{2j+1})$ contains exactly one point $\lambda_j$.

To prove the last assertion let us suppose that some zero $P_i$ of $d\Omega_1(P)$ coincides with one of $d\Omega_3(P)$. Then there exists a real constant $\xi$ such that the differential

$$d\tilde{\Omega}(P) = (\xi d\Omega_1 + d\Omega_3)(P)$$

has double zeros at the points $P_i^+ = (\lambda_i, \mu_i)$ and $P_i^- = (\lambda_i, -\mu_i)$, $\lambda_i \in \Delta_i^0$. Since $a$-periods of this differential obviously vanish, then each interval $\Delta_i^0$, $i = 1, \ldots, n$, contains its zero (cf. Lemma 3.2 and its proof in Appendix 4). Since

$$d\tilde{\Omega}(\lambda) = i\,\frac{\text{real polynomial of degree } n+1}{\mu}\,d\lambda,$$

then $d\tilde{\Omega}(\lambda)$ has exactly $n+1$ finite zeros and all of them are localized. Therefore, $d\tilde{\Omega}(\lambda)$ has no other zeros (except the double zero $\lambda_i$) in $\Delta_i^0$. But in such a case $\int_{\Delta_i^0} d\tilde{\Omega}(\lambda) \neq 0$, which contradicts the normalization $\oint_{a_i} d\tilde{\Omega} = 0$. □

If the map (A6.1) degenerates at a point $\mathbf{E} = (E_1, \ldots, E_{2n+1})$, then we can construct an analytic deformation $\Gamma(\tau) = \Gamma(\mathbf{E}(\tau))$ of the initial curve $\Gamma$ (i.e. a deformation $\mathbf{E}(\tau)$ with $\mathbf{E}(0) = \mathbf{E}$), such that for the vectors $\boldsymbol{V}(\tau)$ and $\boldsymbol{W}(\tau)$, $c(\tau)$ we have

$$\boldsymbol{V}(\tau) = \boldsymbol{V} + O(\tau^2), \qquad \boldsymbol{W}(\tau) = \boldsymbol{W} + O(\tau^2), \qquad c(\tau) = c + O(\tau^2), \tag{A6.2}$$

and the vector of branching points $\mathbf{E}(\tau)$ has a non-zero $\tau$-derivative at $\tau = 0$. Below we prove that such a deformation $\Gamma(\tau)$ cannot exist: the relations (A6.2) imply that $\mathbf{E}'_\tau(0) = 0$.

We define the Abel integrals $\Omega_j(P, \tau)$, $j = 1, 3$ as follows. Let $\gamma_P$ be any path in $\Gamma(\tau)$ from $\sigma P$ to $P$, where $\sigma$ is the involution

$$\sigma : (\lambda, \mu) \mapsto (\lambda, -\mu).$$

We set

$$\Omega_j(P, \tau) = \frac{1}{2}\int_{\gamma} d\Omega_j(P, \tau), \qquad j = 1, 3.$$

Each integral $\Omega_j$ is multi-valued (it is defined up to half-periods of the differential $d\Omega_j$) and

$$\Omega_j(E_r(\tau), \tau) \ni 0, \quad \forall j = 1, 3, \ \ \forall r = 1, \ldots, 2n+1. \tag{A6.3}$$

Let $E_*$ be any finite branching point of $\Gamma(\tau)$ and $\gamma_0$ be a path from $E_*$ to $P$. We can take $\gamma_P = -\sigma\gamma_0 \cup \gamma_0$. Since the differentials $d\Omega_j$ are odd with respect to $\sigma$ (this readily follows from (3.11)), then we have

$$\Omega_j(P, \tau) = \frac{1}{2}\Bigg(\int_{\gamma_0} - \int_{\sigma\gamma_0}\Bigg) d\Omega_j = \int_{\gamma_0} d\Omega_j, \quad \text{where } \gamma_0 \text{ is a path from } E_* \text{ to } P. \tag{A6.4}$$

In particular, the differential of $\Omega_j$ equals $d\Omega_j$.

Let $P = (\lambda, \mu)$ be any point in $\Gamma$ outside the branching points. Then we can identify $P$ with its projection $\lambda$. For $\tau$ small enough the point $\lambda$ lies outside the branching points of $\Gamma(\tau)$. So for $j = 1, 3$ we can define the function $\partial_\tau \Omega_j(\lambda, \tau)\big|_{\tau=0}$.

**Lemma A6.1.** *The functions*

$$P = (\lambda, \mu) \mapsto \partial_\tau \Omega_j(P) := \partial_\tau \Omega_j(\lambda, \tau)|_{\tau=0}, \quad j = 1, 3, \tag{A6.5}$$

*may be extended to meromorphic functions on the curve $\Gamma$. These functions are regular outside the finite branching points $E_1, \ldots, E_{2n+1}$, where they have first-order poles with*

$$Res_{P=E_m} \partial_\tau \Omega_j(P) = x^j_{-1}(m) \partial_\tau E_m(0), \quad j = 1, 3, \quad m = 1, \ldots, 2n+1,$$

*and $x^1_{-1}(m)$, $m = 1, \ldots, 2n+1$, are non-zero constants. The functions (A6.5) are regular at infinity and vanish there. Moreover, for $j = 1$ the function (A6.5) is $O(|u|^3)$ as $u = \lambda^{-1/2}$ tends to zero.*

*Proof.* Owing to the relations (A6.2), the $b$-periods of the differentials $d\Omega_j(P, \tau)$, $j = 1, 3$, are constant up to $O(\tau^2)$. Since their $a$-periods vanish, then different branches of the Abel integrals $\Omega_j(P, \tau)$ differ by const $+O(\tau^2)$, and hence the functions (A6.5) are well defined and analytic outside the branching points.

In the vicinity of any finite branching point $E_m$ of $\Gamma$, not $\lambda$ but $(\lambda - E_m)^{1/2}$ is an analytic coordinate. Using (3.11) we expand the differentials $d\Omega_1$ and $d\Omega_3$ as follows:

$$d\Omega_j(\lambda, \tau) = \sum_{k=-1}^{\infty} (\lambda - E_m)^{k/2} x^j_k(E_m, \tau)\, d\lambda, \quad j = 1, 3. \tag{A6.6}$$

Owing to the first statement of the proposition the coefficients $x^1_{-1}(E_m, 0)$, $m = 1, \ldots, 2n+1$, are non-zero.

From (A6.4) with $E_* = E_m$ and (A6.6) we obtain that near $E_m$ the function $\partial_\tau \Omega_j$ can be written as

$$\partial_\tau \Omega_j(\lambda, 0) = \sum_{k=1}^{\infty} \Big( \frac{2}{k} \partial_\tau x^j_{k-2}(E_m, 0)(\lambda - E_m)^{k/2} + x^j_{k-2}(E_m, 0)(\lambda - E_m)^{(k-2)/2} \partial_\tau E_m \Big).$$

The r.h.s. of the above formula defines near $E_m$ a meromorphic function with a first-order pole at $E_m$.

For $P = (\lambda, \mu)$ with $\lambda$ large enough we shall define $\Omega_j$ using (A6.4), where $\gamma_P$ is the lift to $\Gamma(\tau)$ of the circle in $\mathbb{C}_\lambda$ of the radius $|\lambda|$, cut at the point $\lambda$. Since near infinity we have

$$d\Omega_3 = 3iu^{-4}\, du + d\Omega_3^0, \quad u = \lambda^{-1/2},$$

where the differential $d\Omega_3^0(u,\tau)$ is regular for sufficiently small $u$ (see (3.10)), then

$$\Omega_3(P,\tau) = -iu^{-3} + \frac{1}{2}\int_{\gamma_P} d\Omega_3^0(u,\tau).$$

Hence, the function $\partial_\tau\Omega_3(P) = \frac{1}{2}\int_{\gamma_P}\partial_\tau d\Omega_3^0(u,0)$ is analytic near infinity and vanishes at infinity.

For $j = 1$ we have by (3.10):

$$\Omega_1(P,\tau) = iu^{-1} + icu + O(|u|^3),$$

so $\partial_\tau\Omega_1 = O(|u|^3)$ by (A6.2) and the lemma is proved. □

Since all the numbers $x_{-1}^1(m)$ are non-zero, we have as a consequence of the lemma:

**Corollary.** *To prove the theorem it is sufficient to check that*

$$\partial_\tau\Omega_1(P) \equiv 0. \tag{A6.7}$$

To prove (A6.7), we construct a function $\dot{\Omega}_3$ equal to the '$\tau$-derivative of $\Omega_3$ with $\Omega_1$ fixed'. To do it we fix a point $P \in \Gamma$ such that

$$d\Omega_1(P,0) \neq 0, \tag{A6.8}$$

and consider the following equation for a point $P(\tau) \in \Gamma(\tau)$:

$$\Omega_1(P(\tau),\tau) = \Omega_1(P,0). \tag{A6.9}$$

Owing to (A6.9) and the Implicit Function Theorem, eqn (A6.9) may be uniquely solved for small $\tau$.

We define the function $\dot{\Omega}_3$ as

$$\dot{\Omega}_3(P) := \frac{d}{d\tau}\Omega_3(P(\tau),\tau)|_{\tau=0}. \tag{A6.10}$$

Owing to the theorem's assumptions, replacement of the branch of the integral $\Omega_1$, used in (A6.9), will change the curve $P(\tau)$ by $O(\tau^2)$, and replacement of the branch of $\Omega_3$ in (A6.10) will change $\Omega_3(P(\tau),\tau)$ by const $+ O(\tau^2)$ and will not change the r.h.s. in (A6.10). So the function $\dot{\Omega}_3$ is single-valued.

**Lemma A6.2** *The function $\dot{\Omega}_3$ extends to a meromorphic function on $\Gamma$.*

*Proof.* We claim that

$$\dot{\Omega}_3(P) = \partial_\tau\Omega_3(P) - \partial_\tau\Omega_1(P)\frac{d\Omega_3(P,0)}{d\Omega_1(P,0)} \tag{A6.11}$$

outside the branching points of $\Gamma$ and zeros of $d\Omega_1$. Indeed, identifying a point $P(\tau) = (\lambda(\tau),\mu(\tau)) \in \Gamma(\tau)$ such that $P(0)$ is not a branching point of $\Gamma$ with its projection $\lambda$ (we can do this if $\tau$ is sufficiently small), we write $d\Omega_1$ as $\partial_\lambda\Omega_1 d\lambda$ and get from (A6.9) that $\partial_\tau\lambda(0) = -\partial_\tau\Omega_1(\lambda,0)/\partial_\lambda\Omega_1(\lambda,0)$. Now (A6.11) follows.

The formula (A6.11) proves the lemma since by Lemma A6.1 its r.h.s. extends to a meromorphic function. □

By assertion (1) of the proposition, (A6.9) holds at the points $E_j$, $j = 1, \ldots, 2n+1$. By (A6.3) the solution $P(\tau)$ of (A6.9) with $P = E_j$ is $P(\tau) = E_j(\tau)$ and $\Omega_3(E_j(\tau), \tau) \equiv 0$. So we have

$$\dot{\Omega}_3(E_j, 0) = 0, \quad \forall\, j = 1, \ldots, 2n+1, \tag{A6.12}$$

and the function $\dot{\Omega}_3$ has $2n+1$ zeros in the finite branching points of $\Gamma$.

By (A6.11), (A6.12), and Lemma A6.1, the only possible finite poles of $\dot{\Omega}_3$ lie in the $2n$ zeros of $d\Omega_1$. To study $\dot{\Omega}_3$ near infinity let us observe that there

$$\partial_\tau \Omega_3 = O(|u|), \qquad \partial_\tau \Omega_1 = O(|u|^3),$$

by Lemma A6.1, and $d\Omega_3/d\Omega_1 = O(|u|^{-2})$ by (3.10). So $\dot{\Omega}_3(\infty) = 0$. Altogether the function $\dot{\Omega}_3$ has at least $2n+2$ zeros and no more then $2n$ poles. Hence, $\dot{\Omega}_3 \equiv 0$ (see Springer 1981, p. 175) and

$$\partial_\tau \Omega_3 \, d\Omega_1 = \partial_\tau \Omega_1 \, d\Omega_3. \tag{A6.13}$$

All the poles of $\partial_\tau \Omega_1$ lie in the finite branching points. So by statement (2) of the proposition the r.h.s. of (A6.13) has at least $2n$ zeros outside the branching points. The differential $d\Omega_3(\lambda)$ has one more zero $\lambda_{n+1} \in \mathbb{C}$. To complete the proof we should distinguish two cases:

(a) $\lambda_{n+1}$ lies outside the branching points. Then the r.h.s. in (A6.13) has $2n+2$ zeros in $\Gamma \setminus \{E_1, \ldots, E_{2n+1}\}$. The zeros of $d\Omega_1$ lie outside them by statement (3) of the proposition. Thus, the function $\partial_\tau \Omega_3$ vanishes at these points. So $\partial_\tau \Omega_3$ has $2n+2$ finite zeros, the zero at infinity and no more than $2n+1$ poles. Hence, it vanishes identically, $\partial_\tau \Omega_1 \equiv 0$ by (A6.13), and the theorem is proved.

(b) $\lambda_{n+1} = E_{j*}$ for some $1 \le j_* \le 2n+1$. Then the r.h.s. is regular in $E_{j*}$. Since $d\Omega_1(E_{j*}) \neq 0$, then the function $\partial_\tau \Omega_3$ also is regular in $E_{j*}$. So it has no more than $2n$ poles. This function vanishes at the $2n$ zeros of $d\Omega_1$ and at infinity. Thus, $\partial_\tau \Omega_3 \equiv 0$, $\partial_\tau \Omega_1 \equiv 0$ by (A6.13), and the proof is completed. □

The scheme to prove non-degeneracy of the map (A6.1) presented above is rather general: if for a given integrable equation and its finite-gap solutions we take the statements of the proposition for granted, we can proceed just as above to construct the functions $\partial_\tau \Omega_1$, $\partial_\tau \Omega_3$, and $\dot{\Omega}_3$ which are meromorphic on the spectral curve of the solution. If the vector of additional parameters $c(\tau)$ is chosen in such a way that the function $\dot{\Omega}_3$ vanishes at the infinite points of the spectral curve provided that (A6.2) holds, then the vector $(\boldsymbol{V}, \boldsymbol{W}, c)$ gives the parameterization we look for. (Observe that in the given proof the function $\dot{\Omega}_3$ vanishes at infinity due to the last statement of Lemma A6.1 and, finally, due to the 'clever' choice of the parameter $c$.)

Our proof of the proposition applies to equations with self-adjoint $\mathcal{L}$-operators (for these equations the vectors $\boldsymbol{E}$ of the branching points are real). For some integrable equations with non-self-adjoint $\mathcal{L}$-operators an analogy of the proposition can be obtained if the corresponding potential $u(x)$ is small (this happens, for example, to the SG equation; see Section 4.3). In this case the arguments above prove the following local version of the Theorem: 'the map (A6.1) is non-degenerate at points $\mathbf{E}$ such that the corresponding gaps $|E_{2j+1} - E_{2j}|$ are sufficiently small'. This weaker version of the result still implies the non-degeneracy lemma.

# 4

# The Sine–Gordon equation

In this chapter we consider the SG equation under periodic and even periodic boundary conditions (see Example 2.3 in Section 2.1). The results are parallel to the KdV case and our presentation is much shorter. Missing details can be found in McKean (1981), Belokolos *et al.* (1994), Bikbaev and Kuksin (1992), and Bobenko and Kuksin (1995*a*).

## 4.1 The $L, A$ pair

We recall that the SG equation can be written in a Hamiltonian form both in the variables $(u, v = \dot{u})$ and in the variables $(u, w = (-\partial^2/\partial x^2 + 1)^{1/2}\dot{u})$; see Section 2.1. In the variables $(u, v)$ the SG equation takes the form

$$\dot{u} = -v, \qquad \dot{v} = -u_{xx} + \sin u. \tag{4.1}$$

Equation (4.1) is Lax-integrable and can be written in the Lax form

$$\dot{\mathcal{L}} = [\mathcal{A}, \mathcal{L}],$$

where $\mathcal{L} = \mathcal{L}_{(u,v)}$ and $\mathcal{A} = \mathcal{A}_{(u,v)}$ stand for the following differential operators:

$$\begin{aligned} \mathcal{L} &= -\begin{pmatrix} J & 0 \\ 0 & 0 \end{pmatrix} \frac{\partial}{\partial x} + \begin{pmatrix} \tilde{A} & \tilde{B} \\ \tilde{B} & 0 \end{pmatrix}, \\ \mathcal{A} &= \begin{pmatrix} -E & 0 \\ 0 & E \end{pmatrix} \frac{\partial}{\partial x} - 2\begin{pmatrix} 0 & J\tilde{B} \\ \tilde{B}J & 0 \end{pmatrix}. \end{aligned} \tag{4.2}$$

Here $E$ is the identity $2 \times 2$-matrix, $J = \begin{pmatrix} 0 & 1 \\ -1 & 0 \end{pmatrix}$, and $\tilde{A}$ and $\tilde{B}$ stand for the operators

$$\tilde{A} = \frac{i}{4}(v + u'_x)\begin{pmatrix} 0 & 1 \\ 1 & 0 \end{pmatrix}, \qquad \tilde{B} = \frac{1}{4}\begin{pmatrix} e^{\frac{i}{2}u} & 0 \\ 0 & e^{-\frac{i}{2}u} \end{pmatrix},$$

see McKean (1981) and Faddeev and Takhtajan (1986). The operators $\mathcal{L}$ and $\mathcal{A}$ act on vector-functions, valued in $\mathbb{C}^4$, under $2\pi$-periodic/antiperiodic or $4\pi$-periodic boundary conditions. For the scale $\{\mathfrak{Z}_s\}$ we take one of the corresponding scales of Sobolev vector-functions.

For any smooth $2\pi$-periodic functions $u(t,x)$ and $v(t,x)$ and any smooth complex vector-function $\xi_0(x)$ which is $2\pi$-periodic/antiperiodic or $4\pi$-periodic, the corresponding boundary-value problem for the equation

$$\dot{\xi} = \mathcal{A}\xi, \qquad \xi(0,x) = \xi_0(x)$$

has a unique smooth solution (Pazy 1983). So by the general results described in Section 2.3, the set of eigenvalues of the operator $\mathcal{L}_t = \mathcal{L}_{(u(t,\cdot),v(t,\cdot))}$ under the boundary conditions as above is $t$-independent if $(u, v)$ is a solution for (4.1).

The four-dimensional eigenvalue problem $\mathcal{L}f = \mu f$ can be reduced to a two-dimensional one since denoting $f = \begin{pmatrix} f_- \\ f_+ \end{pmatrix}$, $f_\pm \in \mathbb{C}^2$, we have:

$$-J\frac{\partial}{\partial x}f_- + \tilde{A}f_- + \tilde{B}f_+ = \mu f_-, \qquad \tilde{B}f_- = \mu f_+,$$

so that

$$-J\frac{\partial}{\partial x}f_- + (\tilde{A} + \tilde{B}^2\mu^{-1})f_- - \mu f_- = 0, \tag{4.3}$$

if $\mu \neq 0$.

Now let $M(x), 0 \le x \le 2\pi$, be a *monodromy matrix* for the linear equation (4.3), i.e.

$$\frac{\partial}{\partial x}M + J(\tilde{A} + \tilde{B}^2\mu^{-1})M - \mu JM = 0, \qquad M(0) = E. \tag{4.4}$$

Since this is a traceless linear equation, then $\det M(x) = 1$ for every $x$. So a complex number $m$ is an eigenvalue of $M(2\pi)$ if

$$0 = \det\left(mE - M(2\pi)\right) = m^2 - 2\Delta m + 1, \qquad \Delta = \tfrac{1}{2}\mathrm{tr}\, M(2\pi).$$

The function $\Delta = \Delta(\mu; u, v)$ is a *discriminant* of the spectral problem we discuss.

A complex number $\mu \neq 0, \infty$ is a periodic (antiperiodic) eigenvalue of $\mathcal{L}$ if eqn (4.3) has a non-trivial $2\pi$-periodic (antiperiodic) solution. That is, if $m = 1$ $(m = -1)$ is an eigenvalue of the matrix $M(2\pi)$, or, equivalently, if $\Delta = 1$ (respectively $\Delta = -1$). Finally, $\mu \neq 0, \infty$ is a periodic/antiperiodic eigenvalue if

$$\Delta^2(\mu; u, v) = 1. \tag{4.5}$$

Since the $f_-$-components of the corresponding (vector) eigenfunctions $f = \begin{pmatrix} f_- \\ f_+ \end{pmatrix}$ satisfy (4.3), then the eigenfunctions are smooth.

This implicit description of the spectrum will provide us with the inverse spectral information which in the KdV case (Section 3.1) we extracted from the classical theory of the Sturm–Liouville operator. The 'discriminant approach' is general and applies to other integrable equations (including the KdV, see McKean and Trubowitz 1976).

The potentials $(u, v)$ we consider in this section are assumed to be bounded:

$$\|u\|_{C^1} + \|v\|_{C^0} \le C_\sharp,$$

where $C_\sharp$ is any fixed real constant. Moreover, for technical reasons we denote

$$\lambda = 16\mu^2$$

and below use the spectral parameter $\lambda$ as well as $\mu$.

Elementary analysis of eqn (4.4) (see McKean 1981) shows that the set of periodic/antiperiodic eigenvalues of $\mathcal{L}_{(u,v)}$ is invariant under the symmetry

$$\mu \mapsto -\mu\,; \tag{4.6$_1$}$$

if the potential $(u, v)$ is real, then under the complex conjugation

$$\mu \mapsto \bar{\mu}, \tag{4.6$_2$}$$

and if the potential is even or odd, then under the inversion

$$\lambda \mapsto \frac{1}{\lambda}\,. \tag{4.6$_3$}$$

The first symmetry explains the advantages of the $\mu$-coordinate compared with $\lambda$: using the former we factorize the symmetry (4.6$_1$).

To investigate the periodic/antiperiodic eigenvalues of the $\mathcal{L}$-operator, i.e. the roots of eqn (4.5), we first compute them for the zero potential $u = v = 0$. In this case eqn (4.4) simplifies to

$$\frac{\partial}{\partial x} M = \left(\mu - \frac{1}{16\mu}\right) J M, \quad M(0) = E.$$

So $M(x) = \exp((\mu - 1/16\mu)xJ)$ and $M(2\pi) = \pm E$ if $(\mu - 1/16\mu)$ is a half-integer number. This happens, if $\mu = \pm\mu_k^0$ for some $k$, where

$$\mu_k^0 = \frac{k + k^*}{4}, \quad k \in \mathbb{Z}, \tag{4.7}$$

and

$$k^* \equiv \sqrt{k^2 + 1}.$$

All these roots of (4.5) are real and double since for any $\pm\mu_k^0$ as in (4.7) both eigenvalues of the matrix $M(2\pi)$ equal $+1$ or $-1$. Corresponding eigenfunctions form the bases of the spaces of periodic and antiperiodic functions. In the $\lambda$-presentation the eigenvalues are $l_k$, $k \in \mathbb{Z}$, where

$$l_k = (4\mu_k^0)^2 = (k + k^*)^2.$$

We note that

$$l_k \cdot l_{-k} \equiv 1 \tag{4.8}$$

and

$$l_j = \begin{cases} 4j^2 + 1 + O(j^{-2}), & j \to \infty, \\ \frac{1}{4} j^{-2} + O(j^{-4}), & j \to -\infty. \end{cases} \tag{4.9}$$

The eigenvalues $l_j$ accumulate to infinity and to zero.

Now we discuss the periodic/antiperiodic spectrum of the operator $\mathcal{L}$, when the potential $\xi = (u, v)$ is small in the space

$$X = C^1(S^1) \times C^0(S^1), \qquad \|\xi\|_X = \|u\|_{C^1} + \|v\|_{C^0},$$

or in its complexification $X^c$. Applying the classical perturbation theory (see Kato 1966[25]) we get that for $\xi$ in $\mathcal{O}_{\delta_*}(X^c)$ ($\delta_* > 0$ is sufficiently small) the operator $\mathcal{L}_\xi$ has eigenvalues $\mu_j^\pm(\xi)$, $j \in \mathbb{Z}$, which are algebraic functions[26] of $\xi$ such that $\mu_j^\pm(\xi) \to \mu_j^0$ as $\xi \to 0$.

The eigenvalues $\mu_j^\pm$ extend to algebraic functions on the ball $\mathcal{O}_{C_\sharp}(X^c)$. To show this we note that since $\mathcal{L}_\xi$ is a bounded zero-order perturbation of the operator $\mathcal{L}_0$, then due to the asymptotic perturbation theory (Kato 1966), there exists a number $j_1(C_\sharp)$ such that for $|j| > j_1$ the eigenvalues $\mu_j^\pm$ and $\lambda_j^\pm$ are double-valued algebraic functions of $\xi \in \mathcal{O}_{C_\sharp}(X^c)$, different from other eigenvalues:

$$\lambda_j^\pm \neq \lambda_k^\pm \quad \text{if } \max(|j|, |k|) > j_1 \text{ and } \|\xi\|_{X^c} \le C_\sharp. \tag{4.10}$$

The eigenvalues $\mu_j^\pm$ and $\lambda_j^\pm$ are asymptotically close to $\mu_j^0$ and $l_j$, respectively. In particular,

$$\begin{cases} \lambda_j^\pm = l_j + O(j^{-2}) = 4j^2 + 1 + O(j^{-2}), & j \to \infty, \\ \lambda_j^\pm = l_j + O(j^{-4}) = \frac{1}{4} j^{-2} + O(j^{-4}), & j \to -\infty \end{cases} \tag{4.11}$$

(we use (4.9)).

Owing to (4.10), for $\xi \in \mathcal{O}_{C_\sharp}(X^c)$ the eigenvalues $\mu_j^\pm$ with $|j| \le j_1$ form a system of $2j_1 + 1$ solutions for eqn (4.5), isolated from the rest of the solutions. Since the discriminant $\Delta(\mu; \xi)$ is an analytic function, these eigenvalues form a $(2j_1 + 1)$-valued algebraic function.

Finally, we note that due to ($4.6_2$) the branches $\lambda_j^\pm$ form pairs such that for any real $\xi$ either both $\lambda_j^+(\xi)$ and $\lambda_j^-(\xi)$ are real, or these eigenvalues form a conjugation–invariant pair. It turns out (McKean 1981, p. 207) that the second alternative happens:

$$\lambda_j^+ = \overline{\lambda_j^-}, \quad \forall j \tag{4.12}$$

(maybe $\lambda_j^+ = \lambda_j^-$ is a double real eigenvalue). We enumerate the branches $\lambda_j^+$ and $\lambda_j^-$ in each pair in such a way that $\operatorname{Im} \lambda_j^+ \ge 0$ and $\operatorname{Im} \lambda_j^- \le 0$ for each $j$, if $\xi \in X$.

---

[25] The theory has to be applied to the spectral problem for $\mathcal{L}$, rewritten in the form (4.3).

[26] See Appendix 7 where we discuss the algebraic functions of infinite-dimensional arguments.

We have proved the following result:

**Lemma 4.1.** *The double eigenvalues* $\lambda = l_j$, $j \in \mathbb{Z}$, *of the periodic/antiperiodic spectral problem* $\mathcal{L}_0 f = \mu f$, *written in the* $\lambda$*-coordinate* $\lambda = 16\mu^2$, *extend to the algebraic functions* $\lambda_j^{\pm}(\xi)$, $\xi \in \mathcal{O}_{C_\sharp}(X^c)$, *which are periodic/antiperiodic eigenvalues of* $\mathcal{L}_\xi$. *These functions satisfy relations (4.10) as well as the asymptotics (4.11) and* $\lambda_j^{\pm}$ *is a double-value algebraic function if* $|j| \geq j_1$. *For real potentials the eigenvalues satisfy (4.12) and* $\mathrm{Im}\lambda_j^+ \geq 0$ *and* $\mathrm{Im}\lambda_j^- \leq 0$.

Since an algebraic function is analytic outside its branching points, we have the following corollary from Lemma 4.1:

**Corollary.** *If the potential analytically depends on a finite-dimensional parameter* $r$ *and for some* $j$ *we have* $\lambda_j^+(r) \equiv \lambda_j^-(r)$ *and* $\lambda_j^+(r) \neq \lambda_l^{\pm}(r)$ *if* $l \neq k$, *then* $\lambda_j^+$ *is an analytic function of* $r$.

## 4.2 Theta formulas

In complete analogy with the KdV case, a smooth $2\pi$-periodic vector-function $(u(x), v(x))$ is called a *g-gap potential* if the corresponding equation, eqn (4.5), has exactly $2g$ non-double solutions. (In particular, zero is a zero-gap potential.) Finite-gap periodic potentials form several distinct families with rather different properties (Dubrovin and Natanzon 1982). We are concerned with those potentials that can be deformed to zero. In view of Lemma 4.1 this means that we shall discuss families of finite-gap potentials $(u, v)$ such that for some $g$ and some set $\boldsymbol{\Upsilon} = \{\Upsilon_1, \dots, \Upsilon_g\} \subset \mathbb{Z}$, $\Upsilon_1 < \cdots < \Upsilon_g$, we have

$$\begin{cases} \lambda_j^+ = \lambda_j^-, & \text{if } j \in \mathbb{Z}_{\boldsymbol{\Upsilon}}, \\ \lambda_j^+ \neq \lambda_j^-, & \text{if } j \in \boldsymbol{\Upsilon}. \end{cases}$$

These potentials can be written in terms of theta functions, similar to the Its–Matveev formula (3.12). We discuss corresponding formulas below in this section.

All potentials that we consider are assumed to have sufficiently small complex parts. Moreover, to simplify presentation we decrease the family of potentials assuming that

$$\left|\lambda_{\Upsilon_j}^{\pm}\right| < \left|\lambda_{\Upsilon_{j+1}}^{\pm}\right|, \quad \text{for } j = 1, \dots, g-1.$$

The decreased family is assumed to contain all sufficiently small potentials from the original one (this assumption agrees with the last restriction since $\lambda_j^{\pm}(0) = l_j$). For potentials from this family, spiral segments $\gamma_{\Upsilon_j}$ which join $\lambda_{\Upsilon_j}^-$ with $\lambda_{\Upsilon_j}^+$, $j = 1, \dots, g$, do not intersect each other.[27] For the theory of finite-gap solutions of the SG equation which we present below, these segments play the same role as the open gaps $\Delta_{\nu_j}$ play in the KdV theory (cf. Section 3.2).

---

[27] For real potentials we have $|\lambda_j^+| \equiv |\lambda_j^-|$ by (4.12), so each $\gamma_{\Upsilon_j}$ is a segment of a circle.

Let $(u, v)$ be a $g$-gap potential as above and

$$E_{2j-1} = \lambda^{+}_{\Upsilon_j}, \qquad E_{2j} = \lambda^{-}_{\Upsilon_j}, \quad j = 1, \ldots, g,$$

be single eigenvalues of the operator $\mathcal{L}_{(u,v)}$ (abusing language we call a $\lambda$-eigenvalue single if the corresponding $\mu$-eigenvalue, $\mu = \sqrt{\lambda}/4$, is single). By our assumptions, $\operatorname{Im} E_{2j-1} > 0$, $\operatorname{Im} E_{2j} < 0$, and

$$|E_{2j-1}|, |E_{2j}| < |E_{2j+1}|, |E_{2j+2}|, \quad \text{for } j = 1, \ldots, g-1, \tag{4.13}$$

and

$$E_{2j-1} = \overline{E_{2j}}, \qquad E_{2j-1} \neq E_{2j}, \quad \forall j, \tag{4.14}$$

if the potential is real. We denote $\boldsymbol{E} = \{E_1, \ldots, E_g\}$ and view $\boldsymbol{E}$ both as a set and as the complex $g$-vector $(E_1, \ldots, E_g)$.

We restrict ourselves to a bounded part of the family as above and assume that

$$|E_j| < C, \quad \forall j, \tag{4.15}$$

in addition to (4.13). Since we consider potentials with small imaginary parts, then the corresponding vectors $\boldsymbol{E} \in \mathbb{C}^{2g}$ lie in a small neighbourhood of the real subspace, defined by (4.14).

Since the periodic/antiperiodic discrete spectrum of the operator $\mathcal{L}$ is invariant under the SG flow, then the set of $g$-gap potentials with a fixed single spectrum $\{E_1, \ldots, E_{2g}\}$ is flow-invariant as well.

Let $\Gamma = \{(\lambda, z)\}$ be a Riemann surface of genus $g > 0$, defined by the equation

$$z^2 = \lambda \prod_{j=1}^{2g} (\lambda - E_j).$$

We make the cut $\gamma_0 = [0, +\infty)$ and make cuts along the segments $\gamma_{\Upsilon_1}, \ldots, \gamma_{\Upsilon_g}$, defined above. After $\Gamma$ is cut, it falls into two sheets $\Gamma_+$ and $\Gamma_-$. We choose a canonical basis of cycles $(a_j, b_j)$ on $\Gamma$ $(j = 1, \ldots, g)$, so that the cycle $a_j$ goes around the cut $\gamma_{\Upsilon_j}$ (see Fig. 4.1) and the cycles have the canonical intersection matrix:

$$a_i \circ a_j = b_i \circ b_j = 0, \quad a_i \circ b_j = \delta_{ij}.$$

As in Section 3.2 we take a basis $d\omega_1, \ldots, d\omega_g$ of holomorphic differentials on $\Gamma$, normalized by the conditions

$$\langle d\omega_j, a_m \rangle = 2\pi i \delta_{jm}, \quad j, m = 1, \ldots, g.$$

We define the Riemann matrix $B$ as $B_{jk} = \langle d\omega_j, b_k \rangle$ for $j, k = 1, \ldots, g$, and the theta function as

$$\theta(z) = \sum_{s \in \mathbb{Z}^n} \exp\left(\tfrac{1}{2}(Bs, s) + (z, s)\right).$$

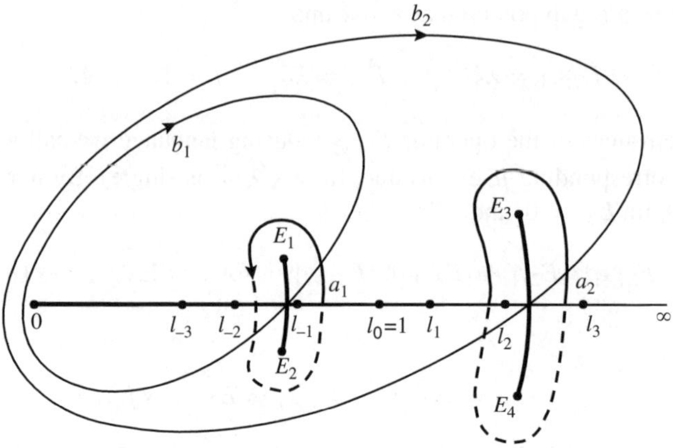


**Fig. 4.1:** The spectral curve with the canonical basis.

We consider the Abelian differentials $d\Omega_1$ and $d\Omega_2$ with zero $a$-periods and such that $d\Omega_1$ has the only pole at the infinity while $d\Omega_2$ has the only pole at zero:

$$\begin{aligned} d\Omega_1(P) &= d\big(\sqrt{\lambda} + \cdots\big), \quad P = (\lambda, z) \to \infty, \\ d\Omega_2(P) &= d\left(\frac{1}{\sqrt{\lambda}} + \cdots\right), \quad P \to 0. \end{aligned} \tag{4.16}$$

Denoting the $b$-periods of $d\Omega_1$ and $d\Omega_2$ as $\mathbf{B}^1$ and $\mathbf{B}^2$, i.e. $\mathbf{B}_j^{1,2} = \langle d\Omega_{1,2}, b_j\rangle$ , we define the wave-number vector $\boldsymbol{V}$ and the frequency vector $\boldsymbol{W}$ as follows:

$$\boldsymbol{V} = \frac{1}{4}(\mathbf{B}^1 - \mathbf{B}^2), \qquad \boldsymbol{W} = \frac{1}{4}(\mathbf{B}^1 + \mathbf{B}^2).$$

Arguments, similar to those used in Section 3.2, show that the vectors $\boldsymbol{V}$ and $\boldsymbol{W}$ are real, provided that (4.14) holds (see Bikbaev and Kuksin 1992, and Bobenko and Kuksin 1995*b*). Let us denote by $i\Delta = i(\pi, \dots, \pi)$ the vector of half-periods of the theta function. Finite-gap solutions of the SG equation with the single spectrum $(E_1, \dots, E_{2g})$ are given by the following theta formula:

$$u(t, x; \boldsymbol{E}, D) = 2i \log \frac{\theta(i(\boldsymbol{V}x + \boldsymbol{W}t + D + \Delta))}{\theta(i(\boldsymbol{V}x + \boldsymbol{W}t + D))}, \tag{4.17}$$

where $D \in \mathbb{T}^g = \mathbb{R}^g/2\pi\mathbb{Z}^g$ is a phase of the solution. On the contrary, for any $D \in \mathbb{T}^g$ and any vector $\boldsymbol{E} \in \mathbb{C}^g$ which satisfies (4.14), both the numerator and the denominator under the log sign in (4.17) do not vanish and the formula (4.17) defines a real solution for the SG equation, see Kozel and Kotlyarov (1976) and Belokolos *et al.* (1994).[28] This

[28] The assumption (4.13) is not needed for this statement to be true since for any vector $\boldsymbol{E}$ as above one can find paths $\gamma_j$ which join $E_{2j}$ with $E_{2j-1}$, are real in the sense that $\overline{\gamma_j} = \gamma_j$, and do not intersect each other. Using these paths instead of the spirals $\gamma_{\Upsilon_j}$ one also gets a real solution for the SG equation. The assumption (4.13) is imposed to choose the paths in a canonical way, continuous in $\boldsymbol{E}$; cf. Remark below.

solution is $2\pi$-periodic if and only if

$$V = V(E) \in \mathbb{Z}^g, \tag{4.18}$$

cf. Appendix 3. A set of all vectors $\boldsymbol{E}$ which satisfy (4.13)–(4.15) form a $2g$-dimensional domain. So a set of all vectors which meet (4.13)–(4.15), (4.18) form a $g$-dimensional[29] algebraic set. Hence, the set of $g$-gap potentials given by formulas (4.17) and (4.18) form a $2g$-dimensional invariant set for the SG equation as in Section 2.1.

*Remark.* Let $\mathcal{E}$ be any connected open bounded subset of the real linear space $\{\boldsymbol{E} \subset \mathbb{C}^g \mid \boldsymbol{E} \text{ satisfies (4.15)}\}$, which contains in its closure the vector $\boldsymbol{l}_\Upsilon = (l_{\Upsilon_1}, l_{\Upsilon_1}, \ldots, l_{\Upsilon_g}, l_{\Upsilon_g})$ and is formed by vectors which meet (4.14). Let us assume that a system of non-intersecting paths $\gamma_1, \ldots, \gamma_g$ can be constructed such that $\gamma_j = \overline{\gamma_j}$, $\gamma_j$ joins $E_{2j}$ with $E_{2j-1}$ $(j = 1, \ldots, g)$ and, first, the paths continuously depend on $\boldsymbol{E} \in \mathcal{E}$ and, second, each path $\gamma_j$ degenerates to the point $l_{\Upsilon_j}$ when $\boldsymbol{E} \to \boldsymbol{l}_\Upsilon$.

The set of finite-gap solutions (4.17) with $\boldsymbol{E} \in \mathcal{E}$ can be used in our constructions instead of the set with vectors $\boldsymbol{E}$ as in (4.13)–(4.15). Clearly, for any given real $g$-gap solution (4.17), corresponding to a vector $\boldsymbol{E}_0$ which satisfies (4.14) and (4.17), a set $\mathcal{E}$ as above can be constructed to contain $\boldsymbol{E}_0$.

## 4.3 Even periodic and odd periodic solutions

Now let us consider the SG equation under the even periodic or odd periodic boundary conditions:

$$u(x) \equiv u(x + 2\pi) \equiv u(-x), \tag{EP}$$
$$u(x) \equiv u(x + 2\pi) \equiv -u(-x). \tag{OP}$$

They imply correspondingly Neumann or Dirichlet boundary conditions on the half-period (see Example 2.3 in Section 2.1). If $(u, v)$ solves (4.1) and $u$ satisfies (EP) or (OP), then $v$ satisfies the same boundary condition in view of the first equation in (4.1).

Elementary arguments based on symmetries of the curve $\Gamma$ (see Bikbaev and Kuksin 1992 and Bobenko and Kuksin 1995*a*, *b*) distinguish among the finite-gap solutions (4.17) those that are even or odd:

**Lemma 4.2.** *The solution (4.17) is even if and only if the set $\boldsymbol{E}$ is symmetric with respect to the inversion $\lambda \mapsto \lambda^{-1}$ and the phase $D \in \mathbb{T}^g$ satisfies $TD = D$, where $T$ is the involution*

$$T(U_1, \ldots, U_g) = (U_g, \ldots, U_1).$$

*The solution is odd if and only if the set $\boldsymbol{E}$ is as above but*

$$TD = D + \Delta \tag{4.19}$$

*($\Delta$ is the same vector as in (4.17)). In both the even and odd cases we have*

$$TW = W, \qquad TV = -V. \tag{4.20}$$

[29] Equation (4.18) form a non-generate system, cf. Lemma 4.3 below.

Owing to the complete analogy between the (OP) and (EP) cases, in what follows we restrict ourselves to the (OP) boundary conditions (for the (EP) case see Bobenko and Kuksin 1995*a*, *b*). The cases of even and odd $g$ have to be treated separately but very similarly. For brevity we consider the even case only, so

$$g = 2n$$

everywhere below.

Comparing the lemma with (4.13) we get that for any even or odd real solution (4.17) the following relations hold:

$$E_{2j} \cdot E_{2(2n-j+1)} = E_{2j-1} \cdot E_{2(2n-j+1)-1} = 1, \quad \forall\, j = 1, \ldots, n. \tag{4.21}$$

By Lemma 4.1, for a small finite-gap potential the corresponding vector $\boldsymbol{E}$ is close to some vector $\boldsymbol{L}_{\Upsilon} = (l_{\Upsilon_1}, l_{\Upsilon_1}, \ldots, l_{\Upsilon_{2n}}, l_{\Upsilon_{2n}})$, where $l_{\Upsilon_i} < l_{\Upsilon_j}$ if $i < j$. If the potential is odd (or even), then we get from (4.21) that $l_{\Upsilon_j} l_{\Upsilon_{2n-j+1}} \equiv 1$, that is

$$T\boldsymbol{\Upsilon} = -\boldsymbol{\Upsilon}$$

(see (4.8)). Since the $l_{\Upsilon_j}$'s are distinct real numbers, $\Upsilon_j \neq 0$ for all $j$. Using (4.13) we get that

$$\Upsilon_1 < \cdots < \Upsilon_n < 0 < \Upsilon_{n+1} < \cdots < \Upsilon_{2n}.$$

Integer $n$-vectors $\boldsymbol{l} = (l_1, \ldots, l_n)$, where $l_j = \Upsilon_{n+j} \in \mathbb{N}$, numerate different families $\mathcal{T}_l^{2n}$ of odd periodic $2n$-gap solutions, contractible to the zero solution. To simplify presentation, we shall discuss only the family formed by finite-gap solutions such that all their first gaps are open. These solutions form the family $\mathcal{T}_l^{2n}$, where $\boldsymbol{l}$ is the vector

$$\boldsymbol{l} = (1, 2, \ldots, n). \tag{4.22}$$

This means that $\Upsilon_1 = -n, \ldots, \Upsilon_{2n} = n$ (and no $\Upsilon_j$ equals zero). We abbreviate this family to $\mathcal{T}^{2n}$.

The set $\mathcal{T}^{2n}$ is a subset of a linear space of potentials $(u(x), v(x))$. By Lemma 4.2 it is a union of $n$-tori, where the finite-gap solutions that fill any torus are parameterized by the reduced phase vector $\tilde{D}$:

$$\tilde{D} = (D_1, \ldots, D_n) \in \mathbb{T}^n$$

(other components of the vector $D$ can be recovered using (4.19)). Finite-gap tori which jointly form the set $\mathcal{T}^{2n}$ are parameterized by the vectors $\boldsymbol{E} \in R$, where

$$R = \{\boldsymbol{E} \in \mathcal{E}_0 \mid \boldsymbol{V}(\boldsymbol{E}) \in \mathbb{Z}^{2n}\},$$

and

$$\mathcal{E}_0 = \{\boldsymbol{E} \in \mathbb{C}^{4n} \mid \boldsymbol{E} \text{ satisfies (4.13)–(4.15) and (4.21)}\}.$$

Owing to (4.21), every $2n$-vector $\boldsymbol{E} \in \mathcal{E}_0$ is uniquely defined by the $n$-vector $\tilde{E} = \tilde{E}(\boldsymbol{E})$, formed by its last $n$ coordinates, and we shall view the set $\mathcal{E}_0$ as a subset of the complex space $\mathbb{C}^{2n}$, formed by the vectors $\tilde{E} = (E_{2n+1}, \dots, E_{4n})$, as well as the subset of $\mathbb{C}^{4n}$. The half-dimension real subspace $L_R \subset \mathbb{C}^{2n}$,

$$L_R = \{\tilde{E} \mid E_{2j-1} = \overline{E_{2j}}, \quad \forall\, j\},$$

is real, i.e. $L_R \cap iL_R = \{0\}$, since the space $iL_R$ is formed by vectors $\tilde{E}$ such that $E_{2j-1} \equiv -\overline{E_{2j}}$. Any vector $\xi \in \mathbb{C}^{2n}$ can be uniquely decomposed as a sum of its real part $\operatorname{Re}\xi \in L_R$ and imaginary part $\operatorname{Im}\xi \in iL_R$. Noting that $\mathcal{E}_0$ is a bounded domain in $L_R$, we define a domain $\Pi^c \subset \mathbb{C}^{2n}$ as

$$\Pi^c = \{\tilde{E} \mid \operatorname{Re}\tilde{E} \subset \mathcal{E}_0,\ |\operatorname{Im}\tilde{E}| < \delta\},$$

where $\delta > 0$ is sufficiently small. Then $\mathcal{E}_0$ is a real part of the complex domain $\Pi^c$, $\mathcal{E}_0 = \Pi^c \cap L_R$, and $R$ is a real part of the corresponding complex analytic set $R^c \subset \Pi^c$.

Let us denote by $\tilde{V}$ and $\tilde{W}$ the vectors formed by the last $n$ components of the vectors $\boldsymbol{V}$ and $\boldsymbol{W}$, respectively. Owing to (4.20), $\boldsymbol{V}(\boldsymbol{E})$ is an integer vector if and only if $\tilde{V}(\boldsymbol{E})$ is too. In particular, the set $R$ is formed by the vectors $\boldsymbol{E} \in \mathcal{E}_0$ such that

$$\tilde{V}(\boldsymbol{E}) \in \mathbb{Z}^n. \tag{4.23}$$

Elements of the set $R$ will be denoted $r$. We treat $R$ as a subset of $\mathbb{C}^{4n} = \{\boldsymbol{E}\}$, or as a subset of $\mathbb{C}^{2n} = \{\tilde{E}\}$.

**Lemma 4.3.** *The set $\mathcal{E}_0 \subset \mathbb{C}^{4n}$ contains in its closure the vector*

$$\boldsymbol{L} = (l_{-n}, l_{-n}, \dots, l_{-1}, l_{-1}, l_1, l_1, \dots, l_n, l_n).$$

*For $r$ sufficiently close to $\boldsymbol{L}$ the map*

$$\mathcal{E}_0 \to \mathbb{R}^{2n}, \quad r \mapsto (\tilde{V}, \tilde{W}), \tag{4.24}$$

*is non-degenerate and $\tilde{V}(r) \equiv \boldsymbol{l} = (1, \dots, n)$.*

The theory of 'small-gap' finite-gap solutions (4.17), i.e. of solutions corresponding to vectors $\boldsymbol{E}$ close to $\boldsymbol{L}$, is very similar to the KdV theory. In particular, replacing for convenience the cuts $\gamma_j$ as they were defined above by the segments $[\lambda_j^-, \lambda_j^+]$, one can use elementary perturbation theory to prove that the differentials $d\Omega_1$ and $d\Omega_2$ are $d\Omega_1 = P_1(\lambda)\mu^{-1}\,d\lambda$ and $d\Omega_2 = P_2(\lambda)\mu^{-1}\,d\lambda$, where $P_1$ and $P_2$ stand for real polynomials of degree $g = 2n$ (cf. Appendices 4 and 5, where similar arguments are used). After this the proof of the non-degeneracy lemma, given in Appendix 6 for the KdV case, applies to the map (4.24) with minor modifications. The relation $\tilde{V} \equiv \boldsymbol{l}$ follows from (4.23) and a small-gap limit for the vector $\tilde{V}(r)$, cf. (A4.2).

For another proof of the lemma, based on direct calculations, see Bobenko and Kuksin (1995*b*).

By the lemma, the system of equations (4.23) has full rank, so the set $R$ is an $n$-dimensional analytic set, smooth near the point $\boldsymbol{L}$. It is unknown if the set $R$ is connected or not. We bypass this subtlety and *replace the set $R$ as it is defined above* *by its connected component* which contains $\boldsymbol{L}$ in its closure. Comparing (4.23) with the last assertion of Lemma 4.3 we see that

$$\tilde{V} \equiv \boldsymbol{l} \text{ in } R.$$

From now on we shall study the SG equation in the $(u, w)$-variables. Accordingly, it takes the form

$$\dot{u} = -\sqrt{A}w, \qquad \dot{w} = \sqrt{A}(u + A^{-1}(\sin u - u)),$$

where $A = -\partial^2/\partial x^2 + 1$, see (2.4). This is a Hamiltonian equation in the symplectic Hilbert scale $(\{Z_s^o\}, \beta_2)$. We recall that the space $Z_s^o$ is a subspace formed by odd periodic vector-functions from the Sobolev space $H^{s+1}(S) \times H^{s+1}(S)$ and that $\beta_2 = \langle \bar{J}(du, dw), (du, dw)\rangle$, where $J(u, w) = (-\sqrt{A}\, w, \sqrt{A}\, u)$ and $\langle\cdot,\cdot\rangle$ signifies the $H^1$-scalar product. Below we consider spaces $Z_s^0$ with $s \geq 0$.

For $r \in R$ let us denote by $\Phi_0(r, \tilde{D})(x)$ the vector-function $(u(x), A^{-1/2}\dot{u}(x))$, where $u(x)$ is the r.h.s. of (4.17) and $\dot{u}(x)$ is its time derivative, calculated for $t = 0$. Now we write the finite-gap solutions, forming the set $\mathcal{T}^{2n}$, as

$$(u, w) = \Phi_0(r, \tilde{D} + \tilde{W}(r)t)(x).$$

The theta map $\Phi_0$ provides global parameterization of $\mathcal{T}^{2n}$:

$$\mathcal{T}^{2n} = \Phi_0(R \times \mathbb{T}^n).$$

This formula shows that $\mathcal{T}^{2n}$ is a union of invariant finite-gap $n$-tori:

$$\mathcal{T}^{2n} = \bigcup_{r \in R} T^n(r), \qquad T^n(r) = \Phi_0(\{r\} \times \mathbb{T}^n).$$

## 4.4 Local structure of finite-gap manifolds

When $r \to \boldsymbol{L}$, the theta function $\theta(z, r)$ converges to 1 (cf. Appendix 5) and the finite-gap solution (4.17) converges to zero. That is, $\Phi_0(r, \tilde{D}) \to 0$ as $r \to \boldsymbol{L}$, for any $\tilde{D}$. By Lemma 4.3, a sufficiently small neighbourhood $R_0$ of $\boldsymbol{L}$ in $R$ is an analytic $n$-manifold. A corresponding part of the set $\mathcal{T}^{2n}$ also is smooth, as well as its closure:

**Lemma 4.4.** *If $\delta > 0$ is sufficiently small, then the set $\mathcal{T}_\delta^{\leq 2n} = \overline{\mathcal{T}^{2n}} \cap \mathcal{O}_\delta(Z_s^o)$, $s \geq 0$, is a $2n$-dimensional analytic submanifold of $Z_s^o$. It passes through the origin and its tangent space there is spanned by the vectors* $(\sin kx, 0)$ *and* $(0, \sin kx)$, $k = 1, \dots, n$. *For any $k \leq n - 1$ and any subset $\{l_1, \dots, l_k\} \subset \{1, \dots, n\}$, a closure of the manifold $\mathcal{T}_{(l_1,\dots,l_k)}^{2k} \cap \mathcal{O}_\delta(Z_s^o)$ is an analytic submanifold of $\mathcal{T}_\delta^{\leq 2n}$.*

In Bobenko and Kuksin (1995*b*) this result is proved by hard direct calculations. In the next section we present another proof, based on the same ideas as in the KdV case (cf. Theorem 3.1′). For the SG case the corresponding arguments are more involved since now the $\mathcal{L}$-operator is not self-adjoint.[30]

[30] This complexifies the proof because for a non-symmetric real $2 \times 2$-matrix there is no linear criterion to check if the matrix has a double eigenvalue, while for a symmetric matrix a criterion exists: the matrix has a double eigenvalue if and only if its deviator vanishes.

Owing to the lemma, a 'small-gap' part $\mathcal{T}_\delta^{2n}$ of a finite-gap set $\mathcal{T}^{2n}$ is smooth. In striking difference with the KdV case, we cannot prove that the whole set $\mathcal{T}^{2n}$ is smooth.[31] Still abusing language we call the sets $\mathcal{T}^{2n}$ finite-gap manifolds.

The finite-gap manifolds $\mathcal{T}^{2n}$ and the corresponding maps $\Phi_0$ satisfy assumptions 1–4 from Section 2.2. Indeed, to prove 1) (analyticity) we recall that $R$ is the real part of the algebraic set $R^c \subset \Pi^c \subset \mathbb{C}^{2n}$. For any vector $\tilde{E} = (E_{2n+1}, \dots, E_{4n}) \subset \Pi^c$ we take the vector $\boldsymbol{E} = (E_1, \dots, E_{4n})$ such that $\tilde{E}(\boldsymbol{E}) = \tilde{E}$ and $E_{2j}E_{2(2n-j+1)} \equiv 1$. The constructions of Section 4.2 assign to this vector $\boldsymbol{E}$ and any point $\tilde{D} \in \mathbb{T}^n$ a complex SG solution $u(t, x)$, given by the formula (4.17), where $D \in \mathbb{T}^{2n}$ satisfies (4.19) and $\tilde{D}$ is the vector formed by its last $n$ coordinates. By Lemma 4.3 the solution $u$ is odd. So denoting $\Psi(\tilde{E}, \tilde{D}) = (u(0, x), \dot{u}(0, x))\,|_{x\in[-\pi,\pi]}$ we get an analytic map $\Psi : \Pi^c \times \{|\mathrm{Im}\,\tilde{D}| < \delta\} \to H_o^s$, where $\delta > 0$ is sufficiently small, $s$ is any integer, and $H_o^s$ stands for a subspace of the Sobolev space $H^s = H^s([-\pi, \pi]; \mathbb{C}^2)$, formed by odd vector-functions. Let $H_{op}^s \subset H^s$ be the subspace, formed by odd periodic functions, and $\pi : H_o^s \to H_{op}^s$ be the corresponding orthogonal projection. The map

$$\Psi_0 = \pi \circ \Psi : \Pi^c \times \{|\mathrm{Im}\,\tilde{D}| < \delta\} \longrightarrow H_{op}^s$$

is analytic and for $\tilde{E} \in R$ it coincides with the map $\Phi_0$, written in terms of the $(u, v)$-variables (rather then $(u, w)$). Hence, for any $s$ the map $\Phi_0$ analytically extends to a map $\Pi^c \times \{|\mathrm{Im}\,\tilde{D}| < \delta\} \longrightarrow Z_s$.

The property 2) holds since by Lemma 4.4 the form $\Phi_0^*\beta_2$ is non-degenerate for $r$ close to $\boldsymbol{L}$; 3) follows from the analyticity of the map $\Phi_0$ and from the formula (4.17). Finally, 4) results from Lemma 4.3.

Owing to Lemma 4.4 we can continue to study small-gap solutions of the SG equation in the same way as in Sections 3.1 and 3.3 we study the KdV equation: since the SG equation has infinitely many integrals of motion (see McKean 1981 and Faddeev and Takhtajan 1986), then due to Vey's theorem the equation restricted to the manifold $\mathcal{T}_\delta^{\le 2n}$ admits analytic at zero Birkhoff coordinates $y_1, \dots, y_{2n}$, where $y = 0$ corresponds to $r = \boldsymbol{L}$, i.e. to the zero solution (cf. Theorem 3.2 and see Appendix 1 in Bobenko and Kuksin (1995a) for another proof of this normal form result). The radii $\mathcal{R}_1, \dots, \mathcal{R}_n$, where

$$\mathcal{R}_j = \sqrt{y_{2j-1}^2 + y_{2j}^2},$$

form a coordinate system on the manifold $R_0$ (which is a small neighbourhood of $\boldsymbol{L}$ in $R$); the form $\beta_2$ restricted to $\mathcal{T}_\delta^{\le 2n}$ equals $\frac{1}{2}\sum d\mathcal{R}_j^2 \wedge dq_j$, where the $q_j$'s are corresponding angles,[32] and the frequency vector $\tilde{W}$ is an analytic at zero vector-function of the actions $I_j = \mathcal{R}_j^2/2$. Repeating arguments from Section 3.3, the coefficients of decomposition of the vector $\tilde{W}$ to series in $I_1, \dots, I_n$ can be calculated. In particular,

$$\tilde{W}_j(0) = \sqrt{j^2 + 1} =: j^* \quad j = 1, \dots, n, \tag{4.25}$$

---

[31] Simply because it is non-smooth. We do not wish here to touch upon the difficult problem of the structure of its singularities.

[32] That is, $q_j = \arg\,(y_{2j-1} + iy_{2j})$.

and linear part of the decomposition is given by the following relations:

$$\left.\frac{\partial W_j}{\partial I_k}\right|_{I=0} = \begin{cases} -16/j^*, & j \neq k, \\ -12/j^*, & j = k. \end{cases} \tag{4.26}$$

We recall that all the first gaps are assumed to be open, see (4.22). Relations similar to (4.25) and (4.26) hold for any finite-gap manifold $\mathcal{T}_I^{2n}$. In particular,

$$W_{n+j}(0) = \tilde{W}_j(0) = l_j^* = \Upsilon_{n+j}^*,$$

where $\boldsymbol{W}$ is a frequency vector, corresponding to this manifold.

Proofs of the asymptotics (4.25), (4.26), based on direct calculations, can be found in Bobenko and Kuksin (1995*b*).

## 4.5 Proof of Lemma 4.4

In this section we abbreviate $\mathcal{O}_\delta(Z_s^o)$ to $\mathcal{O}_\delta$.

Since the $\lambda$-spectrum of the $\mathcal{L}$-operator with an odd periodic potential $(u, w)$ is inversion-invariant (see $(4.6_3)$) and continuously depends on the potential, then to prove that a small odd periodic potential belongs to the finite-gap manifold $\mathcal{T}^{2n}$ we only have to check that $\lambda_j^+(u, w) \neq \lambda_j^-(u, w)$ if $1 \leq j \leq n$ and $\lambda_j^+ = \lambda_j^-$ if $j \geq n+1$.

Let us denote by $L_{(u,w)} = L_{(u,w)}^\mu$ the operator in the l.h.s. of (4.3). Abusing language we say that $\mu_0 \neq 0$ is its eigenvalue if the operator $L_{(u,w)}^{\mu_0}$ has a non-trivial kernel. We have checked (see (4.7)) that the set of $4\pi$-periodic eigenvalues of the operator $L_0$ equals the set of its $2\pi$-periodic/antiperiodic eigenvalues and is $\{\pm\mu_k^0 \mid k \in \mathbb{Z}\}$. Every eigenvalue is double and for any $k \geq 1$[33] eigenvectors, corresponding to the eigenvalue $\mu_k^0$, are $\xi_k^1$ and $\xi_k^2$, where

$$\xi_k^1 = \begin{pmatrix} \sin\dfrac{k}{2}x \\ \cos\dfrac{k}{2}x \end{pmatrix}, \qquad \xi_k^2 = \begin{pmatrix} \cos\dfrac{k}{2}x \\ -\sin\dfrac{k}{2}x \end{pmatrix}.$$

Returning to the operator $\mathcal{L}_0$ we find that its eigenvectors $\Xi_k^1$ and $\Xi_k^2$, are

$$\Xi_{k0}^j = c_k \begin{pmatrix} E \\ 1 \\ \dfrac{1}{\mu_k}\tilde{B} \end{pmatrix} \xi_k^j = c_k \begin{pmatrix} E \\ 1 \\ \dfrac{1}{4\mu_k}E \end{pmatrix} \xi_k^j, \quad j = 1, 2.$$

Here $c_k = \sqrt{\mu_k^0(\pi(4\mu_k^0+1))^{-1}}$ is the normalizing factor, so the vectors have unit norm in the space $L_2 = L_2(\mathbb{R}/4\pi\mathbb{Z};\ \mathbb{C}^k)$.

By Lemma 4.1, for a small potential $(u, w)$ the operator $\mathcal{L}_{(u,w)}$ has two eigenvalues, close to $\mu_k^0$. The corresponding invariant plane $\Pi_k = \Pi_k(u, w) \subset L_2$ is close to the plane

[33] Below we do not use the eigenvalues $-\mu_k^0$ and eigenvalue $\mu_k^0$ with $k \leq 0$.

$\Pi_k^0$, spanned by the vectors $\Xi_{k0}^{1,2}$.[34] The plane $\Pi_k$ analytically depends on the potential $(u, w)$ and is $O(\delta^2)$-close to $\Pi_k^0$ if $\|(u, w)\| = \delta$ (for the same reasons as in the KdV case; cf. the proof of Theorem 3.1′). It has an $L_2$-orthonormal basis $\Xi_k^{1,2}(u, v)$, equal to $\Xi_{k0}^{1,2}$ for $(u, w) = (0, 0)$, continuous in $(u, w)$ and uniquely defined by the following normalization: the vector $\Xi_k^1$ is a vector in $\Pi_k$ which is the closest to the subspace of $L_2$, formed by vector-functions such that their first components are odd functions of $x$.

For $k = 1, 2, \ldots$ let us denote by $M_k(u, w)$ a matrix of the operator $\mathcal{L}\mid_{\Pi_k}$ with respect to the basis $\Xi_k^{1,2}$, and denote by $M_k^D$ the deviator $M_k^D = M_k - \frac{1}{2}(\operatorname{tr} M_k)E$. We consider its matrix elements $(M_k^D)^{ij}$ and abbreviate

$$(M_k^D)^{11} = M_k^1, \qquad (M_k^D)^{12} = M_k^2.$$

Clearly, $M_k^D$ has zero eigenvalues and $M_k$ has a double eigenvalue if $M_k^1 = M_k^2 = 0$.[35] Therefore, $\mathcal{T}^{2n}$ contains the set $\Theta \setminus \Theta_0$, where

$$\Theta = \{(u, w) \in \mathcal{O}_\delta \mid M_k^1 = M_k^2 = 0, \quad \forall k \geq n+1\}$$

and

$$\Theta_0 = \{(u, w) \in \Theta \mid \lambda_j^+ = \lambda_j^-, \quad \text{for some } 1 \leq j \leq n\}.$$

**Lemma 4.5.** *There exists a diffeomorphism $F = F^n : \mathcal{O}_\delta \longrightarrow Z_s^o$ such that*

$$F(0) = 0, \qquad F_*(0) = id, \tag{4.27}$$

*and $\mathcal{O}_{\delta_1}(L) \subset F(\Theta) \subset L$ for some $\delta_1 > 0$, where $L$ is the $2n$-dimensional linear subspace of $Z_s^o$, spanned by the vectors $(\sin jx, 0)$ and $(0, \sin jx)$ with $j = 1, \ldots, n$. Moreover, the set $F(\Theta_0)$ is a closed nowhere dense subset of $F(\Theta)$.*

We prove the lemma at the end of this section. Now we show how this result implies Lemma 4.4. Decreasing the manifold $R_0$ we achieve that $\Phi_0(R_0 \times \mathbb{T}^n) \subset \mathcal{O}_\delta$. Now let us consider the composition

$$G = F \circ \Phi_0 : W_0 = R_0 \times \mathbb{T}^n \to Z_s^o,$$

where $F$ is the map from Lemma 4.5. Since $\mathcal{T}^{2n}$ contains the set $\Theta \setminus \Theta_0$, then range of $G$ contains a domain in the space $L$; we denote it $Q$. We claim that

$$G(W_0) \subset L. \tag{4.28}$$

To prove this assertion we take any system $\psi_1, \psi_2, \ldots$ of vectors in $Z_s^o$ which form an orthogonal complement to $L$ in the Hilbert space $Z_s^o$. We consider all vectors $\psi_j$ such that $\langle G, \psi_j \rangle \not\equiv 0$. If this set of vectors is empty, then (4.28) is proved. Otherwise let us take any vector $\psi_j$ as above and consider the set $K = \{w \in W_0 \mid \langle G(w), \psi_j \rangle = 0\}$.

---

[34] A spectral projector on the plane $\Pi_k$ can be written as a contour integral of a resolvent of the operator $\mathcal{L}$. The resolvent can be expressed in terms of the operator $(L^\mu_{(u,w)})^{-1}$, so it is well defined.

[35] Since $M_k^D$ has zero eigenvalues if and only if its determinant vanishes.

This is a proper analytic subset of $W_0$, so $\operatorname{mes} K = 0$, where $\operatorname{mes} = \operatorname{mes}_{2n}$ stands for the $2n$-dimensional Lebesgue measure. Let us denote by $\Pi$ the orthogonal projection $Z_s^o \longrightarrow L$. Then $Q \subset \Pi \circ G(K)$. The map $\Pi \circ G$ is a Lipschitz mapping of the $2n$-manifold $W_0$ to the $2n$-dimensional space $L$, so it sends zero-measure subsets of $W_0$ to zero-measure subsets of $L$.[36] Hence, $\operatorname{mes} \Pi \circ G(K) = 0$ and $\operatorname{mes} Q = 0$. This contradiction shows that the set of vector $\psi_j$ defined above is empty and (4.28) follows.

We have proved that $F(\mathcal{T}^{2n}) \subset L$. Since $\mathcal{T}^{2n} \supset \Theta \setminus \Theta_0$, then $F(\mathcal{T}^{2n}) \supset F(\Theta) \setminus F(\Theta_0)$ and the closure $\overline{F(\mathcal{T}^{2n})} = F(\overline{\mathcal{T}^{2n}})$ contains the ball $\mathcal{O}_{\delta_1}(L^{2n})$ as in Lemma 4.5 because the set $F(\Theta_0)$ is nowhere dense. That is,

$$\mathcal{O}_{\delta_1}(L) \subset F(\overline{\mathcal{T}^{2n}}) \subset L, \tag{4.29}$$

and the first assertion of the lemma is proven for some new sufficiently small $\delta > 0$. Owing to (4.27) and (4.29), $T_0\overline{\mathcal{T}^{2n}} = F_*(0)^{-1}L^{2n} = L^{2n}$, so the assertion conserning the tangent space follows. To prove the last claim of the lemma we note that $F\left(\mathcal{T}^{2k}_{(l_1,\dots,l_k)} \cap \mathcal{O}_\delta\right)$ is a neighbourhood of the origin in the space $L^{2k}$ which is the subspace of $L$, spanned by the vectors $(\sin l_j x,\ 0)$ and $(0,\ \sin l_j x)$ $(j = 1, \dots, k)$. This follows from (4.29), where the manifold $\mathcal{T}^{2n}$ is replaced by $\mathcal{T}^{2k} = \mathcal{T}^{2k}_{(l_1,\dots,l_k)}$, since the map $F$ restricted to $\mathcal{T}^{2k}$ is exactly the corresponding map $F^k$ for the finite-gap manifold $\mathcal{T}^{2k}$; see construction of $F$ in the proof of Lemma 4.5. □

*Proof of Lemma 4.5.* For $s \geq 0$ let us define a space $\mathfrak{H}^s$ as the set of all sequences $m = (m_1^1, m_1^2, m_2^1, m_2^2, \dots)$ with finite norm $\|m\|_s^2 = \sum j^{2s}\left((m_j^1)^2 + (m_j^2)^2\right)$. Then for any $s \geq 1$ the map

$$\boldsymbol{M}^D : \mathcal{O}_\delta = \mathcal{O}_\delta(Z_s^o) \longrightarrow \mathfrak{H}^s, \qquad \xi = (u, w) \longmapsto (M_1^1(\xi), M_1^2(\xi), M_2^1(\xi), \dots)$$

is well defined and analytic. To calculate the linearized map $\boldsymbol{M}_*^D(0)$, for any $\xi \in (u, w) \in Z_s^o$,

$$u = \frac{1}{\sqrt{\pi}} \sum u_k \sin kx, \qquad w = \frac{1}{\sqrt{\pi}} \sum w_k \sin kx, \tag{4.30}$$

we have to calculate $\frac{d}{d\varepsilon}\boldsymbol{M}^D(\varepsilon\xi)\mid_{\varepsilon=0}$. To do this we argue as in the proof of Theorem 3.1′: since the basis vectors $\Xi_k^1(\varepsilon\xi)$ and $\Xi_k^2(\varepsilon\xi)$ are such that $\Xi_k^{1,2}(\varepsilon\xi) = \Xi_{k0}^{1,2} + O(\varepsilon^2)$, then to calculate the matrix elements of the operator $\mathcal{L}_{\varepsilon\xi} \mid_{\Pi_k(\varepsilon\xi)}$ up to terms $O(\varepsilon^2)$ we can replace the basis $\Xi_k^{1,2}(\varepsilon\xi)$ by $\Xi_{k0}^{1,2}$. Accordingly, denoting the matrix elements by $M_k^{ij}(\varepsilon)$,

[36] This follows, for example, from (A9.2) in Appendix 9, since the $2n$-dimensional Hausdorff measures in $W$ and $L$ are equivalent to the Lebesgue measures; see Falconer (1990) and Federer (1969).

$1 \leq i, j \leq 2$, we have

$$M_k^{11}(\varepsilon) = \int_0^{4\pi} (\mathcal{L}_{\varepsilon\xi}\, \Xi_{k0}^1(x),\, \Xi_{k0}^1(x))\, dx + O(\varepsilon^2)$$

$$= c_k^2 \int_0^{4\pi} \left( \begin{pmatrix} E & \frac{1}{4\mu_k^0} E \end{pmatrix} \cdot \mathcal{L}_{\varepsilon\xi} \cdot \begin{pmatrix} E \\ \frac{1}{4\mu_k^0} E \end{pmatrix} \xi_k^1(x),\, \xi_k^1(x) \right) dx + O(\varepsilon^2).$$

Denoting by $\tilde{A}$ and $\tilde{B}$ the matrices as in (4.2) with the potential $(u, v)$ replaced by $(\varepsilon u, \varepsilon v) = (\varepsilon u, \varepsilon(-\partial^2/\partial x^2 + 1)^{-1/2} w)$, where $u$ and $w$ are defined in (4.30), we calculate the product of the three matrices under the integral sign in the r.h.s. of the last equality. Denoting by const the different $\varepsilon$-independent matrices, we get that the product equals

$$\tilde{A} + \frac{1}{2\mu_k^0}\tilde{B} + \text{const} = \frac{i\varepsilon}{4}(v + u_x') \begin{pmatrix} 0 & 1 \\ 1 & 0 \end{pmatrix}$$

$$+ \frac{i\varepsilon u}{16\mu_k^0} \begin{pmatrix} 1 & 0 \\ 0 & -1 \end{pmatrix} + \text{const} + O(\varepsilon^2).$$

Therefore,

$$M_k^{11}(\varepsilon) = \frac{i\varepsilon c_k^2}{4} \int_0^{4\pi} (v + u_x') 2 \sin \frac{k}{2}x \cos \frac{k}{2}x\, dx$$

$$+ \frac{i\varepsilon c_k^2}{16\mu_k^0} \int_0^{4\pi} u \left( \sin^2 \frac{k}{2}x - \cos^2 \frac{k}{2}x \right) dx + O(\varepsilon^2) + \text{const}$$

$$= \frac{i\varepsilon c_k^2}{2} \int_0^{2\pi} (v + u_x') \sin kx\, dx - \frac{i\varepsilon c_k^2}{8\mu_k^0} \int_0^{2\pi} u \cos kx\, dx + O(\varepsilon^2) + \text{const}.$$

Since $v = (-\partial^2/\partial x^2 + 1)^{1/2} w = (1/\sqrt{\pi}) \sum k^* w_k \sin kx$, then

$$M_k^{11}(\varepsilon) = \frac{i\varepsilon c_k^2 \sqrt{\pi}}{2} k^* w_k + O(\varepsilon^2) + \text{const}.$$

Similar calculations show that

$$M_k^{22}(\varepsilon) = \frac{-i\varepsilon c_k^2 \sqrt{\pi}}{2} k^* w_k + O(\varepsilon^2) + \text{const},$$

$$M_k^{12}(\varepsilon) = \frac{i\varepsilon c_k^2 \sqrt{\pi}}{2} \left( k + \frac{1}{4\mu_k} \right) u_k + O(\varepsilon^2) + \text{const},$$

and $M_k^{21}(\varepsilon)$ equals $M_k^{12}(\varepsilon)$ up to const $+O(\varepsilon^2)$. The deviator $M_k^D$ equals the matrix $M_k$ up to $O(\varepsilon^2)+$ const; hence,

$$\boldsymbol{M}_*^D(0)(u,v) = \frac{i\sqrt{\pi}}{2}(m_1^1, m_1^2, m_2^1, \dots),$$

where

$$m_k^1 = c_k^2 k^* w_k, \qquad m_k^2 = c_k^2\Big(k + \frac{1}{4\mu_k}\Big)u_k, \quad k = 1, 2, \dots.$$

Since $\|(u,v)\|_s^2 = \sum_k \left(1+|k|^{2s+2}\right)(u_k^2 + w_k^2)$, then the map $\boldsymbol{M}_*^D(0)$ defines an isomorphism between $Z_s^o$ and $\mathfrak{H}^s$. Now we define an analytic map $F$ as $F = \boldsymbol{M}_*^D(O)^{-1} \circ \boldsymbol{M}^D$. Then $F$ satisfies (4.27), so by the inverse function theorem $F$ defines a diffeomorphism $\mathcal{O}_\delta \to Z_s^o$. By the construction of this map, $\Theta = F^{-1}(L)$, so $\Theta$ is a $2n$-manifold and $F$ satisfies the first assertion of the lemma.

To prove the last assertion we note that a point $(u,w) \in \Theta$ belongs to $\Theta_0$ if and only if for some $k \le n$ the $2\times 2$-matrix $M_k(u,v)$ has a double eigenvalue. This happens if and only if $\prod_{k=1}^n \det M_k^D = 0$. Calculations of the matrix elements of $M_k(u,v)$, presented above, show that

$$M_k^D(u,v) = M_k^{0D} + \begin{pmatrix} m_k^1 & m_k^2 \\ m_k^2 & -m_k^1 \end{pmatrix} + O(|u,v|^2).$$

Since the vector $(m_1^1, m_1^2, \dots, m_n^2) = m$ forms a cordinate system on $\Theta$ and $O(|u,v|^2) = O(|m|^2)$, then the analytic functions $\det M_k^D$, $1 \le k \le n$, do not vanish identically, as well as their product. Hence, $\Theta_0$ is a proper analytic subset of $\Theta$ and the lemma is proved. □

## Appendix 7. On the algebraic functions of infinite-dimensional arguments

Let $X$ and $X^c$ be a Banach space and its complexification, respectively, and let $O^c$ signify a connected domain in $X^c$. For some $n \ge 1$, let $f_1, \dots, f_n$ be complex functions on $O^c$ such that the set-valued map $x \mapsto \boldsymbol{f} = \{f_1(x), \dots, f_n(x)\}$ is continuous on $O$.

The set $\boldsymbol{f}$ of functions is called an *algebraic function* if for any $m$, any connected complex domain $Q \subset \mathbb{C}^m$, and any analytic map $F : Q \longrightarrow X^c$, the set of functions $\boldsymbol{f} \circ F$ is an algebraic function on $Q$ (for the classical definition of an algebraic function of a finite-dimensional argument see Bochner and Martin (1948) or Definition 5.1 below).

The functions $f_1, \dots, f_n$ are called 'branches of the algebraic function $\boldsymbol{f}$'. Abusing language we also call them algebraic functions.

In non-trivial cases the branches $f_j$ are discontinuous functions[37] and to study them, their sets of discontinuity have to be specified. In this book we are mostly concerned with functions of real arguments and with algebraic functions which are analytic extensions of some continuous functions of real arguments. Accordingly, the branches of analytic functions we consider are continuous on real domains $O^c \cap X$.

[37] Otherwise, by the criterion of analyticity, each $f_j$ is an analytic function.

# 5
# Linearized equations and their Floquet solutions

## 5.1 The linearized equation

Below $(Z, \alpha_2)$ stands for a symplectic space ($Z = Z_d$, $\alpha_2 = \overline{J}\, dz \wedge dz$) with some fixed $d$, as in Chapter 2. We continue to study a quasilinear Hamiltonian equation

$$\dot{u} = J\nabla\mathcal{H}(u) = J(Au + \nabla H(u)) =: V_{\mathcal{H}}(u), \tag{5.1}$$

where $\operatorname{ord} A = d_A$, $\operatorname{ord} \nabla H = d_H < d_A$, $\operatorname{ord} J = d_J$, and $d \geq d_A/2$. The equation is assumed to possess a $2n$-dimensional invariant manifold

$$\mathcal{T}^{2n} \subset Z, \qquad \mathcal{T}^{2n} = \Phi_0(R \times \mathbb{T}^n),$$

with the regular part $\mathcal{T}_0^{2n} = \Phi_0(R_0 \times \mathbb{T}^n)$. We recall (see Chapter 2) that $R$ is a connected $n$-dimensional analytic set which is the real part of a connected complex analytic subset $R^c$ of complex domain $\Pi^c \subset \mathbb{C}^N$, $R_s$ is a proper analytic subset of $R$ which contains its singularities, and $R_0 = R \setminus R_s$. The invariant manifold $\mathcal{T}_0^{2n}$ is analytic and eqn (5.1) defines on $\mathcal{T}_0^{2n}$ a non-degenerate integrable system. Moreover, assumptions 1–4 from Section 2.2 in Chapter 2 have to be satisfied. For convenience we repeat them here:

1. For any $l$, the map $\Phi_0$ extends to an analytic map $\Pi^c \times \{|\operatorname{Im} \mathfrak{z}| < \delta\} \mapsto Z_l^c$.
2. The pull-back form $\Phi_0^*\alpha_2$ is non-degenerate on $R_0 \times \mathbb{T}^n$.
3. The pull-back of eqn (5.1) to $R_0 \times \mathbb{T}^n$ by the map $\Phi_0$ has the form $\dot{r} = 0$, $\dot{\mathfrak{z}} = \omega(r)$, where $\omega$ extends to an analytic map $\Pi^c \mapsto \mathbb{C}^n$.
4. For almost every $r \in R_0$ the tangent map $\omega_*(r) : T_r R_0 \mapsto \mathbb{R}^n$ is non-degenerate.

By 3, any solution $u_0(t)$ of (5.1) in $\mathcal{T}_0^{2n}$ has the form

$$u_0(t) = u_0(t; r_0, \mathfrak{z}_0) = \Phi_0(w_0(t)),$$

where $w_0(t) = (r_0, \mathfrak{z}_0 + t\omega(r_0)) \in R_0 \times \mathbb{T}^n$. We linearize (5.1) about a solution $u_0$ as above to get the non-autonomous linear equation

$$\dot{v} = J(Av + (\nabla H)_*(u_0(t))v) =: JA_t(t)v, \tag{5.2}$$

which is our concern in this section. We recall that linear flow maps of eqn (5.2) (if they exist) are denoted as $S^t_{\tau **}(u_0(\tau))$ (see Definition 1.2), and supplement assumptions 1–4 by

5. For any solution $u_0$ of (5.1) in $\mathcal{T}_0^{2n}$ the flow maps $S^t_{\tau**}(u_0(\tau))$, $-\infty < \tau, t < \infty$, are well defined in the space $Z = Z_d$.

By Theorem 1.3′ the flow maps $S^t_{\tau**}(u)$ $(u \in \mathcal{T}_0^{2n})$ are symplectomorphisms of the symplectic space $(Z, \alpha_2)$.

To study eqn (5.1) near $\mathcal{T}^{2n}$ we shall impose an integrability assumption on the linearized equation (5.2). Roughly speaking, this assumption means that eqn (5.2) has a complete system of time-quasiperiodic Floquet solutions. In Chapter 6 we show how to construct for any Lax-integrable equation as in Chapter 2 an infinite sequence of complex Floquet solutions, naturally parameterized by an index $j \in \mathbb{Z}_n$. It is rather difficult to prove directly the completeness of this system (cf. Krichever 1989, 1991 and Ercolani *et al.* 1987). Instead we shall prove (see Lemma 5.4 below) that a system of Floquet solutions is complete if

(1) when $|j| \to \infty$, these solutions behave as elements of a fixed complex basis of the complexified space $Z^c$ times oscillating exponents;
(2) Floquet exponents of the solutions depend on $r_0$ but not on the angle $\mathfrak{z}_0$. As functions of $r_0$ they do not satisfy identically resonance relations from a list of relevant resonances defined below.

In Chapter 6 we show how to verify properties (1) and (2) for solutions of Lax-integrable equations.

Formal definitions of the properties, given below in Section 5.3, are rather cumbersome since our goal was a friendly easy-to-check definition rather than an elegant and deceptively short one (like on p. 144 of Kuksin 1994).

The time flow of (5.2) is formed by linear symplectomorphisms which preserve tangent spaces to $\mathcal{T}_0^{2n}$. Therefore this flow also defines symplectomorphisms of skew-orthogonal complements $T^{\perp}_{u_0}\mathcal{T}_0^{2n}$ to spaces $T_{u_0}\mathcal{T}_0^{2n}$ in the tangent spaces $T_{u_0}Z \sim Z$.[38]

## 5.2 Floquet solutions

We call a non-zero solution $v(t)$ of the eqn (5.2) a Floquet solution if there exists a section $\Psi$ of the complexified tangent bundle to $Z$, restricted to $\mathcal{T}_0^{2n}$,

$$\begin{array}{ccc} T^cZ|_{\mathcal{T}_0^{2n}} & & \\ \downarrow & \nwarrow^{\Psi} & \\ \mathcal{T}_0^{2n} & \xleftarrow[\sim]{\Phi_0} & R_0 \times \mathbb{T}^n, \end{array}$$

and a complex function $\nu(r)$ such that the solution $v$ has the form

$$v(t) = v(t; r_0, \mathfrak{z}_0) = e^{i\nu(r_0)t}\Psi(w_0(t)), \qquad w_0 = (r_0, \mathfrak{z}_0 + t\omega(r_0)). \tag{5.3}$$

It is assumed that $v(t)$ solves (5.2) for any choice of $r_0 \in R_0$ and $\mathfrak{z}_0 \in \mathbb{T}^n$. We call the function $\nu(r)$ the (*Floquet*) *exponent of a Floquet solution* $v$.

---

[38] A space $T^{\perp}_{u_0}\mathcal{T}_0^{2n}$ is formed by all vectors $\xi \in T_{u_0}Z$ such that $\alpha_2(\xi, \eta) = 0$ for each $\eta \in T_{u_0}\mathcal{T}_0^{2n}$.

A Floquet solution $v(t)$ is called a *skew-orthogonal Floquet solution* if $\Psi$ in (5.3) is a section of the complexified skew-orthogonal bundle $T^{\perp c}\mathcal{T}_0^{2n}$ (its fibres are complexifications of the spaces $T_{u_0}^{\perp}\mathcal{T}_0^{2n}$).

We note that the exponent $\nu(r)$ of a solution $v$ is not uniquely defined since substituting in (5.3) $\Psi = e^{is\cdot\mathfrak{z}}\Psi_1(r,\mathfrak{z})$ with any integer $n$-vector $s$ we write $v$ in terms of the new section $\Psi_1$ as $v = e^{i(\nu(r_0)+\omega(r_0)\cdot s)t}\Psi_1(w_0(t))$. So the exponent $\nu(r)$ is defined up to an element of the $\mathbb{Z}$ module $\omega(r)\cdot\mathbb{Z}^n$, treated as a sub-module of $\mathbb{R}$ (it is dense in $\mathbb{R}$ unless all components of the vector $\omega(r)$ are proportional). The corresponding *factor frequency* $\tilde{\nu}(r)$, equal to the class $\nu(r)+\omega(r)\cdot\mathbb{Z}^n \in \mathbb{R}/(\omega(r)\cdot\mathbb{Z}^n)$, is a well-defined element of the factor module $\mathbb{R}/(\omega(r)\cdot\mathbb{Z}^n)$. Moreover, we show below in Lemma 5.4 that in a non-resonant situation Floquet solutions with the same factor frequency $\tilde{\nu}$ are proportional.

Let us assume that eqn (5.2) has an infinite family of Floquet solutions $v = v_j(t)$ such that different solutions have different exponents. Clearly, if $v_j$ is a Floquet solution, then $\overline{v}_j$ is a solution with the exponent $-\overline{\nu}_j(r)$, corresponding to the section $\overline{\Psi}_j$. We add this solution to the family; if a solution with the exponent $-\overline{\nu}_j(r)$ was already there, we replace it by $\overline{v}_j$. Now the family is invariant with respect to the complex conjugation and the set of all exponents is invariant with respect to the involution $\nu \to -\overline{\nu}$. In addition we *suppose* that the set of exponents is invariant with respect to the complex conjugation $\nu \to \overline{\nu}$ (this assumption holds trivially if all the frequencies are real); hence the set is invariant with respect to the involution $\nu \to -\nu$.

It is convenient to enumerate the Floquet solutions by integers from the set $\mathbb{Z}_n = \{\pm(n+1), \pm(n+2), \dots\}$. We do it in such a way that, first, $\nu_{-j}(p) \equiv -\nu_j(p)$ and, second, $\Psi_{-j} \equiv \overline{\Psi}_j$ if $\nu_j$ is real. So below we consider the following system of Floquet solutions:

$$v_j(t; r_0, \mathfrak{z}_0) = e^{i\nu_j(r_0)t}\Psi_j(r_0, \mathfrak{z}_0 + t\omega(r_0)), \quad j \in \mathbb{Z}_n; \qquad \nu_{-j}(r) \equiv -\nu_j(r). \tag{5.4}$$

For each index $k$ we denote by $\hat{k}$ an index such that $\nu_{\hat{k}} = \overline{\nu_k}$. Clearly, $\hat{\hat{k}} = k$ for any $k$ and $\hat{k} = k$ if $\nu_k$ is real. We note that the hat-map is $r$-independent in any connected sub-domain of $R_0$ where all the functions $\nu_j(r)$ are different.

Let us consider any Floquet solution $v_k$. Then $\overline{v}_k$ is a Floquet solution with the exponent $-\overline{\nu}_k$. A solution with this exponent can be obtained as $v_k \mapsto v_{-k} \mapsto v_{\widehat{-k}}$, or as $v_k \mapsto v_{\hat{k}} \mapsto v_{-\hat{k}}$. These solutions must coincide since the family (5.4) contains no more than one solution with a given exponent; so the hat-map is odd: $\widehat{-k} = -\hat{k}$. Since the two solutions coincide with $\overline{v}_k$, then $\Psi_{-\hat{k}} = \overline{\Psi}_k$. We have obtained that

$$\Psi_{-\hat{k}} = \overline{\Psi}_k \quad \text{and} \quad -\hat{k} = \widehat{-k}, \quad \forall k. \tag{5.4$'$}$$

Now we impose some rather non-restrictive smoothness assumptions on the solutions (5.4). To do this in the right way we note that the sections $\Psi_j$, restricted to a torus $T^n(r)$, are eigenvectors of the linearized time-one shift operator $S^1_{0*}$ which acts on sections of the skew-orthogonal complex bundle $T^{\perp c}\mathcal{T}_0^{2n}|_{T^n(r)}$. Indeed, we have $S^1_{0*}\Psi_j = e^{i\nu_j(r)}\Psi_j$.[39]

[39] The operator $S^1_{0*}$ is a well-known tool to study *hyperbolic* invariant sets (see, for example, Pesin 1989, Section 2.10). The tori $T^n(r)$ we consider usually are *elliptic* and the operator $S^1_{0*}$ has its spectrum in the unit

The operator $S^1_{0*}$ analytically depends on the parameter $r \in R$ and by analogy with classical spectral problems (see Example 5.1 below) it is plausible to assume that its eigenvalues $e^{i\nu_j(r)}$ and their logarithms $i\nu_j(r)$ are algebraic functions of $r$ which might have algebraic singularities at the set

$$\Lambda_j = \{r \mid \nu_j(r) = \nu_p(r) \text{ for some } p \neq j\}.$$

In particular, $\nu_j$ is analytic in $r$ if the set $\Lambda_j$ is empty. The situation becomes too intricate if there are infinitely many non-trivial sets $\Lambda_j$. To avoid this complexification we assume that

(a) There is a point $r \in R_0$ where $\nu_j \neq \nu_k$ if $j \neq k$. Moreover, there exists $j_1$ (depending on $\mathcal{T}^{2n}$) such that $\nu_j(r) \neq \nu_k(r)$ for all $r$, all $k$ and all $j$ such that $|j| \geq j_1$, $j \neq k$.

Since $\nu_{-j} = -\nu_j$, then by this assumption $\nu_j \neq 0$ if $j \geq j_1$.

The exponents $\nu_k$ with $|k| \geq j_1$ are assumed to be real analytic:

(b) For any $k$ such that $|k| \geq j_1$, $\nu_k$ is a real-valued analytic function on $R$ (so $\nu_k \equiv -\nu_{-k}$ and $\Psi_{-k} = \overline{\Psi}_k$). The section $\Psi_k$ extends to an analytic map $\Pi^c \times \{|\operatorname{Im} \mathfrak{z}| < \delta\} \to Z^c$ and $\nu_k$ extends to an analytic function on $\Pi^c$.

In particular, $\hat{k} = k$ if $|k| \geq j_1$.

For sophisticated integrable equations like the SG equation, some exponents $\nu_k(r)$ with $|k| < j_1$ have non-trivial algebraic singularities (see Chapter 6). Recovering later in this section the global properties of the system of Floquet solutions (5.4) we treat them as algebraic functions on the analytic set $R$. Next we cut out of $R$ the set of algebraic singularities to work with the reduced set. Nothing unexpected happens in this way. The reader who trusts this claim, or is not concerned with the 'sophisticated' equations, can assume that all the exponents are analytic functions (i.e. $j_1 = n + 1$) and ignore assumptions (c) and (d) below, where we specify the algebraic singularities.

The assumptions we shall impose now on the exponents $\nu_k$ with $|k| < j_1$ are made *ad hoc*: they are met by Floquet solutions of Lax-integrable equations.

Below, an index $k \in \mathbb{Z}_n$ is called *small* (*big*) if $|k| < j_1$ (respectively $|k| \geq j_1$).

**Definition 5.1.** An $N$-valued continuous complex function $\{\lambda_1, \ldots, \lambda_N\}$ on $\Pi^c$ is called an *algebraic function* if there exists a holomorphic function $F(r, \lambda)$ on $\Pi^c \times \mathbb{C}$ of the form

$$F(r, \lambda) = \lambda^N + f_{N-1}(r)\lambda^{N-1} + \cdots + f_0(r), \tag{5.5}$$

with uniformly in $\Pi^c$ bounded holomorphic coefficients $f_j$, such that the points $\{\lambda_1(r), \ldots, \lambda_N(r)\}$ exhaust all $N$ roots of the equation $F = 0$ and the discriminant $\Delta$ of $F$,

$$\Delta(r) = \prod_{j \neq k} (\lambda_j(r) - \lambda_k(r)),$$

does not vanish identically. The graph of this $N$-valued function is denoted $G_\lambda$, i.e. $G_\lambda = F^{-1}(0) \subset \Pi^c \times \mathbb{C}$.

---

circle. Sections $\Psi_j$ give rise to eigenvectors of $S^1_{0*}$ of the form $e^{is\cdot q}\Psi_j(q)$, $s \in \mathbb{Z}^n$, $j \in \mathbb{Z}_n$. If the system of Floquet solutions is complete (see below), then these vectors form a basis of an appropriate Hilbert space of sections of the bundle. In this case the operator $S^1_{0*}$ has a point spectrum which is dense in the circle.

The functions $\lambda_j$ are called *branches of the algebraic function*, or algebraic functions for short. They are not uniquely defined. Usually the branches of the analytic functions we consider in this book are specified to be continuous on the real domain $\Pi^c \cap \mathbb{R}^N$.

The holomorphic function $F(r, \lambda)$ of the form (5.5) is called a *Weierstrass polynomial.*

Now we specify singularities of the exponents $\nu_k$ with small $k$. We denote $M = j_1 - n - 1$.

(c) The functions $\nu_j$ with small $j$ are continuous in $R$ and are analytic in $R \setminus \Lambda$, where $\Lambda = \bigcup_{|j|<j_1} \Lambda_j$. They have the form

$$\nu_j(r) = \tilde{\nu}(\lambda_j(r), r), \quad j = \pm(n+1), \dots, \pm(n+M),$$

where $\{\lambda_j(r)\}$ is some $2M$-valued algebraic function and $\tilde{\nu}$ is an analytic complex function on $\Pi^c \times \mathbb{C}$, such that $\partial\tilde{\nu}/\partial\lambda \not\equiv 0$.

The functions $\nu_j$ are analytic in $\Pi^c$ outside the discriminant set $D = \Delta^{-1}(0)$. We note that $D \cap R$ is a proper analytic subset of $R$, since, by assumption (a), no two exponents $\nu_j$, $\nu_k$ coincide identically in $R$.

*Remark 1.* The multi-valued map $r \mapsto \{\nu_j(r)\}$, $j = \pm(n+1), \dots, \pm(n+M)$, is analytic bounded outside the discriminant set $D$ and is formed by roots of the polynomial $\prod(\nu - \nu_j(r))$. This polynomial can be written in the form (5.5), where the coefficients are symmetric polynomials of the $\nu_j$'s. So they are holomorphic functions, bounded in $\Pi^c \setminus D$, and their singularities at $D$ can be removed (see Bochner and Martin 1948 and Gunning and Rossi 1965). Thus, the exponents $\nu_j(r)$ with small $j$ form the $2M$ roots of a Weierstrass polynomial. We could treat $\{\nu_j\}$ as a $2M$-valued algebraic function, but do not do this since in applications the multi-valued functions $\{\lambda_j(r)\}$ appear naturally (as eigenvalues of the corresponding $\mathcal{L}$-operator) and since the corresponding sections, the $\Psi_j$'s, also are functions of the $\lambda_j$'s, see item (d) below. □

*Remark 2.* Let us take any two connected components $O_1$ and $O_2$ of $R \setminus D$ and a smooth path from $O_1$ to $O_2$ in $R^c \setminus (D \cup R_s^c)$ (it exists since the codimension of $D \cup R_s^c$ in $R^c$ is at least two; see Bochner and Martin 1948 and Gunning and Rossi 1965). For any small $j$ we analytically continue the functions $\lambda_j$ and $\lambda_{-j}$ along the path from $O_1$ to $O_2$. Since the relation $\nu_j + \nu_{-j} \equiv 0$ is preserved by this continuation, we get in $O_2$ functions $\nu_{j'}$ and $\nu_{-j'}$ with some small $j'$. This means that the exponents $\nu_j$ and the functions $\lambda_j$ form pairs, invariant under the monodromy. □

The set $D \cap R^c$ contains algebraic singularities of the Floquet exponents and is contained in the set $\Lambda$, defined in (c). The latter is a proper analytic subset of $R^c$ since it is formed by zeros of the non-trivial analytic function $\prod(\nu_j - \nu_k)$ (the product is taken over all small $j \neq k$). We note that $\Lambda$ contains zeros of the exponents $\nu_j$ since they are odd in $j$. We add $\Lambda$ to the singular set $R_s^c$:

$$R_s^c := R_s^c \cup (\Lambda \cap R^c), \qquad R_s := R_s \cup (\Lambda \cap R),$$

and modify the regular set $R_0 = R \setminus R_s$ accordingly.

**Example 5.1.** Eigenvalues $\{\lambda_j\}$ of a real matrix $B(a)$ which analytically depend on a real vector-parameter $a$ are zeros of the characteristic equation $\det(B(a)-\lambda E)=0$ and are algebraic functions of $a$. A priori they have singularities at the sets $\Lambda_{jk}=\{\lambda_j=\lambda_k\}$. Some of these singularities can be removed by re-enumerating the eigenvalues before or behind the sets $\Lambda_{jk}$. In particular, if the matrix $B(a)$ is symmetric, then under proper enumeration the eigenvalues have no singularities at all (this is Rellich's theorem). However, if $\lambda_j$ and $\lambda_k$ are real 'before' $\Lambda_{jk}$ and have non-trivial imaginary parts 'behind' $\Lambda_{jk}$, then a singularity at this set is unremovable. For example, eigenvalues of the matrix $\begin{pmatrix} 1 & -a \\ 1 & -1 \end{pmatrix}$ are real for $a<1$ and are complex for $a>1$. At $a=1$ they have unremovable algebraic singularities. □

Now we pass to smoothness of the sections $\Psi_j$ with $|j|<j_1$:

(d) There exists an analytic map $\widetilde{\Psi}:\Pi^c\times\{|\mathrm{Im}\,\mathfrak{z}|<\delta\}\times\mathbb{C}\to Z^c$, such that $\Psi_j(r,\mathfrak{z})=\widetilde{\Psi}(r,\mathfrak{z};\lambda_j(r))$ for $(r,\mathfrak{z})\in R_0\times\mathbb{T}^n$ and all small $j$. The range of the map $\tilde{\Psi}$ is contained in $Z^c_\infty$ and $\tilde{\Psi}$ is analytic as a map, valued in any space $Z^c_s$.

This assumption agrees with the smoothness of the eigenvectors in finite-dimensional spectral problems:

**Example 5.1, continuation.** Let us denote by $B^j$ the $n\times n$ matrix $B^j(a)=B-\lambda_j(a)E$, so $B\xi=\lambda_j\xi$ if $B^j\xi=0$. Let us assume that $\mathrm{rk}\,B^j(a)=n-1$ for $a\notin\Lambda_j=\bigcup_k\Lambda_{jk}$. Then for $a\notin\Lambda_j$ some $(n-1)\times n$-submatrix of $B^j$ also has rank $n-1$. Assuming for simplicity that this rank has the matrix formed by the first $n-1$ lines, we denote by $\xi_m(a)$, $1\le m\le n$, an algebraic complement to the element $B^j_{nm}(a)$ in the matrix $B^j$. Then the vector $\xi=(\xi_1,\dots,\xi_n)$ is non-zero for $a\notin\Lambda_j$ and $\sum_m B^j_{lm}\xi_m=0$ since: for $l=n$ the sum equals $\det B^j=0$ and for $l\ne n$ it vanishes by an elementary linear algebra. The vector $\xi$ is an eigenvector of $B$, $B\xi=\lambda_j(a)\xi$. It is a polynomial in the eigenvalue $\lambda_j$ and in elements of the matrix $B$. It vanishes at $\Lambda_j$. □

Since the exponents $\lambda_j(r),\lambda_{-j}(r)$ with small $j$ form monodromy-invariant pairs, then the function

$$b(r,\mathfrak{z})=-\prod_{j=n+1}^{n+M}\left(\alpha_2[\widetilde{\Psi}(r,\mathfrak{z};\lambda_j(r)),\widetilde{\Psi}(r,\mathfrak{z},\lambda_{-j}(r))]\right)^2$$

is well defined, bounded, and analytic in $\Pi^c\times\{|\,\mathrm{Im}\,\mathfrak{z}|<\delta\}$ outside the branching set $D\times\mathbb{C}^n$. Since the discriminant set $D$ is a proper analytic subset, the singularity at $D$ may be removed (see Bochner and Martin 1948, section VIII.5 or Gunning and Rossi 1965) and $b$ extends analytically to the whole domain $\Pi^c\times\{|\,\mathrm{Im}\,\mathfrak{z}|<\delta\}$. We use this function in Section 5.3 below.

## 5.3 Complete systems of Floquet solutions

Let us take any basis $\{\varphi_j\mid j\in\mathbb{Z}_0\}$ of the Hilbert scale $\{Z_s\}$ as in the beginning of Chapter 2 and assume that the basis is symplectic, i.e.

$$\alpha_2[\varphi_j,\varphi_{-k}]=\langle\overline{J}\varphi_j,\varphi_{-k}\rangle=\delta_{j,k}\mu_j,\quad \text{for all } j\in\mathbb{N},\,k\in\mathbb{Z}_0, \tag{5.6}$$

where $\mu_j$ are some positive real numbers. For $-j < 0$ we set $\mu_{-j} = -\mu_j$, so now the numbers $\mu_j$ are defined for $j \in \mathbb{Z}_0$. Since $\{\varphi_j\}$ is a Hilbert basis, then $\overline{J}\varphi_k = \mu_k\varphi_{-k}$ for every $k \in \mathbb{Z}_0$. Denoting

$$\nu_j^J = \mu_j^{-1}$$

and using that $\overline{J}$ is an isomorphism of the scale $\{Z_s\}$ of order $-d_J \le 0$, we get:

$$C_1^{-1} j^{d_J} \le \nu_j^J \le C_1 j^{d_J}, \quad \forall j \ge 1,$$

with some $C_1 \ge 1$.

Given the basis $\{\varphi_j\}$ we define a complex Hilbert basis $\{\psi_j \mid j \in \mathbb{Z}_0\}$ as follows:

$$\psi_j = \frac{1}{\sqrt{2}}(\varphi_j - i\varphi_{-j}), \qquad \psi_{-j} = \bar{\psi}_j = \frac{1}{\sqrt{2}}(\varphi_j + i\varphi_{-j}), \quad \forall j \in \mathbb{N}.$$

Owing to (5.6), for any $j$ and $k$ we have

$$\alpha_2[\psi_j, \psi_{-k}] = i\delta_{j,k}\mu_j. \tag{5.7}$$

Since $\bar{J}\varphi_k = \mu_k\varphi_{-k}$ for every $k$, then the operators $\bar{J}$ and $J$ are diagonal in this basis:

$$\bar{J}\psi_j = i\mu_j\psi_j, \qquad J\psi_j = i\nu_j^J\psi_j. \tag{5.8}$$

For any real $s$ we denote by $Y_s$ the following subspace of $Z_s$ of codimension $2n$:

$$Y_s = \overline{\text{span}}\{\varphi_j \mid j \in \mathbb{Z}_n\} \subset Z_s\,.$$

The spaces $\{Y_s, \alpha_2\mid_{Y_s}\}$ form a symplectic Hilbert scale with the basis $\{\varphi_j \mid j \in \mathbb{Z}_n\}$.

**Example.** If $\{\varphi_j\}$ is the trigonometric basis as in (1.1), i.e. $\varphi_k = \pi^{-1/2}\cos kx$ and $\varphi_{-k} = -\pi^{-1/2}\sin kx$, then the complex basis $\{\psi_k\}$ is the exponential basis $\psi_k = (2\pi)^{-1/2}e^{ikx}$. □

Let $\{v_j\}$ be a system of Floquet solutions as in Section 5.2 and let $\{\Psi_j\}$ be the corresponding sections. For any $(r, \mathfrak{z}) \in R_0 \times \mathbb{T}^n$ we denote by $\Phi_1(r, \mathfrak{z})$ a complex-linear map from $Y^c = Y_d^c$ to $Z^c$ which identifies $\psi_j$ with $\Psi_j$:

$$\Phi_1(r, \mathfrak{z}) : Y^c \to Z^c, \qquad \psi_j \mapsto \Psi_j(r, \mathfrak{z}), \quad \forall j \in \mathbb{Z}_n. \tag{5.9}$$

The map $\Phi_1$ will be used to formulate an important notion of completeness of a system of Floquet solutions. Before we do this we cut out the set $R_0$ a 'neighbourhood of infinity' and a neighbourhood of the singular set $R_s$ to get an open domain $R_1$:

$$R_1 \Subset R_0 = R \setminus R_s.$$

Possibly, $R_1$ is disconnected. To simplify notation we assume that the domain $R_1$ belongs to a single chart of the analytic manifold $R_0$ and treat $R_1$ as a bounded domain in $\mathbb{R}^n$. We fix any bounded complex domain $R_1^c$ which contains $R_1$ with its complex $\delta$-neighbourhood and does not intersect the singular set $R_s^c$. We denote by $W_1$ the set

$$W_1 = R_1 \times \mathbb{T}^n$$

and denote by $W_1^c$ its complex neighbourhood ,

$$W_1^c = R_1^c \times \{|\text{Im}\,\mathfrak{z}| < \delta\}.$$

**Definition 5.2.** A system of Floquet solutions (5.4) which satisfies the analyticity assumptions (a)–(d) is called *complete* (in the space $Z = Z_d$) if :

(0) it is formed by skew-orthogonal Floquet solutions,

and for any $(r, \mathfrak{z}) \in R_0 \times \mathbb{T}^n$ we have:

(1a) the functions $\beta_j = -i\alpha_2[\Psi_j(r, \mathfrak{z}), \Psi_{-j}(r, \mathfrak{z})]$, $j \in \mathbb{Z}_n$, are $\mathfrak{z}$-independent: $\beta_j = \beta_j(r)$,

(1b) there is a non-empty sub-domain of $R_0$ where no function $\beta_j(r)$ vanishes identically,

(1c) the vectors $\{\Psi_j(r, \mathfrak{z})\}$ form a skew-orthogonal system in the space $T^{\perp c}_{\Phi_0(r,\mathfrak{z})} T^{2n}$, that is:

$$\alpha_2[\Psi_j, \Psi_{-k}] = i\beta_j(r)\delta_{j,k}, \quad \forall j, k. \tag{5.10}$$

(2) The vectors $\{\Psi_j(w)\}$, $w = (r, \mathfrak{z}) \in W_1^c$, are analytic in $w$ and are uniformly asymptotically close to the complex basis $\{\psi_j\}$ and the exponents $\nu_j(r)$ are close to constants. Namely,

(i) the linear map $\Phi_1(w)$ analytically depends on $w \in W_1^c$ as an operator $Y^c \to Z^c$ and equals the natural embedding $\iota : Y^c \hookrightarrow Z^c$ up to a $\Delta$-smoothing operator, $\Delta > 0$:

$$\|\Phi_1(w) - \iota\|_{d,d+\Delta} \le C_1, \quad \text{for all } w \in W_1^c; \tag{5.11}$$

(ii) for large $j$ the functions $\beta_j(r)$ in (5.10) are analytic in $R_1^c$ and are there close to the constants $\mu_j$, defined in (5.6) (cf. (5.8)):

$$|\beta_j(r) - \mu_j| \le C_2|j|^{-d_J-\Delta}, \quad \text{for } r \in R_1^c; \tag{5.12}$$

(iii) the exponents $\nu_j$ analytically extend to $R_1^c$ and are there 'asymptotically close to constants'. Namely, for any $r \in R_1^c$ we have $|\nu_j(r)| \le C_3|j|^{d_A+d_J}$ and

$$|\nabla\nu_j(r)| \le C_4|j|^{\widetilde{\Delta}}, \tag{5.13}$$

with some real $\widetilde{\Delta} < d_A + d_J$.

The constants $C_1$–$C_4$ in this definition may depend on the domain $R_1$ but not on $j$.

Since the vectors $\Psi_j$ analytically extend to $W_1^c$ by item (2), then the functions $\beta_j$ are analytic in $R_1^c$ and relation (5.10) holds in $W_1^c$.

Since $\Psi_{-j}(w) = \overline{\Psi}_j(w)$ for real $w$ and big $j$, then the corresponding functions $\beta_j$ are real and $\beta_{-j} \equiv -\beta_j$. Since $\mu_j \ge C^{-1}j^{-d_J}$, then by assumption (5.12) we have

$$|\beta_j(r)| \ge \tfrac{1}{2}\mu_j, \quad \text{for } r \in R_1^c \text{ and } j > j_2, \tag{5.14}$$

with some new constant $j_2$. We consider the product

$$\widetilde{b}(r) = \prod_{j=n+1}^{j_2} \beta_j^2(r) = b(r) \prod_{j=j_1}^{j_2} \beta_j^2(r),$$

where the function $b = \prod_{j=n+1}^{j_1-1} \beta_j^2$ was introduced at the end of Section 5.2 and was shown to be analytic; now it is $\mathfrak{z}$-independent due to assumption (1a). The functions $\beta_j$

with big $j$ also are $\mathfrak{z}$-independent analytic. So $\tilde{b}$ is analytic in $\Pi^c$ and due to (1b) a set of its zeros is a proper analytic subset of $R^c$. We add it to the complex singular set $R_s^c$,

$$R_s^c := R_s^c \cup \tilde{b}^{-1}(0),$$

and accordingly modify the sets $R_s$ and $R_0$. If it is necessary, we also decrease the domain $R_1$ so that the inclusion $R_1 \Subset R_0 \setminus R_s$ still holds true.

*Remark 3.* The set $R_s$ as it is defined now is the final singular set for our constructions. It comprises: (1) the singular part of the algebraic set $R$, (2) the set of degeneracies of the pull-back symplectic form $\Phi_0^*\alpha_2$, (3) algebraic singularities of the Floquet exponents, and (4) points where any two of them coincide. Finally, it contains (5) the zero set of the function $\tilde{b}$ we have just constructed. The last set is a set of degeneracies of the system $\{\Psi_j\}$ since a vector $\Psi_j(r, \mathfrak{z})$ is skew-orthogonal to the tangent space $T_u\mathcal{T}^{2n}$ and to all the vectors $\Psi_k(r, \mathfrak{z})$ as soon as $\beta_j(r) = 0$ (see (5.10)). At the same time, in Lemma 5.1 below we prove that the vectors $\{\Psi_j\}$ form a basis for the skew-orthogonal space $T_u^{\perp c}\mathcal{T}^{2n}$ if $r \notin R_s$. □

The set $R_1 \Subset R \setminus R_s$ may be chosen to occupy most of $R_0$ in the sense of measure: if $\tilde{R}$ is a bounded chart of the manifold $R_0$, $\text{mes}_n$ is the Lebesgue measure in $\tilde{R}$, and $\gamma$ is any positive number, then $R_1$ can be chosen in such a way that

$$\text{mes}_n(\tilde{R} \setminus R_1) \le \gamma. \tag{5.15}$$

Let us denote by $\mathcal{T}^{2n,c}$ the $2n$-dimensional complex manifold $\Phi_0(W_1^c)$ and for $u = \Phi_0(w) \in \mathcal{T}^{2n,c}$ define the space $T_u^{\perp}\mathcal{T}^{2n,c}$ as the set of all vectors $z \in Z^c$ such that $\alpha_2[z, \xi] = 0$ for every $\xi \in T_u\mathcal{T}^{2n,c}$. For any real $u = \Phi_0(w) \in \mathcal{T}^{2n}$ we have

$$T_u^{\perp}\mathcal{T}^{2n,c} = T_u^{\perp c}\mathcal{T}^{2n}.$$

A complete system of skew-orthogonal Floquet solutions spans the skew-orthogonal spaces $T_u^{\perp}\mathcal{T}^{2n,c}$:

**Lemma 5.1.** *For any $w \in W_1^c$ and for $u = \Phi_0(w)$ the map $\Phi_1(w)$ defines an isomorphism of the spaces $Y^c$ and $T_u^{\perp}\mathcal{T}^{2n,c}$, as well as of $Y_{d+\Delta}^c$ and $T_u^{\perp}\mathcal{T}^{2n,c} \cap Z_{d+\Delta}^c$. In particular, the vectors $\{\Psi_j(w)\}$ form a skew-orthogonal basis of the space $T_u^{\perp}\mathcal{T}^{2n,c}$.*

*Proof.* By (5.11) the map $\Phi_1(w)$ is a compact perturbation of the embedding $\iota$ : $Y^c \to Z^c$, so $\text{ind}_{\mathbb{C}}\, \Phi_1(w) = \text{ind}\, \iota = 2n$. Since the range of $\Phi_1$ lies in $T_u^{\perp}\mathcal{T}^{2n,c}$, then $\dim_{\mathbb{C}} \text{Coker}\, \Phi_1 \ge 2n$. So if we can show that $\text{Ker}\, \Phi_1 = \{0\}$, then the range of $\Phi_1$ equals $T_u^{\perp}\mathcal{T}^{2n,c}$ and the assertion concerning the spaces $Y^c$ and $T_u^{\perp}\mathcal{T}^{2n,c}$ will follow. Suppose that the kernel is non-trivial. Then it contains a non-zero vector $\xi = \sum y_j\psi_j$ and we have

$$0 = \Phi_1\xi = \sum y_j\Psi_j(w).$$

By (5.10), the skew-product of the right-hand side with any vector $\Psi_{-j}(w)$ equals $i\, y_j\beta_j(r)$. Thus, $y_j \equiv 0$ since $\beta_j \not\equiv 0$ outside $R_s^c$ (we recall that this set contains the zero set of the function $\tilde{\beta}$). So $\xi = 0$, a contradiction.

The assertion concerning the spaces $Y^c_{d+\Delta}$ and $T_u^{\perp}\mathcal{T}^{2n,c} \cap Z^c_{d+\Delta}$ follows by the same arguments, since due to (5.11) the map $\Phi_1(w)$ is a compact perturbation of the embedding $\iota : Y^c_{d+\Delta} \to Z^c_{d+\Delta}$. □

Decreasing, if necessary, the complex neighbourhood $W_1^c$ of $W_1$ we get the following result:

**Lemma 5.2.** *For any $s \in [-d - d_J - \Delta, d + \Delta]$ the operator $\Phi_1(w) : Y_s^c \to Z_s^c$ analytically depends on $w \in W_1^c$ and is uniformly bounded. Moreover, for any $s$ as above, the map $\Phi_1(w) - \iota : Y_s^c \to Z^c_{s+\Delta}$ is analytic in $w \in W_1^c$ as well.*

*Proof.* We consider the linear space $\mathbb{R}^{2n} = \operatorname{span}\{\varphi_j \mid j = \pm 1, \ldots, \pm n\} \subset (Z, \alpha_2)$ and provide it with the induced symplectic structure. Next, for any point $w$ in the closure of $W_1$ we take its complex neighbourhood $O^c \subset W_1^c$ and choose a linear symplectomorphism $\Psi_0 = \Psi_0(w) : \mathbb{C}^{2n} \to T_w\mathcal{T}^{2n,c} \subset Z^c_{d+\Delta}$ which is real for real $w$ and analytically depends on $w \in O^c$. (It can be constructed using any analytic Darboux coordinates in the vicinity of $\Phi_0(w)$ in $\mathcal{T}^{2n}$.) By Lemma 5.1 the linear map

$$\Psi(w) : Z^c_{d+\Delta} = \mathbb{C}^{2n} \oplus Y^c_{d+\Delta} \to Z^c_{d+\Delta}, \quad (z, y) \mapsto \Psi_0(w)z + \Phi_1(w)y,$$

defines a symplectomorphism, analytic in $w \in O^c$; the inverse map $\Psi(w)^{-1}$ also is bounded and analytic in $w$. By Proposition 1.3′, applied to the linear maps $\Psi(w)$, the operators $\Psi(w) : Z^c_\theta \to Z^c_\theta$, $-d - d_J - \Delta \leq \theta \leq d + \Delta$, are bounded and analytic in $w \in O^c$.

Since $\Psi \mid_{\{0\}\oplus Y^c} = \Phi_1$, then the map $\Phi_1(w) : Y^c_s \to Z^c_s$ analytically depends on $w \in O^c$. To prove the first assertion of the lemma it remains to cover $W_1$ by a finite system of domains $O^c$ as above and choose a new complex neighbourhood $W_1^c$ which is contained in the union of these domains.

The second assertion follows from Proposition 1.4, applied to the map $\Psi$ (see the remark made after the Proposition 1.4). □

**Example 5.3.** (Birkhoff-integrable systems; see Kuksin 1989, Kappeler 1991, and Bättig *et al.* 1996). Let $Z = Z_0$ be a space of sequences $\xi = (x_1, y_1; x_2, y_2; \ldots)$, given the $l_2$-norm and given the 'usual' symplectic structure by means of the 2-form $J\, d\xi \wedge d\xi$, where $J(x_1, y_1; \ldots) = (-y_1, x_1; \ldots)$. We do not specify the scale $\{Z_s\}$ and the orders of operators involved in the constructions below.

Let us denote $p_j = (x_j^2 + y_j^2)/2$, $q_j = \arg(x_j + iy_j)$ and consider an analytic Hamiltonian $h(p_1, p_2, \ldots)$. The subspace $\mathcal{T}^{2n} \subset Z$, formed by all vectors $\xi$ such that $0 = x_{n+1} = y_{n+1} = \ldots$, is invariant for the Hamiltonian vector field $V_h$, and the system restricted to $\mathcal{T}^{2n}$ obviously is integrable. Let us abbreviate $(p_1, \ldots, p_n) = p^n$ and $(q_1, \ldots, q_n) = q^n$ and denote by $\nu_j$ the functions

$$\nu_j(p^n) = \frac{\partial h(p^n, 0, \ldots)}{\partial p_j}, \quad j \geq 1.$$

We shall identify any $p^n$ with the vector $(p^n, 0, \ldots)$.

The manifold $\mathcal{T}^{2n}$ is filled with solutions

$$\xi(t) = \{p^n = \text{const}\,, q^n = t\nu^n(p^n) + \varphi^n;\ p_r = 0 \text{ for } r > n\},$$

where $\varphi^n \in \mathbb{T}^n$ and $\nu^n = (\nu_1, \dots, \nu_n)$. For any $j > n$ let us consider a smooth variation $\xi(t, \varepsilon)$ of a solution $\xi(t)$, which changes no action $p_l$ except $p_j$ and makes the latter equal to $\varepsilon^2$. That is, $\xi = (x_1, y_1; \dots)$, where

$$p^n(t) = p^n, \qquad q^n(t) = t\nu^n(p^n) + q_0^n(\varepsilon) + O(\varepsilon^2);$$
$$x_l(t) = y_l(t) = 0, \quad \text{if } l > n,\ \ l \neq j,$$

and

$$x_j(t) = \varepsilon\cos(t\nu_j(p^n) + \varphi(\varepsilon)), \qquad y_j = \varepsilon\sin(t\nu_j(p^n) + \varphi(\varepsilon))\,.$$

Here $q_0^n(\varepsilon) \in \mathbb{T}^n$ and $\varphi(\varepsilon) \in S^1$ are phases of the solution $\xi(t, \varepsilon)$. The curve $\tilde{v}_j = \xi'_\varepsilon(t, 0)$ is a solution of the equation, linearized about the solution $\xi(t)$. It equals

$$\tilde{v}_j(t, \varphi) = \{\delta p^n = 0, \delta q^n = (q_0^n)'_\varepsilon(0);\ \delta x(t), \delta y(t)\},$$

where $\delta x_l(t) = \delta y_l(t) = 0$ if $l > n, l \neq j$ and

$$\delta x_j = \cos(t\nu_j(p^n) + \varphi), \qquad \delta y_j = \sin(t\nu_j(p^n) + \varphi), \qquad \varphi = \varphi(0).$$

The curve $v_{triv}(t) = \{\delta p^n = 0, \delta q^n = q_{0\,\varepsilon}^{n\prime}(0);\ \delta x = \delta y = 0\}$ is a trivial solution of the linearized equation (it may be obtained using the variation of $\xi(t)$, corresponding to a shift of the phase-vector $\varphi^n$). An appropriate complex linear combination of the solutions $\tilde{v}_j(t, 0)$ and $\tilde{v}_j(t, \pi)$ and the trivial solution as above takes the Floquet form

$$v_j(t) = \mathrm{e}^{i\nu_j(p^n)t}\Psi_j, \qquad \Psi_j = (0, \dots; i, 1; 0, \dots)$$

(the pair $(i, 1)$ stands on the $j$th place).

Let us suppose that $|\nu_j| \leq Cj^{d_A}$ for some $d_A$ and that (5.13) holds. Then the system of Floquet solutions $\{v_j, \overline{v_j} \mid j \geq n+1\}$ is complete in the sense of Definition 5.2. □

This example illustrates well the definition but it is too simple and too restrictive: to be Birkhoff-integrable a finite-dimensional system has to have dim $Z/2$ integrals of motion, but to have a complete system of Floquet solutions for the equations linearized about solutions in $\mathcal{T}^{2n}$ it needs only $n$ of them (see a version of Floquet's theorem in Section 5.4 below).

To be useful in analytical studies of eqn (5.1) and its perturbations, a system of Floquet solutions should be complete and *non-resonant*:

**Definition 5.3.** A system of Floquet exponents $\{\nu_j(r) \mid j \in \mathbb{Z}_n\}$ satisfying assumptions (a)–(c) from Section 5.2 is called *non-resonant* if:

(3) there exists a domain $O \subset R_0$ such that for all $s \in \mathbb{Z}^n$ and all $j, k \in \mathbb{Z}_n$, $j \neq -k$, we have

$$\omega(r) \cdot s + \nu_j(r) \not\equiv 0 \quad \text{in } O, \tag{5.18}$$
$$\omega(r) \cdot s + \nu_j(r) + \nu_k(r) \not\equiv 0 \quad \text{in } O \tag{5.19}$$

The system of Floquet solutions with non-resonant exponents is also called non-resonant.

The functions in the left-hand side of (5.18) and (5.19) are called *resonance functions*, or *resonances*. We note that assumptions (5.18) and (5.19) admit a compact reformulation in terms of the factor-frequencies $\widetilde{\nu}_j$, introduced in Section 5.2:

$$\widetilde{\nu}_j(r) \not\equiv 0 \text{ and } \widetilde{\nu}_j(r) \not\equiv \widetilde{\nu}_k(r) \text{ in } O \text{ for any } j \text{ and each } k \neq j.$$

The zero set of any resonance is nowhere dense:

**Lemma 5.3.** *If a system of Floquet exponents is non-resonant, then each resonance function as in* (5.18) *and* (5.19) *is non-zero almost everywhere.*

*Proof.* Let $f$ be any resonance as in (5.19). Since the function $f$ is analytic, we should only check that it does not vanish identically in any connected component $O_1$ of the set $R_0$. Let us assume the opposite: $f \equiv 0$ in $O_1$. Since $R_s^c$ is a proper analytic subset of $R^c$, then we can find a smooth path in $R_0^c$ from $O_1$ to $O$ and analytically extend $f$ along this path (see Bochner and Martin 1948). In $O$ we get the relation $\omega \cdot s + \nu_{j'} + \nu_{k'} \equiv 0$, where $\nu_{j'}$ and $\nu_{k'}$ are analytic continuations of $\nu_j$ and $\nu_k$, respectively. By Remark 2 in Section 5.2, $j' \neq -k'$. So the obtained relation contradicts (5.19).

By the same arguments the lemma's assertion also holds true for any resonance as in (5.18). □

Finally we give

**Definition 5.4.** A system of Floquet solutions (5.4) satisfying (a)–(d) is called complete non-resonant if it satisfies assumptions (0)–(3) from Definitions 5.2 and 5.3.

It turns out that assumptions (0), (1a), and (1c) follow from (3):

**Lemma 5.4.** *Any non-resonant system of Floquet solutions satisfy assumptions (0), (1a), and (1c) from Definition* 5.2.

*Proof.* To check (1c) we should prove that for any $j \neq -k$ the function $F(r, \mathfrak{z}) = \alpha_2[\Psi_j, \Psi_k]$ vanishes identically. To do this let us consider the auxiliary function $f(t; r, \mathfrak{z})$,

$$f := \alpha_2[v_j(t), v_k(t)] = e^{i(\nu_j + \nu_k)t} \alpha_2[\Psi_j(w(t)), \Psi_k(w(t))] = e^{i(\nu_j + \nu_k)t} F,$$

where $w(t) = (r, \mathfrak{z} + t\omega(r))$. Since the skew-product of any Floquet solutions $v_j$ and $v_k$ is time independent (see Theorem 1.3′ and assumption 5 from Section 5.1), then

$$0 = \left.\frac{df}{dt}\right|_{t=0} = i(\nu_j + \nu_k)F + \nabla_{\mathfrak{z}} F \cdot \omega.$$

Let us expand $F$ in Fourier series, $F = \sum e^{is\cdot\mathfrak{z}} \widehat{F}(r, s)$. From the previous identity we get that

$$\widehat{F}(r, s)(\nu_j + \nu_k + s \cdot \omega(r)) = 0$$

for all $s$ and all $r$. By Lemma 5.3 the second factor is non-zero for almost all $r$, so $\widehat{F}(r, s) \equiv 0$ and $F(r, \mathfrak{z}) \equiv 0$.

To check (1a) we note that for $k = -j$ we have

$$\text{const} \equiv \alpha_2[v_j(t), v_{-j}(t)] = \alpha_2[\Psi_j(w(t)), \Psi_{-j}(w(t))].$$

Because of (2.5), the curve $w(t)$ is dense in the tori $\{r\} \times \mathbb{T}^n$ for almost all $r$. So $\alpha_2[\Psi_j, \Psi_{-j}]$ is a $\mathfrak{z}$-independent function for almost all $r$. By continuity, it is $\mathfrak{z}$-independent for all $r$, as stated in (1a).

To check (0) we take any variations $\delta r$ and $\delta\mathfrak{z}$ of the initial conditions for the curve $w(t)$ and get the corresponding solutions $V_1$ and $V_2$ for eqn (5.2):

$$V_1(t) = \Phi_{0*}(w(t))(\delta r, 0), \qquad V_2(t) = \Phi_{0*}(w(t))(0, \delta\mathfrak{z}).$$

We claim that $F(r, \mathfrak{z}) := \alpha_2[\Psi_j, \Phi_{0*}(\delta r, 0)] \equiv 0$ for any $j$. Indeed, since

$$\text{const} \equiv \alpha_2[v_j(t), V_1(t)] = e^{i\nu_j t} F,$$

then the claim follows by the same arguments as above if we use relation (5.18) instead of (5.19). Thus, $\Psi_j$ is skew-orthogonal to each vector $\Phi_{0*}(\delta r, 0)$. Using the solution $V_2(t)$ rather than $V_1(t)$ we get that $\Psi_j$ also is skew-orthogonal to each vector $\Phi_{0*}(0, \delta\mathfrak{z})$. Hence, this is a skew-orthogonal solution. □

**Corollary.** *A system of Floquet solutions (5.4) which meets assumptions (a)–(d) from Section 5.2 as well as the assumptions (2) and (3) from Definitions 5.2 and 5.3 is skew-orthogonal to $T^{2n}$ and is complete non-resonant, provided assumption* (1b) *holds. The latter happens, for example, if there exists a point $r_* \in \overline{R}$ such that $\Psi_j(r, \mathfrak{z}) \to \psi_j$ as $r$ tends to $r_*$, for each $j$. Here $\overline{R}$ signifies the closure of $R$ in $\mathbb{R}^N$ where $R$ is a subset.*

In practice the point $r_*$ corresponds to the zero solution of eqn (5.1) (or another trivial solution).

This result simplifies the verification of completeness for a system of Floquet solutions since it is much easier to check the non-resonance relations (5.18) and (5.19) than the completeness (1a)–(1c).

The transformation $\Phi_1$ integrates the linearized equation (5.2): it sends the curves $y_j = e^{i\nu_j(r_0)t}\psi_j$ to solutions $v_j(t)$ of (5.2). It is convenient to have this transformation symplectic and real. To this end, the sections $\{\Psi_j\}$ have to be properly reordered and normalized by multiplying by some analytic functions; simultaneously the basis $\{\psi_j\}$ also has to be transformed by a linear symplectomorphism which changes finitely many of its components only. In this way the following result can be proven:

**Proposition 5.1.** *Given any complete system of Floquet solutions (5.4) we can normalise the sections $\{\Psi_j\}$ and the complex basis $\{\psi_j\}$ is such a way that the new basis still meets* (5.7). *For $|j| \geq j_1$ the functions $\{\psi_j\}$ are orthonormal and*

$$J\psi_j = i\nu_j^J \psi_j, \quad |j| \geq j_1.$$

*The new system of Floquet solutions still is complete. Moreover,*

*(a) for any $(r, \mathfrak{z}) \in W_1^c = R_1^c \times U(\delta)$ the map $\Phi_1(r, \mathfrak{z})$ defines a symplectic isomorphism of $Y^c$ and the skew-orthogonal space $T^{\perp}_{\Phi(r,\mathfrak{z})} T^{2n,c}$, which analytically depends on $(r, \mathfrak{z}) \in W_1^c$;*

*(b) the non-autonomous linear map $\Phi_1(r, \mathfrak{z} + t\omega(r))$ sends solutions $y(t)$ of the autonomous Hamiltonian equation*

$$\dot{y} = JB(r)y, \quad y \in Y^c, \tag{5.20}$$

*to solutions of the linearized equation (5.2), skew-orthogonal to the manifold $\mathcal{T}^{2n}$. The operator $B(r)$ defines a self-adjoint morphism of the scale $\{Y_s^c\}$ of order $d_A$, analytic in $r \in R_1^c$, and*

$$\operatorname{ord} \nabla_r B(r) \le \widetilde{\Delta} - d_J. \tag{5.21}$$

*The operator $JB(r)$ is diagonal in the basis $\{\psi_j\}$ and its eigenvalues are the Floquet exponents of the solutions (5.4): $JB(r)\psi_j = i\nu_j(r)\psi_j$ for each $j$.*

We note that the basis $\{\psi_j\}$ may depend on a connected component of the set $R_1$.

*Proof.* To prove the theorem we replace the sets $R_1$ and $R_1^c$ by any connected components $R_1^0 \subset R_1$ and $R_1^{0c} \subset R_1^c$, where $R_1^0 = R_1^{0c} \cap R_1$, and denote $\mathcal{T} = \Phi_0(R_1^0 \times \mathbb{T})$ and $\mathcal{T}^c = \Phi_0(R_1^{0c} \times \{|\operatorname{Im} z| < \delta\})$. We consider sections $\Psi_j$ with big and small indexes $j$ separately:

(1) $j$ is big. Now the functions $\beta_j(r)$, $r \in R_1^0$, are real non-zero and odd in $j$. For $|j| > j_2$ the function $\operatorname{sgn} j \cdot \beta_j$ is positive by (5.14). If for some $j_1 \le |j| \le j_2$ this function is negative, then we interchange the Floquet solutions $v_j$ and $v_{-j}$. After this transposition every function $\beta_j(r)\nu_j^J$ is positive (we recall that the map $j \mapsto \nu_j^J$ is odd in $j$ and is positive for positive $j$) and we replace each section $\Psi_j$ by $(\nu_j^J \beta_j(r))^{-1/2}\Psi_j$. Then (see (5.7) and (5.10)) for big $j$ we have achieved:

$$\alpha_2[\psi_j, \psi_{-j}] \equiv \alpha_2[\Psi_j(w), \Psi_{-j}(w)], \quad w \in R_1^{0c} \times \{|\operatorname{Im} \mathfrak{z}| < \delta\}. \tag{5.22}$$

In the space $\overline{\operatorname{span}}\{\psi_j \mid j \in \mathbb{Z}_{j_1}\} \subset Y^c$ we consider a linear operator $B(r)$, $r \in R^{0c}$, such that $B(r)\psi_j = (\nu_j(r)/\nu_j^J)\,\psi_j$ for every $j$. That is,

$$JB(r)\psi_j = i\nu_j(r)\psi_j, \quad \forall\, j \in \mathbb{Z}_{j_1}.$$

Obviously, the operator $B(r)$ is symmetric.

(2) $j$ is small. Since the set of Floquet exponents is invariant under the involutions $\nu \mapsto \overline{\nu}$ and $\nu \mapsto -\nu$, and since the exponents do not coincide and do not vanish in $R_1^0$ (see Remark 3), then real (for real $r$) exponents $\nu_j(r)$ do not vanish as well as purely imaginary exponents, and complex ones never take real or purely imaginary values. If a function $\nu_j$ is real, then we normalize $\psi_{\pm j}$ and $\Psi_{\pm j}$ as in the first case and extend the domain of definition of the operator $B(r)$ accordingly. The rest of the exponents $\nu_j$ are either purely imaginary or complex. We consider the more involved complex case only.

If an exponent $\nu_k$ with small $k$ is complex, then the set $\mathcal{K} = \{k, -k, \hat{k}, \widehat{-k}\}$ consists of four different numbers. We take the space $\mathbb{C}^4 = \operatorname{span}_{\mathbb{C}}\{\psi_{\pm k}, \psi_{\pm \hat{k}}\} \subset Y^c$ and choose a new basis $\{\tilde{\psi}_{\pm k}, \tilde{\psi}_{\pm \hat{k}}\}$ such that

$$\tilde{\psi}_{-\hat{k}} = \overline{\tilde{\psi}_k}, \qquad \tilde{\psi}_{\hat{k}} = \overline{\tilde{\psi}_{-k}}, \tag{5.23}$$

$$0 = \alpha_2[\tilde{\psi}_{\pm k}, \tilde{\psi}_{\pm \tilde{k}}], \qquad 1 = \alpha_2[\tilde{\psi}_k, \tilde{\psi}_{-k}] = \alpha_2[\tilde{\psi}_{-\hat{k}}, \tilde{\psi}_{\hat{k}}]. \tag{5.24}$$

We add this space to the domain of definition of the operator $B(r)$ and extend $B(r)$ in the following way:

$$JB(r)\tilde{\psi}_{\pm k} = i\nu_{\pm k}(r)\tilde{\psi}_{\pm k}, \qquad JB(r)\tilde{\psi}_{\pm\hat{k}} = i\nu_{\pm\hat{k}}(r)\tilde{\psi}_{\pm\hat{k}}.$$

The extended operator is symmetric since, first, $\langle B\tilde{\psi}_l, \psi_j\rangle = \langle B\psi_j, \tilde{\psi}_l\rangle = 0$ for any $l \in \mathcal{K}$ and any vector $\psi_j$ as above, and, second,

$$\langle B\tilde{\psi}_{l_1}, \tilde{\psi}_{l_2}\rangle = -\omega_2(JB\tilde{\psi}_{l_1}, \tilde{\psi}_{l_2}) = -i\nu_{l_1}\omega_2(\tilde{\psi}_{l_1}, \tilde{\psi}_{l_2}), \quad \forall l_1, l_2 \in \mathcal{K};$$

so $\langle B\tilde{\psi}_{l_1}, \tilde{\psi}_{l_2}\rangle \equiv \langle B\tilde{\psi}_{l_2}, \tilde{\psi}_{l_1}\rangle$ due to (5.24).

For any $u \in \mathcal{T}^c$ and any $k \in \mathcal{K}$, because of (5.10) we have

$$\alpha_2[\Psi_j(u), \Psi_l(u)] = 0, \quad \forall l \neq -j.$$

Since the function $\tilde{b}(r) = \prod_{l=n+1}^{j_2} \beta_l(r)$ does not vanish in the domain $R_0^c$ (see earlier in this section), then $\beta_j \neq 0$ in $R_0^c$ and $|\beta_j| \geq C^{-1} > 0$ in $R_1^c$ for every $j \in \mathcal{K}$. Using (5.4′) we get that for real $r$ the functions $\beta_k$ and $\beta_{\hat{k}}$ are complex conjugated:

$$\bar{\beta}_k = i\alpha_2[\overline{\Psi}_k, \overline{\Psi}_{-k}] = i\alpha_2[\Psi_{-\hat{k}}, \Psi_{\hat{k}}] = \beta_{\hat{k}}.$$

Next we redefine the vectors $\Psi_k$ and $\Psi_{-\hat{k}}$:

$$\Psi_k := \frac{1}{i\beta_k(r)}\Psi_k, \qquad \Psi_{-\hat{k}} := \frac{1}{i\beta_{-\hat{k}}(r)}\Psi_{-\hat{k}},$$

keeping $\Psi_{-k}$ and $\Psi_{\hat{k}}$ unchanged. The redefined vectors still meet (5.4′). Moreover,

$$\alpha_2[\Psi_k, \Psi_{-k}] = 1 = \alpha_2[\Psi_{-\hat{k}}, \Psi_{\hat{k}}]. \tag{5.25}$$

The transformation of vectors $\Psi_j$, described in step (2), changes the map $\Phi_1$ on a finite-dimensional subspace only. The transformations described in step (1) changes $\Phi_1$ to $D \circ \Phi_1$, where $D$ is the diagonal operator with diagonal elements, equal to $\sqrt{\nu_j^J \beta_j}$ for big $j$. Using (5.12) we get that the new map still satisfies (5.11). It sends one symplectic basis to another (see (5.22)–(5.25)), so it is symplectic. This map is real since it commutes with the complex conjugation; it sends solutions of (5.20) to solutions of (5.2).

Since for $|k| < j_1$ and $|j| \geq j_1$ we have $B(r)\tilde{\psi}_k = -i\nu_k(r)\bar{J}\tilde{\psi}_k$ and

$$B(r)\psi_j = -i\nu_j(r)\bar{J}\psi_j = (\nu_j(r)/\nu_j^J)\psi_j$$

(see (5.8)), where the functions $\nu_l(r)$ are analytic and $|\nu_l/\nu_l^J| \leq C|l|^{d_A}$ by item (2c) of Definition 5.2, then $B(r)$ defines a morphism of the scale of order $d_A$, analytic in $r$. This morphism is self-adjoint since the linear map $B(r)$ is symmetric; see Section 1.2.

The estimate (5.21) follows from (5.13), so the proposition is proved. □

The *leading Lyapunov exponent* of linear equation (5.2) in $Z_d$ is a number $a$ equal to the supremum over all real numbers $a'$ such that

$$\overline{\lim}_{t\to\infty} e^{-a't}\|v(t)\|_d = \infty, \quad \text{for some solution } v(t) \subset Z_d \text{ of (5.2).}$$

A solution $u_0(t)$ of (5.1) is called *linearly stable* if the leading Lyapunov exponent of the corresponding linearized equation (5.2) vanishes.

A direct consequence of Proposition 5.1 is the following:

**Corollary.** *If the linearized equation (5.2) has a complete system of Floquet solutions, then the leading Lyapunov exponent of the equation corresponding to a solution* $u_0 = u_0(t; r, \mathfrak{z})$ *with* $r \in R_1$ *equals* $\nu^I(r) = \max\{\operatorname{Im} \nu_j(r) \mid n < |j| < j_1\}$.[40]

*Proof.* By Proposition 5.1 any variation $u'(t)$ of a solution $u_0(t)$ can be written as $\Phi_{0*}(u_0)(r', \mathfrak{z}') + \Phi_1(u_0)y'$ and in terms of the prime variables eqn (5.2) reads as

$$\dot{r}' = 0, \qquad \dot{\mathfrak{z}}' = \omega_*(r)r', \qquad \dot{y}' = JB(r)y'. \tag{5.26}$$

Decomposing $y'(0)$ in the basis $\{\psi_j\}$ we find that $e^{-at}\|u'(t)\|_s \to 0$ as $t$ grows, if $a > \nu^I(r)$. If $a < \nu^I(r)$ and $\psi_j$ is an eigenvector of $JB(r)$ with the eigenvalue $\nu_j$ such that $\operatorname{Im} \nu_j = a$, then $y'(t) = e^{-i\nu_j t}\psi_{-j}$ is the $y'$-component of a solution of (5.26). A norm of this solution grows with $t$ faster than $e^{at}$. □

In Section 5.4 we quote a result from Kuksin (1992) which states that a *finite-dimensional* system (2.1) which satisfies 1–4 and has $n$ integrals of motion has a complete system of Floquet solutions —this is a version of the classical Floquet theorem (see, for example, Hartman 1964) for multi-dimensional time. For *infinite-dimensional* systems the Floquet theorem is unknown. However, for Lax-integrable equations Floquet solutions can be constructed in at least two different ways. The first one was explained in Proposition 3.1, where we $\varepsilon$-opened any closed gap of the $\mathcal{L}$-operator to obtain an $(n+1)$-gap solution and then differentiated it in $\varepsilon$ at $\varepsilon = 0$ to get a solution of the linearized equation. The second way is to construct Floquet solutions as quadratic forms of eigenfunctions corresponding to closed gaps. We discuss this and use it in Chapter 6.

## 5.4 Lower-dimensional invariant tori of finite-dimensional systems and Floquet's theorem

Let $O$ be a domain in the Euclidean space $\mathbb{R}^{2N}$, given the usual symplectic structure. Let $H_1, \ldots, H_n$, $1 \le n < N$, be a system of commuting Hamiltonians, defined and analytic in $O$. Let $T^n \subset O$ be a torus, analytically embedded in $O$, which is invariant for all $n$ Hamiltonian vector fields $V_{H_j}$. The vector fields are assumed to be linearly independent at any point of the torus.

[40] We recall that the functions $\nu_j(r)$ with $|j| \ge j_1$ are real valued, by assumption (b).

Under mild non-degeneracy assumptions on the system of Hamiltonians (see Nekhoroshev 1994), the torus $T^n$ can be proved to belong to an $n$-dimensional family of invariant $n$-tori $T_r^n$:

$$T^n \subset \mathcal{T}^{2n} = \bigcup_{r \in R} T_r^n, \quad 0 \in R \Subset \mathbb{R}^n; \qquad T^n = T_0^n,$$

where $\mathcal{T}^{2n}$ is an analytic $2n$-dimensional submanifold of $O$. Moreover, the symplectic form, restricted to $\mathcal{T}^{2n}$, is non-degenerate and $\mathcal{T}^{2n}$ admits analytic coordinates $(r, \mathfrak{z})$, $\mathfrak{z} \in \mathbb{T}^n$, such that for every $j = 1, \ldots, n$ the vector field $V_{H_j}$, restricted to $\mathcal{T}^{2n}$, takes the form $\sum_l \omega_j^l(r)\partial/\partial \mathfrak{z}_l$ (the functions $\omega_j^l(r)$ all are analytic).

Instead of presenting here the non-degeneracy assumptions, we just *assume* the existence of a family of invariant $n$-tori as above. Then for any $r$ there exist linear combinations $K_1, \ldots, K_n$ of the original Hamiltonians $H_j$ such that for every $j$ the vector field $V_{K_j}$ restricted to the torus $T_r^n$ equals $\partial/\partial \mathfrak{z}_j$. Accordingly, at any point $(r, \mathfrak{z}) \in T_r^n$ every vector field $V_{K_j}$ defines $N - n$ Floquet multipliers $e^{i\lambda_l^j(r)}$, $l = 1, \ldots, N - n$, corresponding to directions transversal to $\mathcal{T}^{2n}$.[41] For simplicity we assume that $\mathcal{T}^{2n}$ is a linearly stable invariant set of every vector field $V_{K_j}$ (so also of every $V_{H_j}$). Then all the functions $\lambda_l^j(r)$ are real.

The following result is a version of the Floquet theorem 'for multidimensional time'. For a proof, see Kuksin (1992).

**Proposition 5.2.** *Under the assumptions given above, every vector field $V_{H_j}$, linearized about its solutions in $\mathcal{T}^{2n}$, has a complete system of $N - n$ skew-orthogonal Floquet solutions with real exponents $\nu_j(r)$.*

We note that in the finite-dimensional situation which we were discussing, item (2) of Definition 5.2 becomes trivial.

[41] The multipliers are defined as eigenvalues of the linearized time-$2\pi$ flow map of the vector field $V_{K_j}$, restricted to a skew-orthogonal component of the space $T_{(r,\mathfrak{z})}\mathcal{T}^{2n}$. They are $\mathfrak{z}$-independent; see Kuksin (1992).

# 6
# Linearized Lax-integrable equations

## 6.1 Abstract setting

If (5.1) is a Lax-integrable equation, then its $\mathcal{L}$, $\mathcal{A}$ pair can be used to construct solutions of the linearized equation (5.2) as quadratic expressions of eigenfunctions of the $\mathcal{L}$-operator and its adjoint. Below we present the construction, mostly following Krichever (1991).

Let $u(t)$ be a smooth solution of a Lax-integrable equation (5.1) = (2.9). For any smooth vector $w \in Z_\infty$ we denote

$$\mathcal{L}'_t(w) = \mathcal{L}'_{u(t)}(w) = \frac{\partial}{\partial \varepsilon} \mathcal{L}_{u(t)+\varepsilon w}\Big|_{\varepsilon=0}$$

(by assumption (2.10) the operators $\mathcal{L}'_t(v)$ are well-defined morphisms of order $d'$ of the scale $\{\mathfrak{Z}_s\}$), and similarly define operators $\mathcal{A}'_t(v)$. Let $v(t)$ be a smooth solution for the linearized equation (5.2). Then the curve $u(t) + \varepsilon v(t)$ satisfies eqn (5.1) = (2.9) up to a smooth curve $O(\varepsilon^2)$. Differentiating this relation in $\varepsilon$ at $\varepsilon = 0$, we get a Lax representation for the linearized equation (5.2):

$$\frac{d}{dt}\mathcal{L}'_t(v(t)) = [\mathcal{A}'_t(v(t)), \mathcal{L}_t] + [\mathcal{A}_t, \mathcal{L}'_t(v(t))],$$

where $\mathcal{A}_t = \mathcal{A}_{u(t)}$ and $\mathcal{L}_t = \mathcal{L}_{u(t)}$. Let us consider smooth eigenvectors of the operator $\mathcal{L}_0 = \mathcal{L}_{u_0}$ and of its conjugate operator $\mathcal{L}_0^*$, corresponding to the same eigenvalue $\lambda$:

$$\mathcal{L}_0\chi_0 = \lambda\chi_0, \qquad \mathcal{L}_0^*\xi_0 = \lambda\xi_0.$$

We assume that the following initial-value problems,

$$\dot\chi(t) = \mathcal{A}_t\chi(t), \quad \chi(0) = \chi_0, \qquad \dot\xi(t) = -\mathcal{A}_t^*\xi(t), \quad \xi(0) = \xi_0, \tag{6.1}$$

have unique smooth solutions $\chi(t)$ and $\xi(t)$. Then for any $t$ we have $\mathcal{L}_t\chi(t) = \lambda\chi(t)$ and $\mathcal{L}_t^*\xi(t) = \lambda\xi(t)$ (see Lemma 2.3 for the proof of the first relation; the proof of the second is identical).

We claim that

$$\frac{d}{dt}\langle\mathcal{L}'_t(v(t))\chi, \xi\rangle = 0. \tag{6.2}$$

Indeed, abbreviating $\mathcal{L}_t'(v(t))$ to $\mathcal{L}'$ and $\mathcal{A}_t'(v(t))$ to $\mathcal{A}'$, we write the l.h.s. of (6.2) as

$$\begin{aligned}&\langle \mathcal{L}'\chi, \dot{\xi}\rangle + \langle \dot{\mathcal{L}}'\chi, \xi\rangle + \langle \mathcal{L}'\dot{\chi}, \xi\rangle \\ &\quad = \langle \mathcal{L}'\chi, -\mathcal{A}^*\xi\rangle + \langle ([\mathcal{A}', \mathcal{L}] + [\mathcal{A}, \mathcal{L}'])\chi, \xi\rangle + \langle \mathcal{L}'\mathcal{A}\chi, \xi\rangle \\ &\quad = \langle [\mathcal{A}', \mathcal{L}]\chi, \xi\rangle = \langle \mathcal{A}'\mathcal{L}\chi, \xi\rangle - \langle \mathcal{A}'\chi, \mathcal{L}^*\xi\rangle = (\lambda - \lambda)\langle \mathcal{A}'\chi, \xi\rangle = 0.\end{aligned}$$

Since $\mathcal{L}_t'(w)$ linearly depends on $w \in Z_{s'}$ as an operator from $\mathfrak{Z}_{s'}$ to $\mathfrak{Z}_{s'-d}$ (see (2.10)), then

$$\langle \mathcal{L}_t'(w)\chi, \xi\rangle_{\mathfrak{Z}} = \langle w, q_t(\chi, \xi)\rangle_Z, \quad \forall\, w, \tag{6.3}$$

where $q_t(\chi, \xi) = q_{u(t)}(\chi, \xi)$ is an $Z_{-s'}$-valued quadratic form of $\chi, \xi \in \mathfrak{Z}_{s'}$, which is $C^1$-smooth in $t$. Hence, we can rewrite (6.2) as

$$\frac{d}{dt}\langle v(t), q_t(\chi, \xi)\rangle \equiv 0. \tag{6.4}$$

For the moment let us denote $q_t(\chi, \xi) = w$. Then

$$\langle v, A_t J w\rangle = -\langle J A_t v, w\rangle = -\langle \dot{v}, w\rangle = \langle v, \dot{w}\rangle, \tag{6.5}$$

where the last equality follows from (6.4). At this point we assume that the flow maps $S^t_{\tau**}(u(\tau))$ of the linearized equation (5.2) preserve the space $Z_\infty$. Then the set $\{v(t)\}$ formed by values at time $t$ of all smooth solutions of eqn (5.2) equals $Z_\infty$, so $\dot{w} = A(t)Jw$ since (6.5) holds for any $t$ and for all solutions $v(\cdot)$. Therefore $J\dot{w} = JA(t)Jw$, i.e. the curve $Jw(t) = J(q_t(\chi(t), \xi(t)))$ satisfies eqn (5.2).

Thus, linearized Lax-integrable equations have solutions that can be obtained as bilinear forms of eigenfunctions of the $\mathcal{L}$-operator and its adjoint:

**Theorem 6.1.** *If flow maps of the linearized equation (5.2) preserve the space $Z_\infty$ and the curves $\chi(t)$ and $\xi(t)$ are smooth solutions of equations (6.1), then the function $J(q_t(\chi(t), \xi(t)))$ with $q_t$ defined in (6.3) solves the linearized equation (5.2).*

Remarkably, for 'classical' Lax-integrable PDEs the solutions of a linearized equation, given by the theorem, are Floquet solutions which jointly form a complete nondegenerate family. Below we check this property for the KdV and SG equations.

## 6.2 Linearized KdV equation

Now we consider the KdV equation and take for the invariant manifold $\mathcal{T}^{2n}$ a bounded part of any finite-gap manifold $\mathcal{T}_V^{2n}$ of the form

$$\mathcal{T}^{2n} = \bigcup_{r \in R} T_V^n(r), \quad R = \{r \in \mathbb{R}_+^n \mid 0 < r_j < K\ \forall j\} \tag{6.6}$$

with some fixed $K > 0$. We have already checked that this invariant manifold satisfies assumptions 1–4 (see Section 5.1).

For any $n$-gap solution $u_0(t,\cdot) \in T^n_V(r)$ the equation linearized about $u_0$ takes the form

$$\dot{v} = \frac{1}{4} v_{xxx} + \frac{3}{2}\frac{\partial}{\partial x}(u_0(t,x)v). \tag{6.7}$$

Since $u_0(t,x)$ is a smooth function, then this equation is well defined in Sobolev spaces $H^d_0$ with $d \geq 1$; see Example 1.6 or Pazy (1983). Thus, assumption 5 on the invariant manifold also is satisfied.

Equation (6.7) has trivial solutions $\partial\Phi_0/\partial\mathfrak{z}_j$ and $\partial\Phi_0/\partial r_j$, $j = 1,\dots,n$ (see (3.17)). It also has non-trivial Floquet solutions of the form (5.4). We begin with illuminative and elementary constructions of these solutions in the small-amplitude case $|r| \leq \delta \ll 1$, assuming for simplicity that $\boldsymbol{V} = (1,\dots,n)$. We fix any $m \geq n+1$ and $\varepsilon$-open the $m$th gap to get an $(n+1)$-gap solution $u_\varepsilon \in \mathcal{T}^{2n+2}_{(1,\dots,n,m)}$, smooth in $\varepsilon$. By Proposition 3.1, the function $\tilde{v}_m(t,\mathfrak{z}_m) = \partial u_\varepsilon/\partial\varepsilon\mid_{\varepsilon=0}$ solves (6.7), where $\mathfrak{z}_m$ stands for the $(n+1)$th phase of the solution $u_\varepsilon$. Now we use local (near the origin) Darboux coordinates $(y_1,\dots,y_{2n+2})$ on the manifold $\mathcal{T}^{\leq 2n+2}$, constructed in Theorem 3.2 (one has to choose there $n := n+1$). Using the calculations from Example 5.3 (Section 5.3) we get that an appropriate complex linear combination of trivial solutions as above and the solutions $\tilde{v}_m(t,0)$, $\tilde{v}_m(t,\pi)$, written in the $y$-coordinates, has the form $\exp(itW_m(r))(0,\dots,i,1)$, where $W_m(r)$ is the last component of the $(n+1)$-vector $\boldsymbol{W}^{(n+1)}(r,0)$. Since the map $U_*$ sends solutions of the linearized equation, written in the $y$-coordinates, to solutions of (6.7), then for any $m \geq n+1$ we get a Floquet solution $v_m(t) = U(y_0(t))_*(0,\dots,i,1)$:

$$v_m(t,\cdot;r) = \mathrm{e}^{itW_m(r)}U_*(y_0(t))(0,\dots,i,1).$$

To study these solutions for large $r$ we have to write the $(n+1)$-gap solution $u_\varepsilon$ using the Its–Matveev formula and examine the function $u_\varepsilon$ at the degenerate limit $\varepsilon \to 0$. Corresponding calculations can be carried out but they are rather technical (see Krichever 1991). It is easier to construct Floquet solutions using Theorem 6.1. We do this later in this section.

We recall that the $\mathcal{L}$-operator of the KdV equation is the Sturm–Liouville operator $\mathcal{L} = -\partial^2/\partial x^2 - u_0(t,x)$ and consider its complex eigenfunction $\chi(x;\lambda)$ with an eigenvalue $\lambda$, satisfying the Floquet–Bloch boundary conditions:

$$\mathcal{L}\chi(x;\lambda) = \lambda\chi(x;\lambda), \qquad \chi(x+2\pi;\lambda) = \mathrm{e}^{i\rho}\chi(x;\lambda), \quad \rho = \rho(\lambda).$$

This is a periodic (antiperiodic) eigenfunction if $\rho = 0 \bmod 2\pi$ ($\rho = \pi \bmod 2\pi$). Taking the function $\chi(x,\lambda)$ for an initial condition $\chi_0$, we solve the first equation in (6.1) under the same Floquet–Bloch boundary condition $\chi(x+2\pi;\lambda) = \mathrm{e}^{i\rho}\chi(x;\lambda)$ and denote the solution $\chi(t,x;\lambda)$.

Let $\Gamma = \Gamma(r) = \{P = (\lambda,\mu)\}$ be the Riemann surface, defined in Section 3.2. One of the most important and elegant properties of the KdV equation (and of the whole class of Lax-integrable equations) is that $\chi$ as a function of $P$ is meromorphic in $\Gamma\setminus\infty$ and can be normalized to have at infinity the singularity $\exp i\sqrt{\lambda}\,x$ (so $\chi$ is a double-valued function of the spectral parameter $\lambda \in \mathbb{C}$). An eigenfunction $\chi$ which depends on the spectral parameter $P \in \Gamma$ in this specific way is called a *Baker–Akhieser function* (see Baker 1897, Belokolos *et al.* 1994, Dubrovin *et al.* 1976, and Zakharov *et al.* 1980). The Baker–Akhieser function admits a representation in terms of the same theta function $\theta$ and the same vectors $\boldsymbol{V}$, $\boldsymbol{W}$, and $\mathfrak{z}$ as in Section 3.2. The representation is given by the following formula, also due to Its–Matveev; see in Dubrovin *et al.* (1976), Dubrovin (1981), and Belokolos *et al.* (1994):

$$\chi(t, x; r, \mathfrak{z}; P) = e^{\Omega_1(P)x+\Omega_3(P)t}\,\frac{\theta(A(P) + i(\boldsymbol{V}x + \boldsymbol{W}t + \mathfrak{z}))\theta(i\mathfrak{z})}{\theta(A(P) + i\mathfrak{z})\theta(i(\boldsymbol{V}x + \boldsymbol{W}t + \mathfrak{z}))},$$
$$P = (\lambda, \mu) \in \Gamma.$$

Here $A(P)$ is the Abel transformation, the same as in Section 3.2, and $\Omega_1$ and $\Omega_3$ are the Abel integrals of the differentials $d\Omega_1$ and $d\Omega_3$. The integrals are defined modulo periods of the differentials. For $P = (\lambda, \mu)$ with real $\lambda$ we normalize the integrals in the following way:

$$\Omega_{1,3}(\lambda, \mu) = \int_{[E_1,\lambda]} d\Omega_{1,3}, \quad \text{for } (\lambda, \mu) \in \Gamma_+,\ \lambda \in \mathbb{R},$$

where $[E_1, \lambda]$ stands for the path in $\Gamma_+$ through the upper edges of the cuts. We denote by $\sigma$ the holomorphic involution of $\Gamma$ which transposes the sheets:

$$\sigma(\lambda, \mu) = (\lambda, -\mu).$$

Denoting, for any $P = (\lambda, \mu)$ with real $\lambda$, by $\gamma_P$ the path from $\sigma(P)$ to $P$ through $E_1$, equal to $\gamma_P = \sigma(-[E_1, \lambda]) \cup [E_1, \lambda]$, we get that

$$\Omega_{1,3}(P) = \frac{1}{2}\int_{\gamma_P} d\Omega_{1,3},$$

since $\sigma^* d\Omega_j = -d\Omega_j$ due to (3.11).

In a similar way we can normalize the integrals $\Omega_{1,3}(\lambda, \mu)$ when $\lambda$ is a complex number which is prohibited from rotating around any branching point of $\Gamma$. In particular, when $\lambda$ is such that $\operatorname{Re}\lambda \in K$ and

$$K \Subset (E_1, E_2) \cup (E_3, E_4) \cup \cdots \cup (E_{2n+1}, \infty] \tag{6.8}$$

(we recall that $[E_1, E_2], \ldots, [E_{2n+1}, \infty]$ are the cuts on $\Gamma$). Namely, we define $\Omega_{1,3}$ by the same formulas as above, where $[E_1, \lambda]$ stands for the continuous path $[E_1, \operatorname{Re}\lambda] \cup [\operatorname{Re}\lambda, \lambda]$ and $[E_1, \operatorname{Re}\lambda]$ is a segment in $\Gamma_+$ as above, while $[\operatorname{Re}\lambda, \lambda]$ is a (uniquely defined) path in $\Gamma$ such that its projection $\pi([\operatorname{Re}\lambda, \lambda])$ is the segment $[\operatorname{Re}\lambda, \lambda]$ in the $\lambda$-plane. The functions $\Omega_{1,3}$ are well defined and analytic if $\operatorname{Re}\lambda \in K$. Moreover, the same formulas apply when $\Gamma$ has complex branching points $\{E_j\}$ with small imaginary parts. In this case $\Omega_{1,3}$ as functions of $\boldsymbol{E} = (E_1, \ldots, E_{2n+1})$ analytically extend to a small complex neighbourhood of a real vector $\boldsymbol{E}$. A radius of this neighbourhood depends on the compact set $K$.

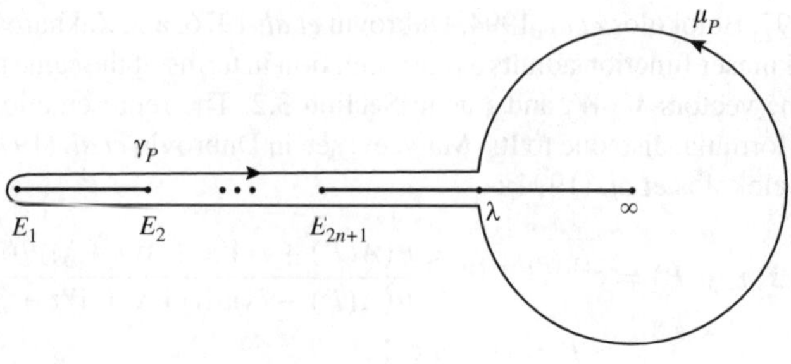


**Fig. 6.1**

Now we take a point $P = (\lambda, \mu)$, close to infinity, and denote by $\mu_P$ the path from $\sigma(P)$ to $P$ equal to a lift to $\Gamma$ of the circle in $\mathbb{C}_\lambda$ centred at infinity, which passes through $\lambda$ and is cut there (see Fig. 6.1). The loop $\gamma_P - \mu_P$ is contractible in $\Gamma \setminus \infty$ since it envelops all the cuts, so $\int_{\gamma_P - \mu_P} d\Omega_j = 0$ and $\Omega_j(P) = \frac{1}{2}\int_{\mu_P} d\Omega_j$. Using this equality and (3.10) with $c = 0$ we get the following asymptotics:

$$\Omega_1(P) = k + O(k^{-2}), \qquad \Omega_3(P) = k^3 + O(k^{-1}), \tag{6.9}$$

where $k = i\sqrt{\lambda}$ (the functions $\Omega_{1,3}$, originally defined for $\operatorname{Re}\lambda \gg 1$, analytically in $k$ extend to a neighbourhood of the infinity).

When the branching points $E_j$ are complex, sufficiently close to the real line, the asymptotics (6.9) hold for the same trivial reasons. Since the vector $\boldsymbol{E}$, formed by the single periodic/antiperiodic eigenvalues, analytically depends on the vector $r$, then (6.9) holds for $r$ from a suitable complex neighbourhood of $\mathbb{R}^n_+$ in $\mathbb{C}^n$[42] and for $k$ from a neighbourhood of infinity in the complex plane.

*Remark.* Strictly speaking, in the Its–Matveev formula for $\chi$ we should use the Abel transformation $A(P)$ with the same initial point $P_0 = E_1$ as in the integral for $\Omega_j$, not $P_0 = \infty$ as in Section 3.2. To replace the integrating from $\infty$ by integrating from $E_1$ in the formula for $A(D)_j$, we have to add the correction $I_j = n\int_\infty^{E_1} dw_j$. Since $\sigma^* dw_j = -dw_j$ (it follows, for example, from (3.9)), then $I_j = \frac{1}{2}n\int_\gamma dw_j$, where $\gamma = [\infty, E_1] \cup (-\sigma[\infty, E_1])$. Since the cycle $\gamma$ envelops all the cuts on the surface $\Gamma$ (see Fig. 6.1), then it is contractible. Hence, $I_j = 0$ and we can use $P_0 = E_1$ as an initial point for the Abel transformation. □

Let us denote

$$f(U; r, \mathfrak{z}; P) = \frac{\theta(A(P) + iU + i\mathfrak{z})\theta(i\mathfrak{z})}{\theta(A(P) + i\mathfrak{z})\theta(iU + i\mathfrak{z})}$$

and rewrite $\chi$ as

$$\chi(t, x; r, \mathfrak{z}; P) = \mathrm{e}^{\Omega_1(P)x + \Omega_3(P)t} f(\boldsymbol{V}x + \boldsymbol{W}t; r, \mathfrak{z}; P). \tag{6.10}$$

[42] We analytically extend the map $r \mapsto \boldsymbol{E} = (E_1, \ldots, E_{2n+1})(r)$ to this neighbourhood.

By the Riemann theorem (see Dubrovin 1981 and Belokolos *et al.* 1994) the first term of the denominator in the formula for $f$ as a function of $P$ has exactly $n$ zeros which form the poles of the function $P \mapsto f$ and lie in the ovals $a_1, \dots, a_n$ (see Appendix 3(ii) for a discussion of eqn (A3.2)). Since $|\theta(i\xi)| \geq C(r) > 0$ for every real vector $\xi$ (see (3.13)) and $A(\infty) = 0$, then the function $f(U; r, \mathfrak{z}; P)$, $P = (\lambda, \mathfrak{z})$ is analytic and bounded for $r$ from an appropriate complex neighbourhood of any compact subset of the set $R$, defined in (6.6), and for $(\lambda, \mathfrak{z}, U)$ from the complex domain

$$\{\,|\mathrm{Im}\,\lambda|, |\mathrm{Im}\,\mathfrak{z}|, |\mathrm{Im}\,U| < \delta, \ \ \mathrm{Re}\,\lambda \in K(r)\}, \tag{6.11}$$

where $\delta > 0$ is sufficiently small and the compact set $K$ satisfies (6.8).

We recall that the closed gaps $[\lambda_{2j-1} = \lambda_{2j}]$ are labelled by indices $j \in \mathbb{N}_V = \mathbb{N} \setminus \{V_1, \dots, V_n\}$. They belong to a suitable set $K$ as in (6.8) which can be chosen uniformly in $r$ from a sufficiently small complex neighbourhood of any real $r = r_0$. For any $P = P_{\pm j}$, where $j \in \mathbb{N}_V$ and $P_{\pm j} = (\pm\sqrt{R(\lambda_{2j})}, \lambda_{2j}) \in \Gamma$, the function $\chi(t, x; P_{\pm j})$ must be a periodic/antiperiodic eigenfunction; hence, it is $4\pi$-periodic in $x$. Since $f$ is $2\pi$-periodic, then the exponential function in (6.10) has to be $4\pi$-periodic in $x$ and we should have $\Omega_1(P_{\pm j}) \in (i/2)\mathbb{Z}$. This relation holds identically in $r$. When $r$ tends to zero, $\Omega_1(P_j)$ tends to $ij/2$; see (A4.3). Therefore,

$$\Omega_1(P_j) = \frac{i}{2}j, \quad j \in \mathbb{N}_V. \tag{6.12}$$

Conversely, for any $P$ that meets (6.12), the function (6.10) is $4\pi$-periodic.

Since the operator $\mathcal{A}$ for the KdV equation is anti-self-adjoint, then the second equation in (6.1) coincides with the first and $\xi(t) = \chi(t)$. Now the quadratic form $q$ as in Theorem 6.1 equals $\chi^2$. Finally, since $J = \partial/\partial x$, then the solutions of the linearized equation (5.2) = (6.7), constructed in Theorem 6.1, are the curves $v_j(t) \in Z$ of the form

$$v_j(t, x; r, \mathfrak{z}) = \left(\frac{(2\pi)^{-1/2}}{2\Omega_1(P_j)}\right)\frac{\partial}{\partial x}\left(e^{2(\Omega_1(P_j)x + \Omega_3(P_j)t)} f^2(\mathbf{V}x + \mathbf{W}t; r, \mathfrak{z}; P_j)\right). \tag{6.13}$$

Here $j \in \mathbb{Z}_V$, $P_j = P_j(r)$, and the first factor in the r.h.s. is a convenient normalization.

Thus, we have obtained a system of Floquet solutions of the form (5.4),[43] where the sections $\Psi_j$ of the bundle $T^c H_0^d|_{T^{2n}}$ have the form

$$\Psi_j(r, \mathfrak{z})(x) = \frac{\partial}{\partial x}\left(\frac{e^{2\Omega_1(P_j)x}}{2\sqrt{2\pi}\,\Omega_1(P_j)} f^2(\mathbf{V}x; r, \mathfrak{z}; P_j)\right), \quad j \in \mathbb{Z}_V, \tag{6.14}$$

and the exponents $\nu_j$ are

$$\nu_j(r) = -2i\Omega_3(P_j) = -2i\int_{E_1}^{P_j} d\Omega_3. \tag{6.15}$$

Since the differential $d\Omega_3$ has the form (3.11) and its integrals along the open gaps vanish, then the exponents $\nu_j(r)$ are real for real $r$ and are analytic in $r$ (they have

[43] The set of indices $\mathbb{Z}_V$ which we use now is in obvious one-to-one correspondence with the set $\mathbb{Z}_n$.

no algebraic singularities). We claim that this system satisfies assumptions (a)–(d) (see Section 5.2) and is complete non-resonant. To simplify notation we suppose that $\boldsymbol{V} = (1, \dots, n)$. Now the complex basis $\{\psi_j \mid j \in \mathbb{Z}_0\}$ is the exponential basis $\psi_j = e^{ijx}/\sqrt{2\pi}$ (cf. the Example in Section 5.3).

### 6.2.1 The system of Floquet exponents is non-resonant

To prove the nonresonance we may assume that the vector $r$ is sufficiently small. For any $j \in \mathbb{Z}_V = \mathbb{Z}_n$ we denote by $\boldsymbol{V}^{(n+1)}$ the $(n+1)$-vector $(\boldsymbol{V}, j)$ and view the torus $T^n_{\boldsymbol{V}}(r)$ as a degenerate $(n+1)$-gap torus $T^{n+1}_{\boldsymbol{V}^{(n+1)}}(r, 0)$ (see Theorem 3.1′). Comparing (6.15) with formula (A4.5) from Appendix 4 we get that $\nu_j(r) = W^{(n+1)}_{n+1}(r, 0)$. Since the frequency vector $\omega$ for finite-gap solutions that fill the torus $T^n_{\boldsymbol{V}}(r)$ is $\omega = \boldsymbol{W}$, then the non-resonance relation (5.18) that has to be checked takes the form

$$\sum_{l=1}^{n} W^{(n+1)}_l(r, 0)s_l + W^{(n+1)}_{n+1}(r, 0) \not\equiv 0. \tag{6.16}$$

We can suppose that $s \neq 0$; say, $s_1 \neq 0$. By Lemma 3.4, for $r = (\varepsilon, 0, \dots, 0)$ we have

$$W^{(n+1)}_l = \text{const} + \delta_{l,1}\frac{3}{8V_1}\varepsilon^2 + O(\varepsilon^4).$$

Therefore, the l.h.s. of (6.16) equals $\text{const} + s_1(3/8V_1)\varepsilon^2 + O(\varepsilon^4)$. It does not vanish identically and (6.16) follows. The non-degeneracy relation (5.19) holds true by similar arguments.

### 6.2.2 The system is complete

The assumptions (a)–(d) are checked below. So by the Corollary to Lemma 5.4 we only have to check assumptions (1b) and (2) from Definition 5.2. Because of relation (A5.3) from Appendix 5, the function $f(\cdot\,; r, \mathfrak{z}; P)$ converges to one as $r \to 0$. Therefore $\Psi_j(x)$ converges to the complex exponent $(2\pi)^{-1/2}e^{ijx} = \psi_j(x)$, so (1b) follows and it remains to check item (2).

Given any $\gamma > 0$ we fix a subset $R_1 \Subset R$ such that $\operatorname{mes}(R \setminus R_1) < \gamma$ (see (5.15)). For $r \in R_1$ we shall verify the properties (2a)–(2c).

First we show that the map $\Phi_1$ is close to the embedding $\iota$ up to a smoothing map. Since $\Psi_{-j} = \overline{\Psi}_j$, we have to examine the vectors $\Psi_j$ with $j \in \mathbb{N}_n$ only. Since $\lambda(P_j) = \frac{1}{4}j^2 + O(j^{-1})$ by (3.4), then $k(P_j) = (i/2)j + O(j^{-2})$, where $k = i\sqrt{\lambda}$. Using (6.8) and (6.15) we get that

$$\Omega_3(P_j) = -\tfrac{i}{8}j^3 + O(j^{-1})$$

and

$$\nu_j(r) = -\tfrac{1}{4}j^3 + O(j^{-1}), \tag{6.17}$$

uniformly in $r$ from some complex neighbourhood $R_1 + \delta$ of the set $R_1$.

Since any holomorphic differential $d\omega_j$ has the form (3.9) (also if the branching points are complex), then

$$|A(P_j)| \le C \int_{j^2/4}^{\infty} \frac{\lambda^{n-1}\, d\lambda}{\lambda^{n+1/2}} \le C_1 |j|^{-1} \text{ uniformly in } r \in R_1 + \delta.$$

Therefore for all $U$, $\mathfrak{z}$ as in (6.11), and for $r$ from $R_1 + \delta$ the function $f$ is close to one, if $P = P_j$ and $j$ is big:

$$|f(U; r, \mathfrak{z}; P_j) - 1| \le C|j|^{-1}. \tag{6.18}$$

Using (6.12), (6.18), and the Cauchy estimate we find that the functions $\Psi_j(r, \mathfrak{z})$ defined in (6.14) are close to the complex exponents:

$$\Psi_j(r, \mathfrak{z})(x) = \frac{1}{\sqrt{2\pi}} e^{ijx}(1 + \zeta_j(r, \mathfrak{z})(x)), \tag{6.19}$$

where

$$|\zeta_j(r, \mathfrak{z})(x)| \le Cj^{-1}, \quad \text{for } r \in R_1 + \delta \subset \mathbb{C}^n, \ |\operatorname{Im} \mathfrak{z}| \le \delta, \ |\operatorname{Im} x| \le \delta,$$

with some $j$-independent $\delta$ and $C = C(\delta)$.

To check property (2a) from Definition 5.2 with $\Delta = 1$ we shall show that the linear map

$$\sum a_j e^{ijx} \mapsto \sum a_j \zeta_j(x) \tag{6.20}$$

is 1-smoothing, i.e. for any $r \ge 0$ it sends a space $H_0^r(S^1)$ to the space $H_0^{r+1}(S^1)$. To do it we supply the spaces above with the Hilbert bases $\{(\sqrt{2\pi}\, j^r)^{-1} e^{ijx}\}$ and $\{(\sqrt{2\pi}\, j^{r+1})^{-1} e^{ijx}\}$ respectively, and observe that in these bases the map (6.20) has the matrix $M$ with the entries

$$M_{lj} = \frac{l^{r+1}}{j^r} \int e^{i(j-l)x} \zeta_j(x)\, dx$$

(cf. (A1.1) in Chapter 1). Since for $|\operatorname{Im} x| < \delta$ the function $\zeta_j$ is analytic and bounded by $Cj^{-1}$, then $|M_{lj}| \le C_\delta (l/j)^{r+1} e^{-\delta|j-l|}$ (see, for example, in Appendix 9 in Part II). Therefore the $l_1$-norm of any row and any column of the matrix $M$ is bounded by a constant $C'$. Hence, a norm of the map (6.20) as a map from $H_0^r$ to $H_0^{r+1}$ is bounded by the same constant $C'$ due to the Schur criterion and (2a) follows.

Property (2b) follows from (6.19). Indeed, since $\alpha_2[\Psi_j, \Psi_{-j}]$ equals

$$\frac{i}{\nu_j^J} + \frac{1}{2\pi}(\alpha_2[e^{ijx}, e^{-ijx}\zeta_{-j}] + \alpha_2[e^{ijx}\zeta_j, e^{ijx}] + \alpha_2[e^{ijx}\zeta_j, e^{-ijx}\zeta_{-j}]),$$

then $\beta_j - i/\nu_j^J$ equals

$$-\frac{1}{2\pi} \int [(D^{-1}e^{ijx})e^{-ijx}\zeta_{-j} - e^{ijx}\zeta_j D^{-1}e^{-ijx} + (D^{-1}e^{ijx}\zeta_j)e^{-ijx}\zeta_{-j}]\, dx,$$

where $D = \partial/\partial x$. This equality, estimate (6.19), and the Cauchy estimate jointly imply (5.12) with (say) $\varkappa = 3$.

The property (2c) with $d_A + d_J = 3$ and $\widetilde{\Delta} = 1$ is an immediate consequence of (6.17) and the Cauchy estimate.

### 6.2.3 The system satisfies assumptions (a)–(d)

The first assertion of (a) follows from the convergence

$$\nu_j(r) = -2i\Omega_3(P_j) \to -\tfrac{1}{4}j^3$$

(see (A4.4)) which implies that for small $r$ all the functions $\nu_j$ are distinct. Since $\nu_j(0) = -j^3/4$, then the second assertion follows from item (2c) of Definition 5.2 which is checked already with $d_A + d_J = 3$ and $\widetilde{\Delta} = 1$.

Assumption (b) holds since the exponents $\nu_j$ are real for real $r$ and since the exponents and the sections are analytic, see (6.14) and (6.15). Assumptions (c) and (d) are now empty since all the Floquet exponents are analytic functions.

Finally for the domain $R$ as in (6.6) we proved the following result:

**Theorem 6.2.** *For any $\gamma > 0$ and any n-vector $\mathbf{V}$ there exists a subset $R_1 \Subset R$,* $\text{mes}(R \setminus R_1) < \gamma$, *such that the system of Floquet solutions (6.14) with $j \in \mathbb{Z}_\mathbf{V}$ is complete non-resonant on the n-gap manifold $\Phi_0(R_1 \times \mathbb{T}^n) \subset T_\mathbf{V}^{2n}$ (in any space $H_0^d$, $d \geq 1$).*

**Amplification.** *For any $\widetilde{R} \Subset \mathbb{R}_+^n$ the system of Floquet solutions (6.14) is complete non-resonant on $\Phi_0(\widetilde{R} \times \mathbb{T}^n)$.*

Indeed, $\widetilde{R}$ is a compact part of the set $R$ as above. To get a subset of $\widetilde{R}$ where the system of skew-orthogonal Floquet solutions is complete non-resonant we should cut out $\widetilde{R}$ the vicinity of the singular set $R_s$; see Remark 2 in Section 5.3. The singular part of the analytic set $R$ is clearly empty; the Floquet exponents are analytic so the set of algebraic singularities also is empty. The form $\Phi_0^*\omega_2$ is non-degenerate on $\widetilde{R}$ (see Flaschka and Mclaughlin 1976, Veselov and Novikov 1985, and Bättig *et al.* 1996 where this is proved in three different ways); so the set of degeneracies of the symplectic form is empty. The functions $\beta_j(r)$ do not vanish on $R$—this follows from Krichever (1991) (Theorem 1, Section 1.2) or Bättig *et al.* (1996). Hence, the set of degeneracies of the system of functions (6.14) is empty as well. Thus, $R_s = \varnothing$. So we can choose $R_1 = \widetilde{R}$ and the system (6.14) is complete on $\Phi_0(\widetilde{R} \times \mathbb{T}^n)$. It is non-resonant by Theorem 6.2.

We do not present a complete proof of the Amplification (i.e. we do not prove that the pull-back form $\Phi_0^*\omega_2$ and the system (6.14) are non-degenerate) since Theorem 6.2 is sufficient to obtain our main result—the KAM stability of most finite-gap tori.

We note that the triviality of the singular set $R_s$ is not a general property of integrable PDEs: for the SG equation this set is not empty, as we show in Section 6.4.

## 6.3 Higher KdV equations

The $l$th equation from the KdV hierarchy has an $[\mathcal{L}, \mathcal{A}]$ pair with the same $\mathcal{L}$-operator $\mathcal{L} = -\partial^2/\partial x^2 - u$ and with an $\mathcal{A}$-operator of the form $\mathcal{A} = \mathcal{A}_l = \text{const}\, \partial^{2l+1}/\partial x^{2l+1} + \cdots$ (see Dubrovin *et al.* 1976, McKean and Trubowitz 1976, and Zakharov *et al.* 1980).

Solutions $\chi^l$ of eqn (6.1) with $\mathcal{A} = \mathcal{A}_l$ are given by the Its–Matveev formula (6.7), where the differential $\Omega_3$ should be replaced by an appropriate differential $\Omega_{2l+1}$, and the frequency vector $\boldsymbol{W}$ should be replaced by some vector $\boldsymbol{W}^l$ (see Section 3.4). We get Floquet solutions $v_j^l$ of the linearized $l$th equation,

$$v_j^l(t, x; r, \mathfrak{z}) = \mathrm{e}^{i\nu_j^l(r)t}\Psi_j(r, \mathfrak{z}_0 + \boldsymbol{W}^l(r)t)(x), \quad j \in \mathbb{Z}_V,$$

where $\nu_j^l = 2\Omega_{2l+1}(P_j)$ and $\Psi_j$ is given by (6.14). Using the normalization (3.32) we find that

$$\nu_j^l = 2(i/2)^{2p+1} j^{2l+1} + O(j^{2l-3}), \quad j \in \mathbb{N}_V \tag{6.21}$$

(cf. the asymptotic (6.17) and its proof).

The system of Floquet solutions $\{v_j^l\}$ is complete non-resonant. Indeed, the items of Definition 5.2 from (1) through (2b) describe the properties of the sections $\Psi_j$ which are the same as for the KdV equation, so we have already checked them. Property (2c) with $-\widetilde{\Delta} = -\widetilde{\Delta}^l = 2l - 3$ follows from (6.21). The non-resonance property follows from (3.32) by the same arguments as in the KdV case.

The linearized $l$th equation satisfies assumption 5: its flow maps $S_{\tau**}^t$ are well-defined linear isomorphisms of a space $Z_d, d \geq 1$. Indeed, by Lemma 5.1, outside the singular set $R_s \times \mathbb{T}^n$ the vectors $\{\Psi_j(r, \mathfrak{z})\}$ form an equivalent complex basis of the skew-orthogonal space $T_u^{\perp c}\mathcal{T}^{2n} \subset Z_d$, where $u = \Phi_0(r, \mathfrak{z})$. After we have chosen these bases in the spaces $T_{u_0(\tau)}^{\perp c}\mathcal{T}^{2n}$ and $T_{u_0(t)}^{\perp c}\mathcal{T}^{2n}$, the map $S_{\tau**}^t$ becomes diagonal with the unit diagonal elements $\{\mathrm{e}^{i\nu_j^l(r)(t-\tau)}\}$. So for $r \in R_s$ and any $t$ and $\tau$ the maps $S_{\tau**}^t$ are linear isomorphisms, as stated.

## 6.4 Linearized Sine–Gordon equation

Let us take any odd periodic finite-gap solution $(u, v)$ of the SG equation (4.1) which lies in a finite-gap torus $T^n(r) \subset \mathcal{T}^{2n}$ as in Section 4.3 (the manifold corresponds to the vector $\boldsymbol{l}$ as in (4.22)). In the $(u, v)$-variables the linearized equation for $u$ takes the form

$$\tilde{u}_{tt} - \tilde{u}_{xx} + (\cos u(t, x))\tilde{u} = 0, \tag{6.22}$$

and the $v$-component of a solution recovers as

$$\tilde{v} = -\dot{\tilde{u}};$$

in the $(u, w)$-variables the equation for $u$ should be supplemented by the following equation for $w$:

$$\tilde{w} = -A^{-1/2}\dot{\tilde{u}}.$$

Abusing language, we shall say that $(\tilde{u}, \tilde{v})$ (or $(\tilde{u}, \tilde{w})$) as above is a solution of the linearized equation (6.22).

Since the function $u$ is smooth, then the linearized equation in the $(\tilde{u}, \tilde{w})$-variables is well defined in any space $Z_s^o$, $s \geq 0$. Thus, the invariant manifold $\mathcal{T}^{2n}$ meets the assumption 5 from Section 5.1 (as well as assumptions 1–4, see Section 4.3).

We shall construct Floquet solutions for eqn (6.22), using Theorem 6.1. Since the operator $\mathcal{A}$ is anti-self-adjoint, then in (6.1) $\chi(t) \equiv \xi(t)$, so the vector-function $J(q_t(\chi, \chi))$ satisfies (6.22). Since $J(u, v) = (-v, u)$, then to calculate the $u$-component of $Jq_t$ we have to find the $v$-component of $q_t$ (now in the notation of Section 6.1 we substitute $u := (u, v)$ and $v := (\tilde{u}, \tilde{v})$).

Denoting by $\mathcal{L}^\varepsilon$ the operator $\mathcal{L}$, corresponding to the potential $(u, v) + \varepsilon(u_1, v_1)$, we have

$$\left.\frac{d\mathcal{L}^\varepsilon}{d\varepsilon}\right|_{\varepsilon=0} = \frac{i}{4}(v_1 + u'_{1x}) \begin{pmatrix} \begin{pmatrix} 0 & 1 \\ 1 & 0 \end{pmatrix} & 0 \\ 0 & 0 \end{pmatrix} + \cdots,$$

where the dots stand for an operator, proportional to $u_1$. Therefore the l.h.s. of (6.3) with $\xi = \chi$ equals $(i/2) \int v_1(x)\chi_1\chi_2(t, x)\, dx + \cdots$, so that the $v$-component of $q_t$ equals $(i/2)\chi_1\chi_2$. We have seen that the function

$$\tilde{u}(t, x, \lambda) = \frac{1}{\sqrt{2\pi}}(\chi_1\chi_2)(t, x; \lambda) \tag{6.23}$$

is a $4\pi$-periodic solution for (6.22) if $\chi_1$ and $\chi_2$ are the first two components of the Baker–Akhiezer (vector-) function $\chi(t, x; P) \in \mathbb{C}^4$,

$$\mathcal{L}_{(u(t,\cdot),v(t,\cdot))}\, \chi(t, \cdot; P) = \lambda\chi, \quad P = (\lambda, \mu) \in \Gamma = \Gamma(r),$$

which is $4\pi$-periodic in $x$.

Similar to the KdV case, the function $\chi$ is meromorphic in $P \in \Gamma \setminus \{0, \infty\}$ and can be written as

$$\chi = \chi(t, x; r, \tilde{D}; P) = e^{(i/2)(\kappa(P)x + \nu(P)t)} f(\tilde{V}x + \tilde{W}t; r, \tilde{D}; P),$$

where

$$\kappa(P) = \tfrac{1}{2}(\Omega_1 + \Omega_2)(P), \qquad \nu(P) = \tfrac{1}{2}(\Omega_1 - \Omega_2)(P)$$

and $\Omega_{1,2}$ are integrals of the differentials $d\Omega_{1,2}$ along a path $\gamma_{0P}$ from 0 to $P$.[44] The vector-function $f(q; r, \tilde{D}; P) \in \mathbb{C}^4$ is analytic in $q$, $\tilde{D} \in \mathbb{T}^n$, $r \in R$, and $P \in \Gamma(r) \setminus \{0, \infty\}$; it can be written explicitly in terms of the theta function, defined in Section 4.2 (see Ercolani *et al.* 1987, 1990 and Belokolos *et al.* 1994). The function $\chi$ is $4\pi$-periodic if

$$\kappa(P) \in \tfrac{1}{2}\mathbb{Z} \tag{6.24}$$

(this is a well-defined equation since a change in the path $\gamma_{0P}$ changes its l.h.s. by an integer number). Evoking notation from Section 4.1 we see that (6.24) implies that $\pi(P)$

[44] A change in the path $\gamma_{0P}$ changes the function $f$.

is a point from the $4\pi$-periodic spectrum of the $\mathcal{L}$-operator. That is, $\pi(P)$ equals $\lambda_j^+$ or $\lambda_j^-$ for some $j$ (here $\pi$ stands for the projection $\Gamma \ni (\lambda, \mu) \mapsto \lambda$). Since the potential $(u, v)$ is finite-gap, then

$$\lambda_j^+ = \lambda_j^- =: \lambda_j, \quad \forall\, j \in \mathbb{Z}_n.$$

Using (4.10) we see that the eigenvalues $\lambda_j$ with $|j| > j_1$ are exactly double.[45] Since the $\lambda$-spectrum is invariant with respect to the complex conjugation (see $(4.6_2)$ and (4.12)), then

$$\lambda_k(r) \in \mathbb{R}, \quad \text{if } |k| > j_1. \tag{6.25}$$

On the contrary, the eigenvalues $\lambda_j$ with $|j| \le j_1$ can be complex; see McKean (1981).

Below we are interested in the eigenvalues $\lambda_j$ with $j \in \mathbb{N}_n$. Since they are double, then the Baker–Akhiezer function $\chi$ is $4\pi$–periodic at both points $P_j^\pm \in \pi^{-1}(\lambda_j)$.[46]

Now for $j \in \mathbb{Z}_n$ we determine the solution $\tilde{u}_j$ of (6.22) as follows:

$$\tilde{u}_j = \begin{cases} \tilde{u}(t, x; P_j^+), & \text{if } j \in \mathbb{N}_n, \\ \overline{\tilde{u}_{-j}}, & \text{if } j \in -\mathbb{N}_n. \end{cases}$$

Here the function $\tilde{u}$ is defined as in (6.23) with $\lambda \in \mathbb{C}$ replaced by $P \in \Gamma$, and the hat-map $k \mapsto \hat{k}$ was constructed in Section 5.2. We note that the function $\bar{\tilde{u}}$ is a solution since (6.22) is a real-coefficient equation.

Let us denote by $\Pi$ the projector that sends a periodic (vector-) function $\eta(x)$ to its odd part $\frac{1}{2}(\eta(x) - \eta(-x))$, and denote

$$\xi_j^o = (\tilde{u}_j^o, \tilde{w}_j^o), \quad \text{where } \tilde{u}_j^o(t,x) = \Pi\tilde{u}_j(t,x), \ \ \tilde{w}_j^o(t,x) = -A^{-1/2}\frac{d}{dt}\tilde{u}_j^o.$$

This odd periodic vector-function is a solution of (6.22). Indeed, since $\cos u(t, x)$ is an even function of $x$, then $\Pi(\cos u)\tilde{u} = \cos u\, \Pi\tilde{u}$. Hence, applying $\Pi$ to eqn (6.22) with $\tilde{u} = \tilde{u}_j$ we find that $\tilde{u}_j^o$ also satisfies the equation.

For any $j \in \mathbb{N}_n$ we have $\tilde{u}_j^o(t,x) = e^{i\nu(P_j)t}\Pi\big(e^{i\kappa(P_j)x} f_1 f_2\big)$, where $f_1 f_2$ is the function $f_1 f_2(\tilde{V}x + \tilde{W}t; r, \tilde{D}, P_j)$. Accordingly, if $j \in \mathbb{N}_n$, then

$$\begin{aligned} \xi_j^o = \frac{1}{\sqrt{2\pi}} e^{i\nu(P_j)t}\Pi\big(&e^{i\kappa(P_j)x} f_1 f_2, \\ &- A^{-1/2}\big[e^{i\kappa(P_j)x}(i\nu(P_j) f_1 f_2 + \tilde{W}\cdot\nabla_q f_1 f_2)\big]\big), \end{aligned} \tag{6.26}$$

and $\xi_j^o = \overline{\xi_{-j}^o}$ if $j \in -\mathbb{N}_n$. Thus, we have constructed a system of Floquet solutions for eqn (6.22) of the form (5.4).

By construction, $f_1 f_2$ is an analytic function of all its arguments; $\nu$ and $\kappa$ are analytic functions of $P \in \Gamma$. Since $\pi(P_j^+) = \lambda_j$, then by Lemma 4.1 $\pi(P_j(r))$ is an algebraic

[45] We recall (see Section 4.2) that this means that double is the corresponding eigenvalue $\mu = \sqrt{\lambda}/4$.

[46] We denote by $P_j^\pm$ a point in $\pi^{-1}(\lambda_j)$ which belongs to the sheet $\Gamma_\pm$.

function of $r$. Due to the corollary from that lemma, this function is analytic if $|j| > j_1$. Thus, the solutions $\xi_j^o$ are analytic in $x$ and $\tilde{D}$ and algebraic in $r$. They are analytic in $r$ if $|j| > j_1$.

The wave-number $\kappa(P_j)$ and the exponent $\nu(P_j)$ can be interpreted in terms of $(2n+2)$-gap solutions with two infinitesimal extra gaps, at least for small-gap solutions. Indeed, let $(u, w)(t) = \Phi_0(r, \tilde{D} + \tilde{W}(r)t) \in \mathcal{T}^{2n}$ be a finite-gap solution of the SG equation such that $|r - \boldsymbol{L}| = \rho \ll 1$. Then by the last assertion of Lemma 4.4 (with $n = j$ and $k = n$), for $0 < \varepsilon \ll \rho$ there exists a finite-gap solution $(u_\varepsilon, w_\varepsilon) \subset \mathcal{T}^{2n+2}_{(1,\dots,n,j)}$ which converges to $(u, w)$ when $\varepsilon \to 0$. The corresponding wave vector $\tilde{V}^{(n+1)}$ and the frequency-vector $\tilde{W}^{(n+1)}$ are $(n+1)$-vectors such that

$$\tilde{V}^{(n+1)}_{n+1} \longrightarrow \kappa(P_j), \qquad \tilde{W}^{(n+1)}_{n+1} \longrightarrow \nu(P_j) \quad \text{as } \varepsilon \to 0. \tag{6.27}$$

These limits follow from the same elementary arguments as in the KdV case (see Appendix 5).

Owing to the first limit in (6.27) and the last assertion of Lemma 4.3,

$$\kappa(P_j) = j. \tag{6.28}$$

This relation is proved for $r$ close to $\boldsymbol{L}$. Since $P_j$ is an algebraic function of $r$, then (6.28) holds identically in $r$. It specifies the formula (6.26).

Owing to the second limit in (6.27) and the last formula in Section 4.4,

$$\nu(P_j) \to j^* \equiv \sqrt{1 + j^2} \quad \text{as } r \to \boldsymbol{L}. \tag{6.29}$$

Asymptotic evaluation of the exponents $\nu(P_j)$ (cf. Section 6.2.2) shows that

$$\nu(P_j) = j^* + O(j^{-1}) \quad \text{as } j \to \infty \tag{6.29'}$$

(see Bikbaev and Kuksin 1992).

Arguing as in the KdV case (see Appendix 5 and (6.18)) we can see that the function $f_1 f_2$ in (6.26) is asymptotically close to one:

$$|f_1 f_2 - 1| = o(1) \ \text{ as } r \to \boldsymbol{L} \quad \text{and} \quad = O(j^{-1}) \ \text{ as } j \to \infty, \tag{6.30}$$

where $f_1 f_2 = f_1 f_2(\mathfrak{z}; r, \tilde{D}, P)$ and the estimates hold uniformly in $\mathfrak{z}$ from a complex neighbourhood of the real torus.

The system of Floquet solutions that we have constructed meets assumptions (a)–(d) from Section 5.2. Indeed, (a) follows from (6.25) and (6.29) while (b)–(d) result from previous discussions of the smoothness of the function $f_1 f_2$ and the exponent $\nu$.

The system of Floquet solutions $\{\xi_j^o \mid j \in \mathbb{N}_n\}$ is complete non-resonant:

**Non-resonance.** Using Lemma 4.4 we constructed in Section 4.4 coordinates $\mathcal{R}_1, \dots, \mathcal{R}_n$ on the small-gap part $R_0$ of the algebraic set $R$ such that the point $\boldsymbol{L}$ lies in the closure $\overline{R_0}$ and has coordinates $\mathcal{R} = 0$. As in the KdV case, we have to check the relation (6.16) and a similar relation, equivalent to (5.19).

Let us take any $j \geq n+1$ and consider the resonant function in the l.h.s. of (6.16), where $\tilde{W}_{n+1}^{(n+1)}(r,0) = v_j(r)$. We shall study this function using the coordinates $\mathcal{R}$ and denote it $\eta(\mathcal{R})$. Owing to Lemma 4.4 with $n := n+1$, $\eta$ is an analytic function of the arguments $I_k = \mathcal{R}_k^2/2$, $k = 1, \ldots, n$ (see the discussion at the end of Section 4.3). So if $\eta \equiv 0$, then

$$\eta(0) = 0, \qquad \frac{\partial \eta}{\partial I_l}(0) = 0, \quad \text{for } l = 1, \ldots, n.$$

Abbreviating $\sum_{k=1}^{n}$ to $\sum$ and using (4.25) and (4.26), we rewrite the first equality as

$$\sum k^* s_k + j^* = 0 \tag{6.31}$$

and rewrite the second as

$$-4\left(\sum \frac{4}{k^*} s_k - \frac{s_l}{l^*} + \frac{4}{j^*}\right) = 0, \quad l = 1, \ldots, n. \tag{6.32}$$

In particular, $s_l / l^* = C =$ const for all $l \leq n$. Substituting this relation into (6.31) and (6.32) we get that

$$C \sum k^{*2} + j^* = 0$$

and

$$C(4n-1) + \frac{4}{j^*} = 0.$$

We can eliminate $C$ from these equalities to find that $j^{*2}(4n-1) = 4\sum k^{*2}$. That is,

$$(j^2+1)(4n-1) = 4\sum (k^2+1).$$

We have obtained a contradiction since an integer number $j^2+1$ never can be divided by four. This contradiction proves that $\eta \not\equiv 0$.

Proof of the second relation is similar, see Bobenko and Kuksin (1995*a*).

We note that our arguments use essentially the assumption (4.22).

**Completeness.** The Floquet solutions (6.26) have the form (5.4), where for $j \geq n+1$ the sections $\Psi_j$ equal $(2\pi)^{-1/2}\Pi(\mathrm{e}^{i\kappa(P_j)x} f_1 f_2, \ldots)$. Owing to (6.28), (6.29), and (6.30), for every $j$ we have the following convergence:

$$\Psi_j \longrightarrow \frac{1}{\sqrt{2\pi}}\Pi(\mathrm{e}^{ijx}, -A^{-1/2} i j^* \mathrm{e}^{ijx}) = \frac{1}{\sqrt{2\pi}}(i \sin jx, \sin jx) \quad \text{as } r \to \mathbf{L}.$$

Similarly,

$$\Psi_{-j} \longrightarrow \frac{1}{\sqrt{2\pi}}(-i \sin jx, \sin jx) \quad \text{as } r \to \mathbf{L}.$$

Therefore the corollary from Lemma 5.4 applies and we only have to check assumption (2) from Definition 5.2. To do this we should not study complicated solutions $\xi_j^o$ with complex exponents $\nu$ since they correspond to small $j$ ($|j| \leq j_1$), and large $r$. On the contrary, we have only to consider $|j| \gg 1$ or $r$ close to $\boldsymbol{L}$. In these two asymptotic cases Floquet solutions for the linearized SG equation behave as complex exponents (see (6.29)–(6.30)), so to check the completeness we can argue as in Section 6.2. See Bikbaev and Kuksin (1992) and Bobenko and Kuksin (1995*a*).

# 7
# The normal form

## 7.1 A normal form theorem

We continue to study the Hamiltonian equation (5.1) near an invariant manifold $\mathcal{T}^{2n} = \Phi_0(R \times \mathbb{T}^n)$ which possesses the properties 1–5 as in Section 5.1.

Proposition 5.1 puts the linearized equation (5.2) into a constant coefficient normal form, provided that this equation possesses a complete system of Floquet solutions. In this chapter we show that under this assumption eqn (5.1) itself can be put to a convenient normal form in a neighbourhood of $\mathcal{T}^{2n}$. Namely, we show that the action-angle variables $(p, q)$ on $\mathcal{T}^{2n}$ can be supplemented by a skew-orthogonal to $\mathcal{T}^{2n}$ vector-coordinate $y$ in such a way that in the new coordinate system the symplectic form is $(dp \wedge dq) \oplus \alpha_2^Y$ and the Hamiltonian is

$$h(p) + \tfrac{1}{2}\langle B(p)y, y\rangle + h_3(p, q, y), \quad h_3 = O(\|y\|^3).$$

Here $B(p)$ is the self-adjoint operator from Proposition 5.1 and the term $h_3$ defines a Hamiltonian vector field of the same order as the non-linear part $J\nabla H$ of the original equation (this is a crucial property of the normal form!).

We assume that the linearized equation (5.2) has a complete family of skew-orthogonal Floquet solutions $v_j(t)$ as in (5.4), and define the singular subset $R_s$, $R_s = R_s^c \cap R$ as in Section 5.3 (see there Remark 2). As in Section 5.3, we choose any sub-domain $R_1$ that lies in a compact part of the regular set $R_0 = R \setminus R_s$, i.e. $R_1 \Subset R_0$. A normal form as above will be constructed in the vicinity of the manifold $\mathcal{T}_1^{2n} = \Phi_0(R_1 \times \mathbb{T}^n)$.

By Lemma 2.1 eqn (2.1) is integrable in $\Phi_0(R_0 \times \mathbb{T}^n)$. So we can cover $\Phi_0(R_1 \times \mathbb{T}^n)$ by a finite system of open sub-domains such that in each one the equation admits analytic action-angle variables $(p, q)$ as in (2.6). To simplify notation we suppose that the action angles exist globally in $\Phi_0(R_1 \times \mathbb{T}^n)$. We shall use these coordinates instead of $(r, \mathfrak{z})$. Accordingly, we write $\mathcal{T}_1^{2n}$ as $\mathcal{T}_1^{2n} = \Phi_0(P \times \mathbb{T}^n)$, where $P = \{p\} \subset \mathbb{R}^n$ and $\mathbb{T}^n = \{q\}$.

We denote $W = P \times \mathbb{T}^n$. The map $\Phi_0 : W \to \mathcal{T}^{2n} \subset Z$ analytically extends to a bounded analytic map $W^c \to Z^c$, where $W^c$ is a complex neighbourhood of $W$ of the form $W^c = (P + \delta) \times \{q \in \mathbb{C}^n/2\pi\mathbb{Z}^n \mid |\operatorname{Im} q| < \delta\}$. We treat $W$ and $W^c$ as sub-manifolds of the Hilbert manifolds $\mathcal{Y}$ and $\mathcal{Y}^c$,

$$\mathcal{Y} = \mathcal{Y}_d = \mathbb{R}^n \times \mathbb{T}^n \times Y_d, \qquad \mathcal{Y}^c = \mathcal{Y}_d^c = \mathbb{C}^n \times (\mathbb{C}^n/\mathbb{Z}^n) \times Y_d^c,$$

where

$$Y_d = \overline{\operatorname{span}}\,\{\varphi_j \mid j \in \mathbb{Z}_n\} \subset Z_d.$$

Since $\omega = \nabla h$ (see Lemma 2.2), then we write the skew-orthogonal Floquet solutions $v_j(t)$ as

$$v_j(t;p,q) = e^{i\nu_j(p)t}\Psi_j(p, q + t\nabla h(p)), \quad p \in P,\ q \in \mathbb{T}^n,\ j \in \mathbb{Z}_n. \tag{7.1}$$

The linear in $y$ map $y \mapsto \Phi_1(p, q + t\nabla h)y$ as in (5.9) reduces the linearized equation (5.2) to the constant-coefficient linear equation

$$\dot{y} = JB(p)y, \ldots, \tag{7.2}$$

where the dots stand for components of the linearized equation in directions tangent to $W \times \{0\}$ (see Proposition 5.1). We denote by $\mathcal{S}_\delta = \mathcal{S}_\delta(Y_d)$ the manifold

$$\mathcal{S}_\delta = W \times \mathcal{O}_\delta(Y), \quad Y = Y_d,$$

and denote by $\mathcal{S}_\delta^c$ its complex neighbourhood $\mathcal{S}_\delta^c = W^c \times \mathcal{O}_\delta(Y^c)$. We give $\mathcal{S}_\delta$ a symplectic structure by means of the 2-form $(dp \wedge dq) \oplus \alpha_2^Y$, where $\alpha_2^Y = \alpha_2|_Y$. Since $\alpha_2 = \bar{J}dz \wedge dz$ and the spaces $\{Y_s\}$ are $\bar{J}$-invariant, then

$$\alpha_2^Y = \bar{J}dy \wedge dy.$$

Our goal in this section is to prove the following Normal Form Theorem:

**Theorem 7.1.** *Let the Hamiltonian equation (5.1) and its invariant sub-manifold $\mathcal{T}^{2n}$ satisfy assumptions* 1–5; *let a sub-domain $\mathcal{T}_1^{2n} = \Phi_0(P \times \mathbb{T}^n)$ be as above and (7.1) be a complete system of skew-orthogonal Floquet solutions of the linearized equation (5.2). Then there exists $\delta_1 > 0$ and an analytic symplectomorphism $G : (\mathcal{S}_{\delta_1}, dp \wedge dq \oplus \alpha_2^Y) \to (Z, \alpha_2)$ such that $G(\mathcal{S}_{\delta_1})$ is a neighbourhood of $\mathcal{T}_1^{2n}$ and*

$$\mathcal{H} \circ G = h(p) + \tfrac{1}{2}\langle B(p)y, y\rangle + h_3(p, q, y).$$

*Here $h_3 = O(\|y\|^3)$ is an analytic functional such that its gradient map is of order $\tilde{d} =$* $\max\{d_H, -\Delta - d_J, \widetilde{\Delta} - d_J\}$, *i.e.* $\|\nabla_y h_3(p, q, y)\|_{d-\tilde{d}} \le C\|y\|_d^2$ *for any* $(p, q, y) \in \mathcal{S}_{\delta_1}$.

The proof of the theorem occupies the rest of this section.

To simplify the presentation we suppose below that all the frequencies $\nu_j(p)$ are real and consequently the operator $B(p)$ is diagonal in the $\varphi_j$-basis of the space $Y$:

$$B(p)\varphi_j = \frac{\nu_j(p)}{\nu_j^J}\varphi_j, \quad \forall\, j \in \mathbb{Z}_n.$$

The general case differs from this special case only in more awkward notations since we should treat differently (but in much the same way) the indices $j$, corresponding to real, imaginary, and complex frequencies $\nu_j$.

We start with the affine in $y$ map $\Phi$:

$$\Phi = \Phi_0 + \Phi_1 : \mathcal{S}_\delta^c \to Z^c, \quad (p, q, y) \mapsto \Phi_0(p, q) + \Phi_1(p, q)y.$$

It is real (sends $\mathcal{S}_\delta$ to $Z$), bounded on bounded subsets of $\mathcal{S}_\delta^c$, and is weakly analytic by assumptions (b) and (d). So $\Phi$ is an analytic map by the criterion of analyticity.

By Lemma 5.1 its linearizations at points from $W^c \times \{0\}$ define isomorphisms of $\mathbb{R}^{2n} \times Y$ and $Z$. Thus, by the Inverse Function Theorem the map $\Phi$ defines an analytic isomorphism of $\mathcal{S}^c_{\delta'}$ and a complex neighbourhood of $\mathcal{T}_1^{2n}$ in $Z$, provided that $\delta' \leq \delta$ is sufficiently small.[47]

Next we study the symplectic properties of the map $\Phi$. Since the restriction of $\Phi$ to $W \times \{0\}$ equals $\Phi_0$ and the restriction to any disc $\{w\} \times \mathcal{O}_\delta(Y)$ equals $\Phi_1(w)$ up to a translation, then these restrictions are symplectic. In particular, for any $w \in W$ the map $\Phi_*(w, 0)$ is a linear symplectomorphism. Hence, the pull-back form $\omega_2(w, y)$,

$$\omega_2 := \Phi^* \alpha_2,$$

equals $(dp \wedge dq) \oplus \alpha_2^Y$ for $w = 0$ and these two forms coincide, being restricted to any disc$\{w\} \times \mathcal{O}_\delta(Y)$. This means that the difference

$$\omega_\Delta = \omega_2 - dp \wedge dq \oplus \alpha_2^Y$$

may be written as

$$\omega_\Delta = j_{WW}(w, y)\, dw \wedge dw + j_{WY}(w, y)\, dy \wedge dw + j_{YW}(w, y)\, dw \wedge dy,$$

where $j_{YW}(w, y) = j^*_{WY}(w, y)$ and the linear operators $j_{WW}$, $j_{WY}$, and $j_{YW}$ vanish for $y = 0$ (see Section 1.3 for the notation we use).

In the calculations we carry out below we adopt gradient notations for linearizations of the maps $\Phi$ and $\Phi_1$ in $w$. Namely, we write

$$\Phi_*(w, y)(\delta w, 0) = \sum \nabla_{w_j} \Phi(w, y) \delta w_j =: \nabla_w \Phi \cdot \delta w,$$

where $\nabla_w \Phi = (\nabla_p \Phi, \nabla_q \Phi) \in Z \times \cdots \times Z$ ($2n$ times). Similarly, we write $\Phi_{1*}(\delta w, 0) = \nabla_w \Phi_1 \cdot \delta w$, where any component $\nabla_{w_j} \Phi_1$ is a linear operator $Y \to Z$. In this notation we have:

$$\begin{aligned}
\omega_2[\delta y, \delta w] &= \alpha_2[\Phi_1\, \delta y, \Phi_{0*}\, \delta w + (\nabla_w \Phi_1 \cdot \delta w) y] \\
&= \alpha_2[\Phi_1\, \delta y, (\nabla_w \Phi_1 \cdot \delta w) y] = \langle \bar{J} \Phi_1\, \delta y, \nabla_w \Phi_1 y \rangle \cdot \delta w
\end{aligned}$$

and

$$\omega_2[\delta w, \delta y] = \langle \bar{J}(\nabla_w \Phi_1 \cdot \delta w) y, \Phi_1\, \delta y \rangle.$$

Hence,

$$\begin{aligned}
j_{WY}(w, y)\, \delta y &= \langle \bar{J} \Phi_1(w)\, \delta y, \nabla_w \Phi_1(w) y \rangle, \\
j_{YW}(w, y)\, \delta w &= -j^*_{WY}(w, y)\, \delta \Phi_1(w)^* \bar{J}(\nabla_w \Phi_1 \cdot \delta w) y.
\end{aligned} \tag{7.3}$$

---

[47] To get this result one has to cover the set $W^c \times \{0\}$ by balls $B_w$, $w \in W^c$, such that the Inverse Function Theorem applies to $\Phi$ restricted to each ball; to find a finite system of these balls which cover $W \times \{0\}$ and next choose $\delta' > 0$ so small that $\mathcal{S}^c_{\delta'}$ is contained in the union of these balls.

Abbreviating $(\delta w, \delta y) \in \mathbb{R}^{2n} \times Y$ to $\delta\mathfrak{z}$, we write the form $\omega_\Delta$ as $\omega_\Delta = \bar{J}_\Delta\, d\mathfrak{z} \wedge d\mathfrak{z}$, where $\bar{J}_\Delta$ is the operator matrix:

$$\bar{J}_\Delta = \bar{J}_\Delta(w, y) = \begin{bmatrix} j_{WW} & j_{WY} \\ -j_{WY}^* & 0 \end{bmatrix}. \tag{7.4}$$

The form $\omega_\Delta$ is exact, as well as the forms $\omega_2$ and $dp \wedge dq \oplus \alpha_2^Y$, i.e. $\omega_\Delta = d\omega_1$. Lemma 1.3 represents the 1-form $\omega_1$ as

$$\begin{aligned}\omega_1(w, y) &= \left( \int_0^1 \langle \bar{J}\Phi_1(w)y, \nabla_w \Phi_1(w)ty \rangle\, dt \right) dw \\ &= \tfrac{1}{2}\langle \bar{J}\Phi_1(w)y, \nabla_w \Phi_1(w)y \rangle\, dw = \tfrac{1}{2}\alpha_2[\Phi_1(w)y, \nabla_w \Phi_1(w)y]\, dw.\end{aligned}$$

We sum up the obtained results in the following lemma.

**Lemma 7.1.** *The form* $\omega_2 = \Phi^*\alpha_2$ *equals* $(dp \wedge dq) \oplus \alpha_2^Y + d(L(w, y)\, dw)$, *where the* $2n$*-vector* $L$ *is* $L = \frac{1}{2}\alpha_2[\Phi_1(w)y, \nabla_w \Phi_1(w)y]$.

So far we have examined how the map $\Phi$ transforms the symplectic form $\alpha_2$. Now we calculate how it changes the Hamiltonian $\mathcal{H}$. To begin with we analyse how the non-autonomous linear transformation $\Phi_1$ transforms the quadratic part $\langle Au, u\rangle$ of the Hamiltonian $\mathcal{H}$.

For any $w = (p, q) \in W$ we denote $\Phi^t = \Phi_1(p, q + t\nabla h)$. Since the non-autonomous symplectic linear map $\Phi^t$ sends solutions $y(t)$ of eqn (7.2) to solutions $v(t) = \Phi^t y(t)$ of (5.2), then we have the following equalities:

$$\begin{array}{ccc} \dot{v} & = & \dot{\Phi}^t y + \Phi^t \dot{y} \\ \| & & \| \\ JA_t v & & \dot{\Phi}^t y + \Phi^t JB(p)y \\ \| & & \\ JA_t\Phi^t y. & & \end{array}$$

Thus,

$$JA_t\Phi^t y = \dot{\Phi}^t y + \Phi^t JB(p)y. \tag{7.5}$$

Taking a skew-product of (7.5) with $-v$, we get:

$$\begin{array}{ccc} \langle JA_t\Phi^t y, \bar{J}v\rangle & = & \langle \dot{\Phi}^t y + \Phi^t JB(p)y, \bar{J}v\rangle \\ \| & & \| \\ \langle \Phi^{t*}A_t\Phi^t y, y\rangle & & \langle \dot{\Phi}^t y, \bar{J}\Phi^t y\rangle + \langle B(p)y, y\rangle, \end{array} \tag{7.6}$$

where we use that $\langle \Phi^t JBy, \bar{J}\Phi^t y\rangle = \langle JBy, \bar{J}y\rangle = \langle By, y\rangle$ by the symplecticity of the map $\Phi^t$.

Since for $t = 0$ we have $A_t = A + (\nabla H)_*(\Phi_0(w))$ and $\dot\Phi^t = \nabla_q\Phi_1(w)\cdot\nabla h(p)$, then relation (7.5) with $t = 0$ implies that

$$\Phi_1(w)JB(p) = J(A + (\nabla H)_*(\Phi_0(w)))\Phi_1(w) - \nabla_q\Phi_1(w)\cdot\nabla h(p). \qquad (7.7)$$

Similarly, (7.6) implies that

$$\begin{aligned}&\big\langle(B(p) - \Phi_1(w)^*(A + (\nabla H)_*(\Phi_0(w)))\Phi_1(w))y,\, y\big\rangle\\ &\quad = \langle\Phi_1(w)^*\bar J\big(\nabla_q\Phi_1(w)\cdot\nabla h(p)\big)y,\, y\rangle = \langle\mathfrak{A}(w)y,\, y\rangle,\end{aligned}$$

where $\mathfrak{A}$ stands for the symmetrization of the operator $\Phi_1^*\bar J(\nabla_q\Phi_1\cdot\nabla h)$, i.e.

$$\mathfrak{A}(w) = \tfrac{1}{2}\big(\Phi_1(w)^*\bar J(\nabla_q\Phi_1(w)\cdot\nabla h(p)) - (\nabla_q\Phi_1(w)^*\cdot\nabla h(p))\bar J\Phi_1(w)\big).$$

Since this relation holds for any vector $y \in Y$, then

$$B(p) - \Phi_1(w)^*(A + (\nabla H)_*(\Phi_0(w)))\Phi_1(w) = \mathfrak{A}(w). \qquad (7.8)$$

**Lemma 7.2.** *The operator $\mathfrak{A}$ defines a $(\Delta + d_J)$-smoothing symmetric map $\mathfrak{A} : Y_d^c \to Y_{d+d_J+\Delta}^c$, analytic in $w \in W^c$.*

*Proof.* The operator $\mathfrak{A}$ is symmetric by its construction. It remains to check its smoothness.

Since $\nabla_q\Phi_1 = \nabla_q(\Phi_1 - \iota)$, then by (5.11) and the Cauchy estimate the operator $\nabla_q\Phi_1\cdot\nabla h$ analytically depends on $w \in W^c$ as a map $Y_d^c \to Z_{d+\Delta}$. By Lemma 5.2 the operator $\Phi_1(w)^*\bar J : Z_{d+\Delta}^c \to Y_{d+d_J+\Delta}^c$ also is analytic in $w$. Hence, the first term of the operator $\mathfrak{A}$ defines an analytic in $w \in W^c$ map $Y_d^c \to Y_{d+d_J+\Delta}^c$.

Using Lemma 5.2 once again we find that the operator $\bar J\Phi_1(w) : Y_d^c \to Z_{d+d_J}^c$ is analytic in $w$. Owing to the second assertion of this lemma and the Cauchy estimate, the map $\nabla_q\Phi_1^*\cdot h : Z_{d+d_J}^c \to Y_{d+d_J+\Delta}^c$ is analytic in $w$ as well. Combining these two statements, we find that the second term of $\mathfrak{A}$ also defines an analytic in $w \in W^c$ map $Y_d^c \to Y_{d+d_J+\Delta}^c$. This completes the proof. □

Equalities (7.7) and (7.8) were obtained for real $w$. Since these relations are analytic in $w$, they remain true for any $w \in W^c$.

Now we write the transformed Hamiltonian $\mathcal{H}\circ\Phi$ as

$$\mathcal{H}\circ\Phi = \tfrac{1}{2}\langle A\Phi_0, \Phi_0\rangle + \langle A\Phi_0, \Phi_1 y\rangle + \tfrac{1}{2}\langle A\Phi_1 y, \Phi_1 y\rangle + H(\Phi),$$

and separate its part, affine in $y$:

$$\begin{aligned}\mathcal{H}\circ\Phi =& \big(\tfrac{1}{2}\langle A\Phi_0, \Phi_0\rangle + H(\Phi_0)\big) + \big(\langle A\Phi_0, \Phi_1 y\rangle + \langle\nabla H(\Phi_0), \Phi_1 y\rangle\big)\\ &+ \big(\tfrac{1}{2}\langle A\Phi_1 y, \Phi_1 y\rangle + H(\Phi_0 + \Phi_1 y) - H(\Phi_0) - \langle\nabla H(\Phi_0), \Phi_1 y\rangle\big).\end{aligned}$$

The first term in the r.h.s equals $h(p)$.

By Lemma 7.1 the form $\omega_2 = \Phi^*\alpha_2$ equals $(dp\wedge dq)\oplus\alpha_2^Y$, when $y = 0$. Hence, for $y = 0$ the $y$-component of eqn (5.1), written in the $(p, q, y)$-variables, is $J\nabla_y(\mathcal{H}\circ\Phi)$.

It equals zero since the set $\{y = 0\}$ is invariant for the equation. Thus, the second term vanishes.

By (7.8), $\langle A\Phi_1 y, \Phi_1 y\rangle = \langle By, y\rangle - \langle \nabla H_* \Phi_1 y, \Phi_1 y\rangle - \langle \mathfrak{A} y, y\rangle$. Therefore the third term in the r.h.s. equals $\frac{1}{2}\langle B(p)y, y\rangle + h_2(p, q, y)$, where

$$h_2 = -\tfrac{1}{2}\langle(\nabla H)_*(\Phi_0)\Phi_1 y, \Phi_1 y\rangle - \tfrac{1}{2}\langle \mathfrak{A}(w)y, y\rangle$$
$$+ H(\Phi_0 + \Phi_1 y) - H(\Phi_0) - \langle \nabla H(\Phi_0), \Phi_1 y\rangle.$$

It is easy to see, using Lemmas 5.2 and 5.3, that $h_2$ defines an analytic gradient map $\nabla_y h_2\ \mathbb{R}^{2n} \times Y_d \to Y_{d-\tilde{d}}$.

Thus, the affine in $y$ map $\Phi$ transforms the Hamiltonian $\mathcal{H}$ to

$$\mathcal{H} \circ \Phi = h(p) + \tfrac{1}{2}\langle B(p)y, y\rangle + h_2(p, q, y),$$

where $h_2 = O\|y\|^2$ and $\operatorname{ord} \nabla h_2 = \tilde{d}$.

Our next goal is to normalize the symplectic structure $\omega_2 = \Phi^*\alpha_2$ in $\mathcal{S}_\delta$ by means of the Moser–Weinstein theorem (Lemma 1.4). The theorem states that $\varphi^*\omega_2 = (dp \wedge dq) \oplus \alpha_2^Y$, where $\varphi$ is the time-one shift $S_0^1$ along trajectories of a non-autonomous equation:

$$\dot{\mathfrak{z}} = V^t(\mathfrak{z}), \quad \mathfrak{z} = (w, y).$$

The vector field $V^t : \mathcal{S}_\delta \to \mathbb{R}^{2n} \times Y_d$ is obtained as a solution of the equation

$$-(\bar{J}_0 + t\bar{J}_\Delta)V^t = a(\mathfrak{z}, y), \tag{7.9}$$

where

$$\bar{J}_0(\delta p, \delta q, \delta y) = (-\delta q, \delta p, \bar{J}\,\delta y),$$

the operator $\bar{J}_\Delta$ is as in (7.4), and the map $a$ is such that the differential of the 1-form $a(\mathfrak{z})d\mathfrak{z}$ equals $\omega_2 - (dp \wedge dq) \oplus \alpha_2^Y$. By Lemma 7.1, $a(\mathfrak{z}) = (L(\mathfrak{z}), 0)$, where the $2n$-vector $L(\mathfrak{z})$ is specified in the lemma.

We claim that the map $\varphi$ sends $\mathcal{S}_{\delta_1}^c$ to $\mathcal{S}_\delta^c$ ($\delta_1$ is sufficiently small compared with $\delta$) and transforms $\mathcal{H} \circ \Phi$ to a Hamiltonian of similar form:

**Lemma 7.3.** *The Hamiltonian $\mathcal{H} \circ \Phi \circ \varphi$ is equal to*

$$\mathcal{H} \circ \Phi \circ \varphi = h(p) + \tfrac{1}{2}\langle B(p)y, y\rangle + \tfrac{1}{2}\langle \mathfrak{B}(p, q)y, y\rangle + h_3(p, q, y), \tag{7.10}$$

*where $B(p)$ is the same as in (7.2) and $\mathfrak{B}(p, q)$ is a linear operator of order $\tilde{d}$, analytic in $(p, q)$ ($\tilde{d}$ is the same as in Theorem 7.1). The function $h_3 = O(\|y\|^3)$ has an analytic gradient map of order $\tilde{d}$, $\|\nabla_y h_3(p, q, y)\|_{d+\tilde{d}} \le C\|y\|^2$.*

The statement of the lemma is quite obvious for a finite-dimensional phase space $\mathcal{S}_\delta$, but not in the infinite-dimensional situation. Indeed, the transformation $\varphi$ has the form $\varphi = \mathrm{id} + \widetilde{\varphi}$, where $\widetilde{\varphi}$ is a $\Delta$-smoothing map such that $\widetilde{\varphi} = O(\|y\|)^2$. Thus, the transformed Hamiltonian gets the term

$$\langle B(p)y, \widetilde{\varphi}\rangle, \tag{7.11}$$

which is $O(\|y\|^3)$ with a gradient map of order $d_A - \Delta$. The number $d_A - \Delta$ could be relatively big and the term (7.11) could spoil the forthcoming constructions. Fortunately, (7.11) vanishes up to a smoother term. This is essentially what the lemma states.

We prove the lemma in Section 7.2 and now show how to complete the theorem's proof, given this result. To prove the theorem it remains to check that the operator $\mathfrak{B}$ in (7.10) vanishes. Since $\varphi$ is analytic and $O(\|y\|^2)$-close to the identity, then $\varphi_*(w,0)|_{\{0\}\times Y} = \mathrm{id}$. Denoting $w(t) = (p, q + t\nabla h(p))$ we get that the transformation

$$y(t) \mapsto (\Phi \circ \varphi)_*(w(t), 0)y(t) = (\Phi_*(w(t), 0))y(t)$$

sends solutions of eqn (7.2) to solutions of (5.2).

On other hand, $\Phi \circ \varphi$ is a canonical transformation which transforms solutions of the equation with the Hamiltonian (7.10) to solutions of (5.1). In particular, it sends the curves $w(t)$ to solutions $u_0(t)$ of (5.1). Hence, the linearization $(\Phi \circ \varphi)_*(w(t))$ converts solutions of the linearized equation

$$\dot{y} = J(B(p) + \mathfrak{B}(w(t))y, \ldots \tag{7.12}$$

to solutions of (5.2) and $\varphi_*(w(t))$ converts solutions of (7.12) to solutions of (7.2) (cf. item (b) of Proposition 5.1). Since a $y$-component of the map $\varphi_*(w(t))$ is the identity, then we must have $JB(p)y \equiv J(B(p) + \mathfrak{B}(w(t)))y$. This implies that $\mathfrak{B} \equiv 0$ and the theorem is proved. □

## 7.2 Proof of Lemma 7.3

In this section we denote by $\{\mathcal{Z}_s\}$ a Hilbert scale formed by the spaces $\mathcal{Z}_s = \mathbb{R}^{2n} \times Y_s$ and abbreviate $\mathcal{Z}_d$ to $\mathcal{Z}$. So $T_{\mathfrak{h}}\mathcal{Y} \simeq \mathcal{Z}$ for every $\mathfrak{h}$ in $\mathcal{Y} = \mathcal{Y}_d$.

To study the vector field $V^t$ which defines the transformation $\varphi$ we expand the operator $-(J_0 + t\bar{J}_\Delta)^{-1}$ in the Neumann series:

$$-(\bar{J}_0 + t\bar{J}_\Delta)^{-1} = (\mathrm{id} - tJ_0\bar{J}_\Delta)^{-1}J_0 = \sum_{m=0}^{\infty}(tJ_0\bar{J}_\Delta)^m J_0.$$

The series converges for small $\delta$ since by (7.3) and (7.4) the linear map

$$J_0\bar{J}_\Delta(w, y) : \mathcal{Z}_d \to \mathcal{Z}_{d+\Delta}$$

is analytic in $(w, y) \in \mathcal{S}^c_\delta$ and is proportional to $y$, so its operator-norm is bounded by $C\delta$. Denoting

$$\tilde{a}(w, y) = J_0 a(w, y) = (\omega_2[\Phi_1 y, (-\nabla_q, \nabla_p)\Phi_1 y], 0) \in \mathcal{Z},$$

we solve eqn (7.9) for $V^t$ and find that

$$V^t(w, y) = -(\bar{J}_0 + t\bar{J}_\Delta)^{-1}a(w, y) = \frac{1}{2}\sum_{m=0}^{\infty}(tJ_0\bar{J}_\Delta)^m\tilde{a}(w, y). \tag{7.13}$$

Therefore $V^t = O(\|y\|^2)$ is a $\Delta$-smoothing analytic vector field. In particular, the flow maps $S_0^t$ of the equation $\dot{\mathfrak{z}} = V^t$ with $0 \le t \le 1$ send a domain $\mathcal{S}^c_{\delta_1}$ with sufficiently small $\delta_1$ to $\mathcal{S}^c_\delta$.

Isolating in the r.h.s. of (7.13) the term with $m = 0$ we find that the vector field $V^t$ satisfies the self-similarity identity:

$$V^t(w, y) = \tfrac{1}{2}\tilde{a}(w, y) + \tfrac{1}{2}(tJ_0\bar{J}_\Delta)\sum_{m=0}^{\infty}(tJ_0\bar{J}_\Delta)^m\tilde{a} = \tfrac{1}{2}\tilde{a}(w, y) + tJ_0\bar{J}_\Delta V^t(w, y). \tag{7.14}$$

We begin an analysis of the transformed Hamiltonian with its most complicated term,

$$\tfrac{1}{2}\langle B(p)y, y\rangle \circ \varphi. \tag{7.15}$$

We abbreviate the function $\frac{1}{2}\langle B(p)y, y\rangle \circ S_0^t$ to $\xi_t$, so $\xi_1$ equals (7.15) and $\xi_0 = \langle B(p)y, y\rangle/2$. We have

$$\tfrac{1}{2}\langle B(p)y, y\rangle \circ \varphi - \tfrac{1}{2}\langle B(p)y, y\rangle = \xi_1 - \xi_0 = \int_0^1 \frac{d}{dt}\xi_t\, dt\,.$$

Denoting by $V_p^t$, $V_q^t$, and $V_y^t$ the components of the vector field $V^t$, we get

$$\int_0^1 \frac{d}{dt}\xi_t\, dt = \int_0^1 \Bigg(\underbrace{\frac{1}{2}\Big\langle(\nabla_p B(p)\cdot V_p^t)y, y\Big\rangle}_{q_t} + \underbrace{\Big\langle B(p)y, V_y^t(w, y)\Big\rangle}_{Q_t}\Bigg)\circ S^t dt. \tag{7.16}$$

By (5.21) the function $q_t$ is analytic with a gradient map of order $\widetilde{\Delta} - d_J$. Now we pass to the term $Q_t$. Since the vector $\tilde{a}(w, y)$ has a zero $y$-component, then we get from (7.14) that

$$Q_t = t\langle B(p)y, \Pi_y J_0\bar{J}_\Delta V^t(w, y)\rangle = -t\langle JB(p)y, \Pi_y\bar{J}_\Delta V^t(w, y)\rangle,$$

where $\Pi_y$ stands for the natural projector $\mathcal{Z} \to Y$. Owing to (7.4), $\Pi_y\bar{J}_\Delta V^t = -j_{WY}^* V_w^t$, where we abbreviate $(V_p^t, V_q^t)$ to $V_w^t$. So we get the following formula:

$$Q_t = t\,(j_{WY}(w, y)JB(p)y)\cdot V_w^t(w, y), \tag{7.17}$$

where $\cdot$ stands for the scalar product in $\mathbb{R}^{2n}$. Using (7.3) and (7.7) we write the second factor in the right-hand side of (7.17) as

$$\begin{aligned}\mathbb{R}^{2n} \ni j_{WY}(w, y)JB(p)y &= \langle\bar{J}\Phi_1 JBy, \nabla_w\Phi_1 y\rangle\\ &= -\langle A\Phi_1 y, \nabla_w\Phi_1 y\rangle - \langle(\nabla H)_*(\Phi_0)\Phi_1 y, \nabla_w\Phi_1 y\rangle\\ &\quad - \langle\bar{J}(\nabla_q\Phi_1\cdot\nabla h)y, \nabla_w\Phi_1 y\rangle,\end{aligned} \tag{7.18}$$

where $\Phi_1 = \Phi_1(w)$ and $\Phi_0 = \Phi_0(w)$. Using (7.8) we rewrite the first term in the r.h.s. of (7.18) as

$$\begin{aligned}&-\tfrac{1}{2}\nabla_w\langle A\Phi_1(w)y, \Phi_1(w)y\rangle\\ &= \tfrac{1}{2}\nabla_w\big[-\langle B(p)y, y\rangle + \langle(\nabla H)_*(\Phi_0)\Phi_1(w)y, \Phi_1(w)y\rangle + \langle\mathfrak{A}(w)y, y\rangle\big].\end{aligned}$$

Using (5.20) as well as Lemmas 5.2 and 7.5 we find that this is a quadratic in $y$ form, corresponding to a linear operator of order $\max(\widetilde{\Delta} - d_J, d_H, -\Delta - d_J) = \tilde{d}$. By

Lemma 5.2 the second term in the r.h.s. of (7.18) corresponds to a linear in $y$ operator of order $d_H$, and the third term corresponds to an operator of order $-d_J - 2\Delta$. Thus, $j_{YW}(w, y)JB(p)y$ is a quadratic in $y$ form which we write as $\frac{1}{2}\langle\mathfrak{B}_1(w)y, y\rangle$, where $\operatorname{ord}\mathfrak{B}_1 = \tilde{d}$ and the linear operator $\mathfrak{B}_1$ is analytic in $w \in \mathcal{S}_\delta$ (more precisely, $\mathfrak{B}_1$ is a vector, formed by $2n$ operators).

Now let us consider the third factor in (7.17), $V_w^t = (V_{w_1}^t, \ldots, V_{w_{2n}}^t)$. For $l = 1, \ldots, 2n$ we denote by $\Pi_l$ the projector $\mathcal{Z} \to \mathbb{R}$ which sends $(w, y)$ to $w_l$ (so $\Pi_l V^t = V_{w_l}^t$). Below we write estimates for the vector $\tilde{a}$ and the operator $\Pi_l J_0 J_\Delta$ which follow directly from the formulas for $\tilde{a}$ and the operator $j_{YW}$, taking into account the smoothness of the operators $\Phi_1$ and $\nabla\Phi_1$, specified in Lemma 5.2:

$$
\begin{aligned}
&\|\tilde{a}(w, y)\|_m \leq C_m\|y\|\|y\|_{-d-\Delta-d_J}, && \text{for any } m,\\
&\|\Pi_l J_0 \bar{J}_\Delta(w, y)\|_{\mathcal{Z}_{-d-\Delta-d_J},\mathbb{R}} \leq C\|y\|, && \text{for any } l\,,\\
&\|\Pi_l J_0 \bar{J}_\Delta(w, y)\|_{\mathcal{Z}_d,\mathbb{R}} \leq C\|y\|_{-d-\Delta-d_J}, && \text{for any } l\,.
\end{aligned}
\tag{7.19}
$$

For any $l$ let us consider the function $(w, y) \mapsto \Pi_l t J_0 \bar{J}_\Delta(w, y)\tilde{a}(w, y)$. Since the operator-valued map $y \to \bar{J}_\Delta(w, y)$ is linear in $y$, then linearization of this function, applied to a vector $(0, \eta)$, equals

$$(\Pi_l t J_0 \bar{J}_\Delta(w, y))(\tilde{a}_*(w, y)(0, \eta)) + \Pi_l t J_0 \bar{J}_\Delta(w, \eta)\tilde{a}(w, y).$$

Using (7.19) we bound this number by

$$Ct\|y\|\|\tilde{a}_*(w, y)(0, \eta)\|_{-d-\Delta-d_J} + Ct\|\eta\|_{-d-\Delta-d_J}\|a\| \leq C_1 t\|y\|^2\|\eta\|_{-d-\Delta-d_J}.$$

Hence, $\|\nabla_y(\Pi_l t J_0 \bar{J}_\Delta a)\|_{d+\Delta+d_J} \leq C\|y\|^2$. Similar estimates hold for higher-order in $t$ terms in (7.13). Since $\|\nabla_y \Pi_l \tilde{a}\|_{d+\Delta+d_J} \leq C\|y\|$ due to the last estimate in (7.18), then for $(w, y) \in \mathcal{S}_\delta^c$ we get

$$
\begin{aligned}
\|\nabla_y V_{w_l}^t\|_{d+\Delta+d_J} &\leq \tfrac{1}{2}\|\nabla_y \Pi_l a + \nabla_y(\Pi_l t J_0 \overline{J}_\Delta a) + \cdots\|_{d+\Delta+d_J}\\
&\leq C(\|y\| + t\|y\|^2 + \cdots) \leq C_1\|y\|
\end{aligned}
$$

(we used (7.19) once again). Thus, the functions $V_{w_j}^t$, $j = 1, \ldots, 2n$, are analytic and bounded by $C\|y\|^2$; their gradient maps are $(\Delta + d_J)$-smoothing and are bounded by $C\|y\|$.

We have seen that any function $Q_t$, $0 \leq t \leq 1$, is analytic in the domain $\mathcal{S}_\delta$, where it is bounded by $C\|y\|^4$ and $\|\nabla_y Q_t\|_{d-\tilde{d}} \leq C\|y\|^3$. Because of formula (7.16), the function $\frac{1}{2}\langle By, y\rangle \circ \varphi - \frac{1}{2}\langle By, y\rangle$ has a gradient map of order $\tilde{d}$, which satisfies similar estimates.

By formula (7.13) the map $V^t$ is analytic $\Delta$-smoothing and is bounded by $C\|y\|^2$. Therefore, the map $\varphi - \mathrm{id}$ is $\Delta$-smoothing: $\|\varphi - \mathrm{id}\|_{d+\Delta} \leq C_1\|y\|^2$ and

$$\|\varphi_*(w, y) - \mathrm{id}\|_{s,s+\Delta} \leq C_2\|y\|, \quad \forall s \in [-d - \Delta - d_J, d + \Delta] \tag{7.20}$$

(the estimate for $\varphi-$ id is obvious and (7.20) follows from (1.19)).

Finally, for any $\mathfrak{z} = (w, y) \in \mathcal{S}_{\delta_1}$ we write the transformed Hamiltonian as

$$\begin{aligned}\mathcal{H} \circ \Phi \circ \varphi(\mathfrak{z}) =& (h(p) + \tfrac{1}{2}\langle By, y\rangle) + \tfrac{1}{2}\langle By, y\rangle \circ \varphi(\mathfrak{z})\\ &- \tfrac{1}{2}\langle By, y\rangle + h \circ \varphi(\mathfrak{z}) - h(\mathfrak{z}) + h_2 \circ \varphi(\mathfrak{z})\end{aligned}$$

and denote

$$\tilde{h}_2 = \left(\tfrac{1}{2}\langle By, y\rangle \circ \varphi - \tfrac{1}{2}\langle By, y\rangle\right) + h \circ \varphi - h + h_2 \circ \varphi.$$

Since $\varphi = O\|y\|$ and $\varphi -$ id$= O\|y\|^2$, then $\tilde{h}_2 = O\|y\|^2$. The gradient of the first term was shown to be of order $\tilde{d}$. Since

$$\nabla(h \circ \varphi) = \varphi^*(\nabla h \circ \varphi) = (\varphi^* - \mathrm{id})\nabla h \circ \varphi + \nabla h \circ \varphi$$

and $\nabla h \in \mathbb{R}^{2n} \times \{0\}$, then due to (7.20) $\nabla(h \circ \varphi)(\mathfrak{z}) \in \mathcal{Z}_{d+\Delta+d_J}$. So the gradient map of the second term has the order $-\Delta - d_J$. The gradient map of the last term has the same order as $\nabla h_2$, i.e. $\tilde{d}$ (we use (7.20) once again). We have seen that

$$\tilde{h}_2 = O\|y\|^2, \qquad \operatorname{ord} \nabla\tilde{h}_2 = \tilde{d}.$$

Now we denote the quadratic part of $\widetilde{h}_2$ as $\frac{1}{2}\langle \mathfrak{B}(p, q)y, y\rangle$ and set $h_3 = \widetilde{h}_2 - \frac{1}{2}\langle \mathfrak{B}y, y\rangle$, so

$$\mathcal{H} \circ \Phi \circ \varphi = h(p) + \tfrac{1}{2}\langle By, y\rangle + \tfrac{1}{2}\langle \mathfrak{B}y, y\rangle + h_3(p, q, y).$$

The operator $\mathfrak{B}$ has the order $\widetilde{d}$, as well as $\nabla_y h_3$, and the lemma is proved. □

## 7.3 Examples

(1) *The Korteweg–de Vries equation.* The KdV equation in a Sobolev space $Z_d = H_0^d(S^1)$ with $d \geq 1$, given a symplectic structure by the form $\alpha_2 = \langle(-\partial/\partial x)^{-1}\, du, du\rangle_{L_2}$, takes the form (2.1) (see Example 2.1). Its restriction to a bounded part $\mathcal{T}^{2n}$ of any finite-gap manifold $\mathcal{T}_V^{2n}$ (see (6.6)) satisfies restrictions 1–5 and the corresponding system of Floquet solutions (6.13) is completely non-resonant with

$$\Delta = \widetilde{\Delta} = 1, \qquad d_J = 1, \qquad d_H = 0, \qquad d_A = 2.$$

Therefore Theorem 7.1 provides KdV with a normal form with $\tilde{d} = 0$. To state the result, we find the singular subset $R_s \subset R$ (see (5.15))[48] and choose any domain $R_1 \Subset R \backslash R_s$. We cover $R_1$ up to its zero-measure subset by non-overlapping sub-domains $R_{11}, R_{12}, \ldots$ such that the KdV equation restricted to any manifold $\Phi_0(R_{1j} \times \mathbb{T}^n) = \mathcal{T}_j^{2n}$ admits action-angle variables $(p, q)$ with $p \in P_j \Subset \mathbb{R}^n$ as in (2.6).

48 In the KdV case the set $R_s$ is empty. We ignore this nice specificity of KdV.

For any $s$ we denote by $Y_s \subset H_0^s(S^1)$ the closed subspace spanned by the functions $\{\cos jx, -\sin jx \mid j \in \mathbb{N}_V\}$. Applying Theorem 7.1 we get:

**Theorem 7.2.** *For any $d \geq 3$ there exists a $\delta > 0$ and an analytic symplectomorphism*

$$G : (P_j \times \mathbb{T}^n \times \mathcal{O}_\delta(Y_d), dp \wedge dq \oplus \alpha_2^Y) \to (H_0^d, \alpha_2)$$

*which contains $\mathcal{T}_j^{2n}$ in its range and is such that $G^{-1}$ transforms KdV to the Hamiltonian system*

$$\dot{p} = -\nabla_q \mathcal{H}, \qquad \dot{q} = \nabla_p \mathcal{H}, \qquad \dot{y} = \frac{\partial}{\partial x} \nabla_y \mathcal{H} \tag{7.21}$$

*with a Hamiltonian $\mathcal{H}$ of the form $\mathcal{H} = h(p) + \frac{1}{2}\langle B(p)y, y\rangle + h_3(p, q, y)$. Here $h(p)$ is the KdV Hamiltonian, restricted to $\mathcal{T}_j^{2n}$, $B(p)$ is the linear operator in $Y_d$ with eigenvectors $\cos mx$, $-\sin mx$ and eigenvalues $\nu_m(p)$ $(m \in \mathbb{N}_V)$, and $h_3 = O(\|y\|_d^3)$ is a function with a zero-order analytic gradient map.*

(2) *Higher KdV equations.* Let us take any $l$th equation from the KdV hierarchy. Since the same (as in the KdV case) sections $\Psi_j$ of the skew-orthogonal bundle to a finite-gap manifold $\mathcal{T}_V^{2n}$ give rise to Floquet solutions of the equation, then the same map $\Phi_1$ reduces the linearized $l$th equation to the equation $\dot{y} = JB^l(p)y$ in the space $Y$. Here $J = \partial/\partial x$ and $B^l(p)$ is a linear operator with the eigenvectors $\cos jx$ and $-\sin jx$ corresponding to the eigenvalues $\nu_j^l(p)$ as in (6.21). Therefore, the same map $G$ with $s \geq 2p+1$ reduces the $l$th KdV equation in the vicinity of $\mathcal{T}_j^{2n}$ (the same as above part of $\mathcal{T}_V^{2n}$) to eqn (7.21) with $\mathcal{H} = \mathcal{H}^l(p, q, y) = h_l(p) + \frac{1}{2}\langle B^l(p)y, y\rangle + h_3^l(p, q, y)$. Here $h_l$ is the Hamiltonian of the $l$th equation, restricted to $\mathcal{T}_j^{2n}$ (so $\nabla h_l = \mathbf{W}^{(l)}$, cf. (3.19)) and the operator $B^l(p)$ has the eigenvalues $\nu_j^l$. Now $\Delta = d_J = 1$ as in the KdV case, $d_A = 2l$, $d_H = 2l - 2$, and $\widetilde{\Delta} = 3 - 2l$ by (6.21). So $\tilde{d} = 2l - 2$ and $h_3 = O(\|y\|_d^3)$ has an analytic gradient map of order $2l - 2$.

(3) *The Sine–Gordon equation.* For the SG equation under the odd periodic (OP) boundary conditions in the variables $(u, w) \in Z_s^o$ $(s \geq 0)$,

$$\dot{u} = -\sqrt{A}\, w, \qquad \dot{w} = \sqrt{A}\,(u + A^{-1}(\sin u - u)),$$

let us consider its finite-gap manifold $\mathcal{T}^{2n} = \Phi_0(R \times \mathbb{T}^n)$ as in Section 4.3. We checked that this manifold satisfies assumptions 1–5, and in Section 6.4 we constructed a complete non-resonant system of Floquet solutions for the linearized equation. Accordingly, for any compact subset $R_1$ of a regular part $R_0$ of the algebraic set $R$ we can find a countable system of non-overlapping smooth domains $R_{1j}$ which cover $R_1$ up to a zero-measure subset, such that: for any $j$ in the vicinity of the manifold $\Phi_0(R_{1j} \times \mathbb{T}^n)$ in $Z_s^o$ the SG equation admits the normal form, described in Theorem 7.1. In difference with the KdV case, for some domains $R_{1j}$ the corresponding linear Hamiltonian operators $JB$ have non-imaginary eigenvalues.

To have this normal form result, it is not really essential to consider the SG equation under the OP boundary conditions. Indeed, for any $g \geq 1$ and any integer $g$-vector $\Upsilon$

the theta formula (4.17) subject to (4.18) defines a $2g$-dimensional finite-gap manifold, formed by $g$-dimensional invariant tori of the SG equation under periodic boundary conditions (corresponding arguments are more strightforward compared with the OP case). Arguing as in Section 6.4, for the linearized SG equation we construct a system of Floquet solutions of the form (6.26). Now we have 'twice as many' of them since the solutions are parameterized by an index $(l, \kappa)$, where $l \in \mathbb{Z}_\Upsilon \cup 0$ and $\kappa \in \{+, -\}$. The set of exponents $\nu(P_{l\pm})$ is asymptotically double. Namely,

$$\nu(P_{l\pm}) - l^* \longrightarrow 0$$

both when $l \to \infty$ and when the open gaps $[E_{2j-1}, E_{2j}]$ shrink. This set of Floquet solutions can be checked to be complete non-resonant, but because of the asymptotic degeneracy corresponding arguments become more technical compared with the elementary number-theory ones used in the odd periodic case in Section 6.4.

Accordingly, the SG equation under periodic boundary conditions can also be put to the normal form in the vicinity of its finite-gap manifold.

# II PERTURBED EQUATIONS

# 8

# A KAM theorem for perturbed non-linear equations

## 8.1 The Main Theorem and related results

Let $(\{Z_s\}, \alpha_2)$, $\alpha_2 = \bar{J}\, dz \wedge dz$, be a scale of symplectic Hilbert spaces as in Chapter 1, Section 1.2 (so the operator $\bar{J}$ defines an isomorphism of the scale of order $-d_J \leq 0$) and let $\mathcal{H}$ be a quasilinear Hamiltonian of the form

$$\mathcal{H} = \tfrac{1}{2}\langle Az, z\rangle + H(z),$$

where $A$ is a self-adjoint isomorphism of the scale of order $d_A > -d_J$. We fix any $d \geq d_A/2$ and assume that the function $H$ is analytic in the space $Z_d$ (or in a neighbourhood in $Z_d$ of the manifold $\mathcal{T}_0^{2n}$, see below) and defines an analytic gradient map of order $d_H$, $\nabla H : Z_d \to Z_{d-d_H}$. We have $d_H < d_A$ due to the quasilinearity of the Hamiltonian $\mathcal{H}$. The corresponding Hamiltonian equation takes the form

$$\dot{u} = J\nabla\mathcal{H}(u) = J(Au + \nabla H(u)), \tag{8.1}$$

where $J = (-\bar{J})^{-1}$ defines an isomorphism of the scale of order $d_J \geq 0$.

As in Sections 2.1 and 5.1 we assume that eqn (8.1) has an invariant manifold $\mathcal{T}_0^{2n} = \Phi_0(R_0 \times \mathbb{T}^n)$ filled with quasiperiodic solutions $u_0(t; r, \mathfrak{z}) = \Phi_0(r, \mathfrak{z} + t\omega(r))$ which satisfies assumptions 1–5 in Section 5.1. The manifold $R_0$ is the regular part of an $n$-dimensional real analytic set $R$ (which in its turn is a real part of a complex analytic set $R^c$). By $\widetilde{R}$ we denote any chart on $R_0$ analytically diffeomorphic to a bounded connected subdomain of $\mathbb{R}^n$. We identify $\widetilde{R}$ with this domain and supply it with the $n$-dimensional Lebesgue measure $\mathrm{mes}_n$.

As in Chapter 5, we also consider linearization of eqn (8.1) about a solution $u_0$ as above:

$$\dot{v} = J(Av + (\nabla H)_*(u_0(t))\, v), \tag{8.2}$$

and assume that (8.2) has Floquet solutions $v_j(t)$,

$$v_j(t; r, \mathfrak{z}) = e^{i\nu_j(r)t}\Psi_j(r, \mathfrak{z} + t\omega(r)), \quad j \in \mathbb{Z}_n, \tag{8.3}$$

where $\nu_{-j} \equiv -\nu_j$.

Our concern in this section is a Hamiltonian perturbation of eqn (8.1):

$$\dot{u} = J(Au + \nabla H(u) + \varepsilon \nabla H_1(u)), \tag{8.4}$$

and the behaviour of the solution for (8.4) near the manifold $\mathcal{T}_0^{2n}$. We assume that $H_1$ is an analytic functional such that its gradient map $\nabla H_1$ is analytic of order $d_H$ in a neighbourhood of the manifold $\mathcal{T}_0^{2n}$ in $Z_d$.

By $\tilde{d}$ we denote the real number from Theorem 7.1:

$$\tilde{d} = \max \{d_H,\ -\Delta - d_J,\ -\widetilde{\Delta} - d_J\},$$

where $d_H$ and $d_J$ are as above, $-\Delta$ is the order of the linear operator $\Phi_1 - \iota$ (see (5.11)), and $\widetilde{\Delta}$ is the exponent of growth in $j$ of 'variable parts' of the Floquet exponents $\nu_j(r)$ (see (5.13)).

Let us fix any $\tilde{\rho}$ such that $0 < \tilde{\rho} < 1/3$. Now we state a KAM theorem which is the main result of this book:

**Theorem 8.1 (the Main Theorem).** *Let the invariant manifold $\mathcal{T}_0^{2n}$ satisfy assumptions 1–5 and the system of Floquet solutions (8.3) is completely non-resonant. Moreover, assume that:*

*(1) (spectral asymptotic): $d_1 := d_A + d_J \geq 1$ and*

$$\nu_j(r) = K_1 j^{d_1} + K_1^1 j^{d_1^1} + K_1^2 j^{d_1^2} + \cdots + \tilde{\nu}_j(r),$$

*where $K_1 > 0$, $d_1 > d_1^1 > \cdots$ (the dots stand for a finite sum), and the functions $\tilde{\nu}_j$ analytically extend to $R^c$, where they are bounded by $Cj^{\varkappa}$ with some $\varkappa < d_1 - 1$;*

*(2) (quasilinearity): $\tilde{d} < d_1 - 1$.*

*Then most of the invariant tori $T^n(r)$ of eqn (8.1) persist in (8.4) when $\varepsilon \to 0$ in the following sense: for any chart $\widetilde{R} \subset R_0$ as above, a Borel subset $\widetilde{R}_\varepsilon \Subset \widetilde{R}$ and a Lipschitz embedding $\Sigma^\varepsilon : \widetilde{R}_\varepsilon \times \mathbb{T}^n \longrightarrow Z_d$, analytic in the second variable, can be found such that:*

*(i)* $\text{mes}_n(\widetilde{R} \setminus \widetilde{R}_\varepsilon) \longrightarrow 0$ *as* $\varepsilon \longrightarrow 0$,

*(ii) the map $(\Sigma^\varepsilon - \Phi_0) : \widetilde{R}_\varepsilon \times \mathbb{T}^n \longrightarrow Z_d$ is bounded by $C\varepsilon^{\tilde{\rho}}$, as well as its Lipschitz constant, and is analytic in $q \in \mathbb{T}^n$;*

*(iii) each torus $T_\varepsilon^n(r) := \Sigma^\varepsilon(\{r\} \times \mathbb{T}^n)$, $r \in \widetilde{R}_\varepsilon$, is invariant for eqn (8.4) and is filled with its time-quasiperiodic solutions $u = \mathfrak{h}_\varepsilon(t)$ of the form $\mathfrak{h}_\varepsilon(t) = \mathfrak{h}_\varepsilon(t; r, \mathfrak{z}) = \Sigma^\varepsilon(r, \mathfrak{z} + t\omega_\varepsilon(r))$, where $|\omega_\varepsilon - \omega| + \text{Lip}\,(\omega_\varepsilon - \omega) \leq C\varepsilon^{\tilde{\rho}}$.*

Let $\text{mes}_n^{\mathcal{H}}$ be the $n$-dimensional Hausdorff measure on $R$ (see Falconer 1990 and Appendix 8 below) and let $\mu_n$ be any finite measure, absolutely continuous with respect to $\text{mes}_n^{\mathcal{H}}$. Then the regular set $R_0 \subset R$ is a set of full $\mu_n$-measure since the singular set $R \setminus R_0$ has a positive codimension. Since $\mu_n$ is absolutely continuous with respect to the Lebesgue measure[49] and since the charts $\tilde{R}$ as above jointly cover $R_0$, then by the Main

---

49 Since $\text{mes}_n^{\mathcal{H}}$ is, see Falconer (1990) and Federer (1969).

Theorem most of the tori $T^n(r)$, $r \in R$, persist in the perturbed equation in the sense that those which persist correspond to $r$ from a subset $R_\varepsilon$ such that $\mu_n(R \setminus R_\varepsilon) \to 0$ as $\varepsilon \to 0$. In applications below we use the Main Theorem in this global reformulation.

We note that the theorem's assertions are empty unless $\varepsilon > 0$ is sufficiently small since the set $\widetilde{R}_\varepsilon$ may be empty for non-small $\varepsilon$.

**Amplification.** *The statements (ii) and (iii) of Theorem 8.1 remain true with $\tilde{\rho}$ replaced by any $\rho' < 1$. Moreover, $|\omega_\varepsilon - \omega| \le C\varepsilon$.*

We denote

$$\widetilde{T}^{2n} = \Phi_0(\widetilde{W}), \quad \widetilde{W} = \widetilde{R} \times \mathbb{T}^n \quad \text{and} \quad \widetilde{T}^{2n}_\varepsilon = \Sigma^\varepsilon(\widetilde{W}_\varepsilon), \quad \widetilde{W}_\varepsilon = \widetilde{R}_\varepsilon \times \mathbb{T}^n.$$

The set $\widetilde{T}^{2n}_\varepsilon$ is a remnant of the invariant manifold $\widetilde{T}^{2n}$ in the perturbed equation (8.4).[50]

Since $\widetilde{T}^{2n}$ is a $2n$-dimensional manifold embedded in $Z_d$, then its $2n$-dimensional Hausdorff measure $\text{mes}^{\mathcal{H}}_{2n}\widetilde{T}^{2n}$ is finite and positive: this follows from the estimate (A8.2) applied to the map $\Phi_0 : \widetilde{W} \longrightarrow \widetilde{T}^{2n}$ and to its inverse. The remnant set $\widetilde{T}^{2n}_\varepsilon$ is very irregular (it is totally disconnected), yet still carries most of a measure of the set $\widetilde{T}^{2n}$:

**Proposition 8.1.** *Under the assumptions of Theorem 8.1,*

$$\text{mes}^{\mathcal{H}}_{2n}\widetilde{T}^{2n}_\varepsilon \ge \text{mes}^{\mathcal{H}}_{2n}\widetilde{T}^{2n} - o(1) \quad \textit{as } \varepsilon \to 0.$$

*Proof.* By assertion (i) of the theorem and by the estimate (A8.7) (see Appendix 8) we get that

$$\text{mes}^{\mathcal{H}}_{2n}(\widetilde{W} \setminus \widetilde{W}_\varepsilon) = o(1). \tag{8.5}$$

The map $\Phi_0 : \widetilde{W}_\varepsilon \longrightarrow \Phi_0(\widetilde{W}_\varepsilon) \subset \widetilde{T}^{2n}$ is Lipschitz and has a Lipschitz inverse, so the map $\Sigma^\varepsilon \circ \Phi_0^{-1} : \Phi_0(\widetilde{W}_\varepsilon) \longrightarrow \widetilde{T}^{2n}_\varepsilon$ has the form $\text{id} + L$, where $\text{Lip}\, L \le C\varepsilon^{\tilde{\rho}}$ (we use assertion (ii)). Now estimate (A8.5) implies that $\text{mes}^{\mathcal{H}}_{2n}\widetilde{T}^{2n}_\varepsilon \ge \text{mes}^{\mathcal{H}}_{2n}\Phi_0(\widetilde{W}_\varepsilon) - O(\varepsilon^{\tilde{\rho}})$. Since $\text{mes}^{\mathcal{H}}_{2n}(\Phi_0(W \setminus \widetilde{W}_\varepsilon)) = o(1)$ by (8.5) and (A8.2), then the assertion follows. □

Under the assumptions of Theorem 8.1, a solution $u_0(t; r, \mathfrak{z})$ of (8.1) is linearly stable if all Floquet exponents $\nu_j(r)$ are real (see the Corollary to Proposition 5.1). Let us assume that this is the case for all $r \in \widetilde{R}$. Then the solutions $\mathfrak{h}_\varepsilon(t; r, \mathfrak{z})$ of the perturbed equation (8.4) with $r \in \widetilde{R}_\varepsilon$ also are linearly stable, provided that this equation, linearized about $\mathfrak{h}_\varepsilon$, satisfies some a priori estimate. We recall that by assumption 5 the flow maps $S^t_{\tau**}(\mathfrak{h}_\varepsilon(\tau))$ of the linearized equation are well defined in the space $Z_d$. We say that the linearized equation is *uniformly well defined* (in $Z_d$) if

$$\|S^t_{\tau**}(\mathfrak{h}_\varepsilon(\tau))\|_{d,d} \le C_1 e^{C_2(t-\tau)}, \quad \text{for all } t, \ \tau. \tag{8.6}$$

The solutions $\mathfrak{h}_\varepsilon(\tau) = \mathfrak{h}_\varepsilon(\tau; r, \mathfrak{z})$ lie in the torus $T^n_\varepsilon(r)$ and the map $\mathfrak{z} \mapsto \mathfrak{h}_\varepsilon(0)$ is a diffeomorphism of the standard $n$-torus and $T^n_\varepsilon(r)$. Therefore the assumption (8.6) is

---

[50] By no means do we claim that the invariant tori $T^n(r)$ with $r \in \widetilde{R} \setminus \widetilde{R}_\varepsilon$ really disappear when we switch in the perturbation $\varepsilon J\nabla H_1$, it is just unknown what happens to them, even when the phase-space $Z$ is finite-dimensional. See Moser (1973) and section III 'Beyond the tori' in Lazutkin (1993).

fulfilled if for every phase $\mathfrak{z} \in \mathbb{T}^n$ and $t \geq 0$ the flow map $S^t_{0**}(\mathfrak{h}_\varepsilon(0; r, \mathfrak{z}))$ is a bounded linear operator in $Z_d$, continuously depending on $\mathfrak{z}$ and $t$.

**Theorem 8.2.** *If under the assumptions of Theorem 8.1 all Floquet exponents $\nu_j(r)$ are real for $r \in \widetilde{R}$, then a solution $\mathfrak{h}_\varepsilon(t)$ is linearly stable, provided that eqn (8.4) linearized about this solution is uniformly well defined.*

(The examples we consider in Chapter 9 show that the assumption of uniform well-definedness is quite non-restrictive.)

We prove the two theorems and the amplification, reducing them to similar statements concerning perturbations of parameter-depending linear systems. We present the reduction in next section and prove the theorems on parameter-depending equations in Chapter 9.

## 8.2 Reduction to a parameter-depending case

We perform the reduction in four steps.

**Step 1** (localization). Let us denote by $R_f$ the set of singularities of the frequency map $\omega$, $R_f = \{r \in \widetilde{R} \mid \det \omega_*(r) = 0\}$, and denote $\widetilde{R}_s = (R_s \cap \widetilde{R}) \cup R_f$, where $R_s$ is the singular set, constructed in Section 5.3 (see there Remark 2). By assumption 4, $R_f$ is a proper analytic subset of $\widetilde{R}$. So $\widetilde{R}_s$ also is one, and for any given positive $\gamma_0$ we can find a finite system of $M$ connected sub-domains $\widetilde{R}_l \subset \widetilde{R} \setminus \widetilde{R}_s$ such that $\operatorname{dist}(r_j, r_{j'}) \geq C(\gamma_0) > 0$ if $r_j \in \widetilde{R}_j$ and $r_{j'} \in \widetilde{R}_{j'}$ with $j \neq j$. Moreover,

(a) $\operatorname{mes}(\tilde{R} \setminus \cup \widetilde{R}_l) < \gamma_0$.

(b) The Hamiltonian system restricted to $\Phi_0(\widetilde{R}_l \times \mathbb{T}^n)$ admits analytic action-angle variables $(p, q)$, where $p \in P_l \Subset \mathbb{R}^n$ and $q \in \mathbb{T}^n$. The map $(p, q) \mapsto (r, \mathfrak{z})$ is such that $r = r(p)$ and $\mathfrak{z} = q + \mathfrak{z}_0(P)$. This map, its inverse, and the Hamiltonian $h = h_l(p)$ all are $\delta$-analytic with some positive $\delta = \delta(\gamma_0)$. By Lemma 2.2, $\nabla h(p) \equiv \omega(r(p))$ and $\mathfrak{z} = q + \mathfrak{z}_0(P)$.

(c) For every $l$ the gradient map $p \mapsto \nabla h(p) \equiv \omega(r(p))$ defines a diffeomorphism $P_l \longrightarrow \Omega_l \Subset \mathbb{R}^n$ which is $\delta$-analytic as well as its inverse.

(d) Since each domain $\widetilde{R}_l$ is connected, then the eigenvectors $\psi_j$ of the operator $JB(r)$ are $r$-independent when $r \in \widetilde{R}_l$.

**Step 2** (a normal form theorem). In this step of the proof and in Step 3 we consider any fixed domain $\widetilde{R}_l$ as above and drop the index $l$.

Applying Theorem 7.1 we find an analytic symplectomorphism $G$ such that $G^{-1}$ transforms eqn (8.1) in the vicinity of $\Phi_0(\widetilde{R} \times \mathbb{T}^n)$ to the form given in the theorem. The same symplectomorphism converts the perturbed equation (8.4) to the Hamiltonian system

$$\dot{p} = -\nabla_q \mathcal{H}_\varepsilon, \qquad \dot{q} = \nabla_p \mathcal{H}_\varepsilon, \qquad \dot{y} = J\nabla_y \mathcal{H}_\varepsilon. \tag{8.7}$$

Here $p \in P$, $q \in \mathbb{T}^n$, $y \in \mathcal{O}_\delta(Y_d)$, and

$$\mathcal{H}_\varepsilon = h(p) + \tfrac{1}{2}\langle B(p)y, y\rangle + h_3(p, q, y) + \varepsilon H_1(p, q, y)$$

with $h_3 = O(\|y\|_d^3)$ and ord $\nabla_y h_3 = \tilde{d}$. The operator $B(p)$, the functions $h$, $h_3$, and $H_1$, and their gradients all are $\delta$-analytic in the corresponding domains.

**Step 3** (introducing a parameter). Let us consider the following neighbourhoods of the torus $T_0^n = \{0\} \times \mathbb{T}^n \times \{0\}$ in $\mathcal{Y} = \mathcal{Y}_d$ and $\mathcal{Y}^c = \mathcal{Y}_d^c$:

$$\begin{aligned} Q_\delta &= \mathcal{O}_\delta(\mathbb{R}^n)\times\mathbb{T}^n \times \mathcal{O}_\delta(Y_d) \subset \mathcal{Y}_d = \mathbb{R}^n \times \mathbb{T}^n \times Y_d, \\ Q_\delta^c &= \mathcal{O}_\delta(\mathbb{C}^n)\times\{|\operatorname{Im} q| < \delta\} \times \mathcal{O}_\delta(Y_d^c) \subset \mathcal{Y}_d^c = \mathbb{C}^n \times (\mathbb{C}^n/2\pi\mathbb{Z}^n) \times Y_d^c. \end{aligned} \tag{8.8}$$

In eqn (8.7) we perform a shift of the action $p$:

$$(p, q, y) = (\tilde{p} + a,\ \tilde{q},\ \tilde{y}) =: \operatorname{Shift}_a(\tilde{p},\ \tilde{q},\ \tilde{y}),$$

where $a \in P$ is a parameter of the shift. After this transformation the Hamiltonian $\mathcal{H}_\varepsilon$ becomes an analytic function $\mathcal{H}_\varepsilon(\tilde{p},\ \tilde{q},\ \tilde{y};\ a)$ of the tilde-variables from the domain $Q_\delta^c$. It has the following form:

$$\mathcal{H}_\varepsilon = h(a) + \nabla h(a) \cdot \tilde{p} + \tfrac{1}{2}\langle B(a)\tilde{y},\ \tilde{y}\rangle + \varepsilon H_1(\tilde{p} + a,\ \tilde{q},\ \tilde{y}) + \tilde{h}_3(\tilde{p},\ \tilde{q},\ \tilde{y};\ a),$$

where

$$\tilde{h}_3 = O(\|\tilde{y}\|_d^3 + |\tilde{p}|^2 + |\tilde{p}|\|\tilde{y}\|_d^2), \qquad \|\nabla_y \tilde{h}_3\|_{d-\tilde{d}} = O(\|\tilde{y}\|_d^2 + |\tilde{p}|\|\tilde{y}\|_d)$$

(so ord $\nabla\tilde{h}_3 = \tilde{d}$).

The functions $h$, $H_1$, and $\tilde{h}_3$, and the Floquet exponents $\nu_j$ are analytic bounded functions of the parameter $a \in P + \delta$. Because of property (c) from Step 1, the map $P \ni a \mapsto \omega = \nabla h(a) \in \Omega$ defines an analytic Lipschitz diffeomorphism of $P$ and a bounded domain $\Omega \subset \mathbb{R}$. We drop the tildes and change the parameter $a$ to $\omega$. Now the Hamiltonian $\mathcal{H}_\varepsilon$ reads as

$$\mathcal{H}_\varepsilon(p, q, y; \omega) = h(a) + \omega \cdot p + \tfrac{1}{2}\langle B(\omega)y, y\rangle + \varepsilon H_1(p, q, y; \omega) + h_3(p, q, y; \omega).$$

The operator $JB$ is diagonal in the complex symplectic basis $\{\psi_j\}$, constructed in Proposition 5.1:

$$JB\psi_j = i\nu_j(r)\psi_j, \quad \forall\, j \in \mathbb{Z}_n.$$

Since the Hamiltonian $\mathcal{H}_\varepsilon$ is $\delta$-analytic, then by the Cauchy estimate it is Lipschitz in $a \in P$ as well as in $\omega \in \Omega$. This is all we need from its dependence in the parameters.

In the vicinity of the torus $T_0^n = \{0\} \times \mathbb{T}^n \times \{0\}$ in $Q_\delta$ the Hamiltonian $\mathcal{H}_\varepsilon$ is a perturbation of the $q$-independent Hamiltonian $\mathcal{H}_0$,

$$\mathcal{H}_0 = \omega \cdot p + \tfrac{1}{2}\langle B(\omega)y, y\rangle$$

(we ignore the irrelevant constant $h(a)$). Indeed, $\varepsilon$ is small and the term $h_3$ has a high-order zero on $T_0^n$.

The Hamiltonian equations with the Hamiltonian $\mathcal{H}_\varepsilon(p, q, y; \omega)$ take the form

$$\left.\begin{aligned}\dot{p} &= -\nabla_q(\varepsilon H_1 + h_3), \qquad \dot{q} = \omega + \nabla_p(\varepsilon H_1 + h_3),\\ \dot{y} &= J(B(\omega)y + \varepsilon \nabla_y H_1 + \nabla_y h_3).\end{aligned}\right\} \tag{8.9}$$

We abbreviate $(p, q, y)$ to $\mathfrak{h}$ and rewrite (8.9) as

$$\dot{\mathfrak{h}} = V_{\mathcal{H}_\varepsilon}(\mathfrak{h}).$$

In the context of eqns (8.9), we call the functions $\nu_j(\omega)$ (i.e. the eigenvalues of the operator $JB$, divided by $i$), *frequencies* of the linear equation.

Hamiltonian vector fields with Hamiltonians of the form $\mathcal{H}_\varepsilon$ are studied in Kuksin (1993). Now we break the proof of Theorem 8.1 to present the main theorem from Kuksin (1993) in a form generalized to suit our purposes. After this we make the last step to complete the proof.

## 8.3 A KAM theorem for parameter-depending equations

To state the theorem we need, we relax restrictions on the Hamiltonian $\mathcal{H}_\varepsilon$ as in assumptions (0)–(3) below:

(0) The operator $JB(\omega)$ is diagonal in the complex basis $\{\psi_j \mid j \in \mathbb{Z}_n\}$ as in Proposition 5.1. Namely,

$$JB(\omega)\psi_j = i\nu_j(\omega)\psi_j, \quad \forall j.$$

(1) The complex functions $\nu_j(\omega)$, $j \in \mathbb{Z}_n$, are Lipschitz, are real for $|j| \geq j_1$ with some $j_1 \geq n+1$, and are odd in $j$: $\nu_j \equiv -\nu_{-j}$. For $j \geq n+1$ and for some fixed $\omega_0 \in \Omega$ the following asymptotics hold:

$$\left.\begin{aligned}&|\nu_j(\omega_0) - K_1 j^{d_1} - K_1^1 j^{d_1^1} - K_1^2 j^{d_1^2} - \cdots| \leq K j^{\tilde{d}},\\ &\operatorname{Lip} \nu_j \leq K j^{\tilde{d}},\end{aligned}\right\} \tag{8.10}$$

where $K_1 > 0$, $d_1 \geq 1$, $0 \leq \tilde{d} < d_1 - 1$, and the dots stand for a finite sum with some exponents $d_1 > d_1^1 > d_1^2 > \cdots$.

(2) The functions $h_3$ and $H_1$ are analytic in $(p, q, y) \in Q_\delta^c$ and everywhere in $Q_\delta^c$ satisfy the estimates:

$$\left.\begin{aligned}&|H_1| + \|\nabla_y H_1\|_{d-\tilde{d}+d_J} \leq 1, \quad \forall\omega,\\ &|h_3| \leq K(|p|^2 + \|y\|_d^3), \quad \forall\omega,\\ &\|\nabla_y h_3\|_{d-\tilde{d}+d_J} \leq K(|p|\|y\|_d + \|y\|_d^2), \quad \forall\omega,\\ &\text{the same estimates hold for Lipschitz constants}\\ &\text{in } \omega \in \Omega \text{ of functions } H_1 \text{ and } h_3 \text{ and their gradients.}\end{aligned}\right\} \tag{8.11}$$

(3) $\Omega$ is a bounded Borel set in $\mathbb{R}^n$ of positive Lebesgue measure, such that $\operatorname{diam}\Omega \leq K_2$ and $|\omega| \leq K$ for every $\omega \in \Omega$.

Let us choose any $\rho \in (0, \frac{1}{3})$ and denote by $\Sigma_0$ the map $(q, \omega) \mapsto (0, q, 0) \in Q_\delta$. For eqns (8.9) the following theorem holds, which states that the torus $T_0^n$ persists as an invariant torus of (8.9) for most $\omega$, if $\varepsilon$ is sufficiently small:

**Theorem 8.3.** *Suppose that assumptions (0)–(3) hold. Then there exist integers $j_2 \geq n$ and $M_1$, depending only on $n$, $d_1$, $\tilde{d}$, $K$, $K_1$, $K_2$ and $K_1^1$, $K_1^2$ …, with the following property. If*

$$|s \cdot \omega + l_{n+1}\nu_{n+1}(\omega) + \cdots + l_{j_2}\nu_{j_2}(\omega)| \geq K_3 > 0 \tag{8.12}$$

*for all $\omega \in \Omega$, all integer n-vectors s and all $(j_2 - n)$-vectors l such that*

$$|s| \leq M_1, \quad 1 \leq |l| \leq 2,$$

*then for arbitrary $\gamma > 0$ and for sufficiently small $\varepsilon \leq \bar{\varepsilon}(\gamma)$ $(\bar{\varepsilon} > 0)$, a Borel subset $\Omega_\varepsilon \subset \Omega$ and a Lipschitz embedding $\Sigma_\varepsilon : \mathbb{T}^n \times \Omega_\varepsilon \longrightarrow Q_\delta$, analytic in $q \in \mathbb{T}^n$, can be found with the following properties:*

*(a)* mes $(\Omega \setminus \Omega_\varepsilon) \leq \gamma$;
*(b)* $\|\Sigma_\varepsilon - \Sigma_0\|_{\mathcal{Y}_d}^{\mathbb{T}^n \times \Omega_\varepsilon, \mathrm{Lip}} \leq C\varepsilon^\rho$;
*(c) each torus $\Sigma_\varepsilon(\mathbb{T}^n \times \{\omega\})$, $\omega \in \Omega_\varepsilon$, is invariant for the flow of eqn (8.9) and is filled with its quasiperiodic solutions $\mathfrak{h}(t)$ of the form $\mathfrak{h}(t; q, \omega) = \Sigma_\varepsilon(q + \omega' t, \omega)$, where $\omega' = \omega'(\omega)$ and $|\omega' - \omega| + \mathrm{Lip}\,(\omega' - \omega) \leq C\varepsilon^\rho$.*

**Amplification.** *Assertions (b) and (c) hold with $\rho$ replaced by one. The constant C in (c) is $\gamma$-independent.*

**Theorem 8.4.** *If in Theorem 8.3 all the frequencies $\nu_j$ are real, then any solution $\mathfrak{h}(t)$ is linearly stable provided that eqn (8.9) linearized about this solution is uniformly well defined in the space $\mathbb{R}^{2n} \times Y_d$.*

For proofs of the theorems and of the amplification see Sections 10 and 11.

## 8.4 Completion of the proof of the Main Theorem

**Step 4** (proof of Theorems 8.1 and 8.2, given Theorems 8.3 and 8.4). Now we apply Theorem 8.3 to eqn (8.9) with $\Omega$ equal to a Borel subset $\Omega_l$ of the domain $\widetilde{\Omega}_l = \{\omega(r) \mid r \in \widetilde{R}_l\}$, $l = 1, \ldots, M$, which we construct below.

Assumptions (1)–(3) hold with the constants from $n$ through $d_1^1$, $d_1^2$, … the same as in Theorem 8.1, while the constants $K$ and $K_2$ depend on $\gamma_0$. We take $j_2 = j_2(\gamma_0)$ and $M_1 = M_1(\gamma_0)$ as in Theorem 8.3 and consider all resonances as in (8.12). Since the system of Floquet exponents $\{\nu_j(r)\}$ is non-resonant, then each resonance does not vanish identically. Since these functions are analytic, we can find $K_3 = K_3(\gamma_0)$ and for every $l$ can find a subset $\Omega_l \subset \widetilde{\Omega}_l$ such that mes $(\widetilde{\Omega}_l \setminus \Omega_l) \leq \gamma_0/M$ and (8.12) holds for all $\omega \in \Omega_l$.

For every $l$ we apply Theorem 8.3 with $\gamma = \xi(\gamma_0)$ (the function $\xi > 0$ will be chosen later) to find the subset $\Omega_{l\varepsilon} \subset \Omega_l$, mes $(\Omega_l \setminus \Omega_{l\varepsilon}) < \gamma$, and the map $\Sigma_{l\varepsilon} : \mathbb{T}^n \times \Omega_{l\varepsilon} \longrightarrow Q_\delta$.

Now we are in position to define the set $\widetilde{R}_\varepsilon \subset \widetilde{R}$ and the map $\Sigma^\varepsilon : \widetilde{R}_\varepsilon \times \mathbb{T}^n \longrightarrow Q_\delta$, claimed in Theorem 8.1. We set:

$$\widetilde{R}_\varepsilon = \bigcup_{l=1}^{M} \{r \in R_l \mid \omega(r) \in \Omega_{l\varepsilon}\}, \qquad \Sigma^\varepsilon(r, \mathfrak{z}) = G \circ \mathrm{Shift}_p \circ \Sigma_{l\varepsilon}(q(\mathfrak{z}), \omega(r)),$$

where $q = q(r, \mathfrak{z})$, $p = p(r, \mathfrak{z})$, and the map $(r, \mathfrak{z}) \mapsto (p, q)$ is the action-angle transformation from Step 1.

The set $\widetilde{R}_\varepsilon$ and the map $\Sigma^\varepsilon$ satisfy all the claims of Theorem 8.1. Indeed,

$$\begin{aligned} \mathrm{mes}\,(R \setminus \widetilde{R}_\varepsilon) = \mathrm{mes}\,(\widetilde{R} \setminus \cup \tilde{R}) &+ \sum_{l=1}^{M} \mathrm{mes}\,\{r \in \widetilde{R}_l \mid \omega(r) \in \widetilde{\Omega}_l \setminus \Omega_l\} \\ &+ \sum_{l=1}^{M} \mathrm{mes}\,\{r \in \widetilde{R}_l \mid \omega(r) \notin \Omega_{l\varepsilon}\}. \end{aligned}$$

Denoting $\sup|\det \partial\omega/\partial r|^{-1}$ by $C(\gamma_0)$ we see that $\mathrm{mes}\,(R \setminus \widetilde{R}_\varepsilon)$ is bounded by $\gamma_0 + \gamma_0 + C(\gamma_0)M\xi(\gamma_0)$. This is smaller than $3\gamma_0$, if we choose $\xi(\gamma_0) = \gamma_0/(C(\gamma_0)M)$. This means that we can choose $\gamma_0 = \gamma_0(\varepsilon)$ in such a way that $\mathrm{mes}\,(R \setminus \widetilde{R}) \le 3\gamma_0$ goes to zero with $\varepsilon$, and assertion (a) of Theorem 8.1 holds.

The tori $\Sigma^\varepsilon(\{r\} \times \mathbb{T}^n)$ are invariant for eqn (8.4) and are filled with its quasiperiodic solutions of the form $u = \mathfrak{h}_\varepsilon(t)$, where $\omega_\varepsilon = \omega'(p(r))$.

The estimates for $\Sigma^\varepsilon - \Phi_0$ and $\omega_\varepsilon(r) - \omega(r)$ readily follow from the corresponding estimates in Theorem 8.3.

It remains to majorize the Lipschitz constants of the difference maps as above. Let us take any two points $(r_1, \mathfrak{z}_1)$ and $(r_2, \mathfrak{z}_2)$ in $\widetilde{R}_\varepsilon \times \mathbb{T}^n$. If $r_1$ and $r_2$ belong to the same set $\widetilde{R}_l$, then the estimates for the increments[51] of $\Sigma^\varepsilon - \Phi_0$ and $\omega_\varepsilon - \omega$ follow from the corresponding estimates for the increments of $\Sigma(0, q, 0)$ and $\omega' - \omega$. If $r_1$ and $r_2$ belong to different sets $\widetilde{R}_l$, then $|r_1 - r_2| \ge C(\gamma) > 0$ and the increments of the differences divided by increments of the arguments are bounded by $C_1\varepsilon^\rho/C(\gamma)$. Since we can choose the rate of decay $\gamma(\varepsilon) \to 0$ to be as slow as we wish, then we can achieve $C_1\varepsilon^\rho/C(\gamma) \le C_2\varepsilon^{\tilde{\rho}}$, if we chose for $\rho$ in Theorem 8.3 any number from the interval $(\tilde{\rho}, 1/3)$. Thus, the estimates for the Lipschitz constants are proved.

The last arguments also show that the estimates $|\omega' - \omega| \le C\varepsilon$ and $\mathrm{Lip}\,(\omega' - \omega) \le C\varepsilon$ imply that $|\omega_\varepsilon - \omega| \le C\varepsilon$ and $\mathrm{Lip}\,(\omega_\varepsilon - \omega) \le C_{\rho'}\varepsilon^{\rho'}$ for any $\rho' < 1$. This means that the Amplification to Theorem 8.3 implies the Amplification to Theorem 8.1.

Finally, since linearization of the symplectomorphism which sends solutions $\mathfrak{h}(t)$ of (8.11) to solutions $\mathfrak{h}_\varepsilon(t)$ transforms solutions of the corresponding linearized equations, then Theorem 8.2 follows from Theorem 8.4.

---

51 That is, the estimate $|(\omega_\varepsilon - \omega)(r_1) - (\omega_\varepsilon - \omega)(r_2)| \le C\varepsilon^\rho |r_1 - r_2|$, etc.

## 8.5 Around the Main Theorem

The Main Theorem of this book was first stated in Kuksin (1989) in a less general form, where it was proved with some details missing.[52] The Normal Form Theorem from Section 7.1 was proved in Kuksin (1989) (see there Lemmas 6–8) in the context of analytic functions (rather than algebraic ones). The 'infrastructure' of the Main Theorem, i.e. convenient ways to construct Floquet solutions and to check their completeness through non-degeneracy and non-resonance, was developed later, see Kuksin (1998*a*) and the references in Sections 5 and 6.

The KAM Theorem 8.3 for parameter-depending linear systems and for perturbations, given by *bounded* non-linear operators was first proved in Kuksin (1987, 1988); independently, a similar result was obtained in Wayne (1990). The same theorem for *unbounded* perturbations demands the additional non-trivial step—Theorem 5.1. It was proved in Kuksin (1998*b*) (a preprint of this work arrived in 1995). Finite-dimensional versions of Theorem 8.3 were first proved in Eliasson (1988) and then in Pöschel (1989).

Theorem 8.3 is an important result by itself since it applies to parameter-depending Hamiltonian PDEs with small non-linearities and with a one-dimensional space variable. See Kuksin (1993), where many applications are given.

Theorem 8.3 is proved under the assumption that the unperturbed linear system has a

$$\text{single spectrum} \quad \{\pm i\lambda_j \mid j \in \mathbb{N}\}, \tag{8.13}$$

where

$$\lambda_j = Cj^d + o(j^d), \quad d \geq 1, \tag{8.14}$$

and

$$|\lambda_j - \lambda_{j-1}| \geq C_1^{-1} j^{d-1}, \quad \forall\, j. \tag{8.15}$$

For systems with small *bounded* non-linearities the single-spectrum assumption (8.13) can be replaced by the assumption that the eigenvalues $\lambda_j$ asymptotically have the same multiplicity $m \geq 2$ and the corresponding spectral spaces asymptotically are 'much the same'.[53] This version of Theorem 8.3 is due to Chierchia and You (1999), who observed that the proof of the theorem, given in Kuksin (1993) and Pöschel (1996*a*), generalizes to the asymptotically multiple situation as above if to find the operator $f^{yy}$ from the homological equation (10.21) (see Chapter 10 below) one treats its Hilbert matrix $F$ as a block matrix, formed by $m \times m$ blocks, i.e. as a Hilbert matrix over the ring of $m \times m$ complex matrices rather than a matrix over complex numbers. (These arguments do not apply to systems with small *unbounded* non-linearities since for our

---

[52] The main omission was that a KAM theorem for unbounded perturbations of a parameter-depending linear system (Theorem 8.3 of this book) was given there without a proof.

[53] For the most important case $m = 2$ this means the following: the eigenvalues form pairs $\lambda_j^+, \lambda_j^-$ such that $|\lambda_j^+ - \lambda_j^-| \leq Cj^{-\tilde{d}}$ with a suitable $\tilde{d} > 0$. The linear Hamiltonian operator, restricted to corresponding invariant complex planes in the complexified phase-space, equals $i\lambda_j E + O(j^{-\tilde{d}})$.

proof of Theorem 5.1 it is important that the unknown function $x(q)$ in eqn (5.1) is a scalar-valued—not a matrix-valued—one.)

The version of Theorem 8.3 due to Chierchia and You applies to non-linear wave and non-linear Schrödinger equations under periodic boundary conditions since the linear parts of these equations have asymptotically double spectra which satisfy (8.14) and (8.15).

The assumption (8.15) can be relaxed and replaced by some (rather implicit) restrictions on clusters, formed by the sequence $\{\lambda_j\}$. This follows from another KAM scheme, applicable to the problems we discuss. The scheme is due to Craig and Wayne (1992) and was developed further by Bourgain (1998); for a brief description, see Appendix 10 below. The main advantage of the Craig–Wayne–Bourgain approach is that it applies to non-linear perturbations of the two-dimensional linear Schrödinger equations under periodic boundary conditions: for these equations (8.14) holds with $d = 1$, assumption (8.15) is violated, but control for the clusters is sufficient to prove that most time-quasiperiodic solutions of the linear equation with a potential of a general form withstand small non-linear perturbations (Bourgain 1998). The disadvantages of this approach are that, first, it does not apply to equations with unbounded perturbations and, second, it does not allow one to control the Lyapunov stability of solutions.

Except for the global results concerning KAM persistence most finite-gap solutions and the results on perturbations of parameter-depending linear equations we have just discussed, the 'KAM for PDEs' theory includes the third topic. Namely, theory of *small oscillations in non-linear Hamiltonian PDEs*. Let us consider, for example, a non-linear Klein–Gordon equation with an odd non-linearity:

$$u_{tt} = u_{xx} - mu + \sum_{k=1}^{\infty} a_k u^{2k+1}, \quad m, a_1 > 0, \tag{8.16}$$

$$u(t, 0) \equiv u(t, \pi) \equiv 0.$$

Appropriate positive constants $b_1$ and $b_2$ can be found such that (8.16) can be written as

$$u_{tt} = u_{xx} - b_1 \sin b_2 u + O(|u|^5),$$

i.e. as a high-order perturbation of the SG equation $u_{tt} = u_{xx} - b_1 \sin b_2 u$. Accordingly, most small-amplitude finite-gap solutions of the SG equation persist in eqn (8.16) and a set of the solutions which persist is 'asymptotically dense near the zero solution'.[54] This result follows from a version of the Main Theorem, where the set $R$ has the size $\varepsilon$ to a positive degree (see Kuksin 1993, p. 53). A corresponding theorem was proved in Bobenko and Kuksin (1995*a*). Later it was observed that it is technically easier to treat (8.16) as a perturbation of another integrable system, namely its Birkhoff normal form; see Kuksin and Pöschel (1996) and Pöschel (1996*b*).

---

[54] That is, for any vertexed at the origin open cone in the phase-space $\overset{\circ}{H}{}^1[0, \pi] \times L_2[0, \pi]$ (see item 4 of Example 2.3 in Section 2.1), the set of solutions which persist intersects the cone by an infinite set which has the origin in its accumulation point.

## Appendix 8. Lipschitz analysis and Hausdorff measure

A map $G$ that sends a metric space $Q_1$ to a metric space $Q_2$ is called Lipschitz if its Lipschitz constant, $\operatorname{Lip} G$, is finite, where

$$\operatorname{Lip} G = \sup_{x \neq y} \frac{\operatorname{dist}_2(G(x), G(y))}{\operatorname{dist}_1(x, y)}.$$

(Throughout the book, $Q_1$ and $Q_2$ are subsets of Banach spaces or of the direct product of an $n$-torus with a Banach space.) In particular, if $X^c$ and $Y^c$ are complex Hilbert (or Banach) spaces and a map $F : X^c \supset Q \to Y^c$ admits an analytic extension to a neighbourhood $Q + \delta$, where it is bounded by $C$, then

$$\operatorname{Lip}(F : Q \to Y^c) \leq C\delta^{-1}$$

by the Cauchy estimate.

We recall (see Federer 1969 and Babin and Vishik 1989) that a subset $A \subset Q_1$ has a finite $m$-dimensional *Hausdorff measure* $\operatorname{mes}_m^{\mathcal{H}}(A)$ and the measure is less than $C < \infty$, if for each $\delta > 0$ we can cover $A$ by a countable system $F$ of subsets $S \subset Q_1$ such that $\operatorname{diam} S < \delta$ for every $S$ in $F$ and

$$\boldsymbol{\alpha}(m)2^{-m} \sum_{S \in F} (\operatorname{diam} S)^m < C, \tag{A8.1}$$

where $\boldsymbol{\alpha}(m) > 0$ is a positive constant, equal to the $m$-volume of the unit ball $\mathcal{O}_1(\mathbb{R}^m)$ if $m$ is an integer. Now $\operatorname{mes}_m^{\mathcal{H}}(A)$ is defined in the natural way:

$$\operatorname{mes}_m^{\mathcal{H}}(A) = \inf\{C \mid \operatorname{mes}_m^{\mathcal{H}}(A) < C\}$$

(as usual, $\operatorname{mes}_m^{\mathcal{H}}(A) = \infty$ if the set under inf is empty).

Since a Lipschitz map $G$ as above sends a covering $F = \{S\}$ of a subset $A \subset Q_1$ to the covering $G(F) = \{G(S \cap F)\}$ of the set $G(F) \subset Q_2$ and $\operatorname{diam} G(S) \leq \operatorname{Lip} G \cdot \operatorname{diam} S$, then

$$\operatorname{mes}_m^{\mathcal{H}} G(A) \leq (\operatorname{Lip} G)^m \operatorname{mes}_m^{\mathcal{H}} A. \tag{A8.2}$$

Now let $A$ be a subset of a Banach space $B$ and let $G : A \to B$ be a map of the form

$$G = \mathrm{id} + G_1, \qquad \operatorname{Lip} G_1 \leq \mu < 1. \tag{A8.3}$$

Then the map $G^{-1} : G(A) \to A$ is well defined and

$$\operatorname{Lip} G^{-1} \leq (1 - \mu)^{-1}. \tag{A8.4}$$

Indeed, if $G(x_j) = y_j$ for $j = 1, 2$, then $(x_1 - x_2) + (G_1(x_1) - G(x_2)) = y_1 - y_2$. So

$$\|y_1 - y_2\| \geq \|x_1 - x_2\| - \|G_1 x_1 - G_1 x_2\| \geq (1 - \mu)\|x_1 - x_2\|$$

and (A8.4) follows.

Using (A8.2) with $G = G$ and $G = G^{-1}$ we estimate how a Lipschitz map of the form (A8.3) changes the Hausdorff measures of sets:

$$(1-\mu)^m \mathrm{mes}_m^{\mathcal{H}} A \le \mathrm{mes}_m^{\mathcal{H}} G(A) \le (1+\mu)^m \mathrm{mes}_m^{\mathcal{H}} A. \tag{A8.5}$$

If $A$ is a subset of $\mathbb{R}^m$, then its upper Lebesgue measure, $\mathrm{mes}_m^* A$, is defined in a way similar to (A8.1). Namely, $\mathrm{mes}_m^* A < C'$ if $A$ can be covered by a countable system of balls $B_j = \mathcal{O}_{r_j}(b_j, \mathbb{R}^m)$ such that

$$\alpha(m) \sum r_j^m < C' \tag{A8.6}$$

(the radii $r_j$ can be chosen smaller than any given $\rho > 0$) and $\mathrm{mes}_m^* A$ is the infimum over all $C'$ with this property. Choosing $F = \{B_j \cap A\}$ we get that $\mathrm{mes}_m^{\mathcal{H}} A < C'$. Conversely, given any covering $F = \{S\}$ of $A$, for each $S$ we denote $r_S = \mathrm{diam}\, S$ and choose a point $a_S \in S$. Then the system of balls $B_S = \mathcal{O}_{r_S}(a_S, \mathbb{R}^m)$ covers $A$ and $\alpha(m) \sum r_S^m \le 2^m C$. Thus,

$$\mathrm{mes}_m^{\mathcal{H}} A \le \mathrm{mes}_m^* A \le 2^m \mathrm{mes}_m^{\mathcal{H}} A, \quad \text{for any } A \subset \mathbb{R}^m. \tag{A8.7}$$

If $A$ is a Borel subset of $\mathbb{R}^m$, then $\mathrm{mes}_m^* A = \mathrm{mes}_m A$. Moreover, $\mathrm{mes}_m A = \mathrm{mes}_m^{\mathcal{H}} A$ (see Federer 1969). We shall not use this deep result since the elementary estimates (A8.7) are sufficient for our purposes.

If $A$ is a Borel subset of $\mathbb{R}^m$ and $G : A \to \mathbb{R}^m$ is a Lipschitz map of the form (A8.3), then we can repeat the arguments used to derive (A8.5) to estimate $\mathrm{mes}_m G(A)$ via $(1+\mu)^m \mathrm{mes}_m A$. Indeed, since $A$ is a Borel set, then $\mathrm{mes}_m^* A = \mathrm{mes}_m A$ and for any $C' > \mathrm{mes}_m A$ we can find a covering of $A$ by balls $\mathcal{O}_{r_j}(b_j, \mathbb{R}^m)$, $j = 1, 2, \ldots$, which satisfies (A8.6). Next we extend $G$ to a map $\tilde{G} : \mathbb{R}^m \to \mathbb{R}^m$ with the same Lipschitz constant $\le 1 + \mu$ (this is the Kirszbraun theorem; see Federer 1969). The balls $\mathcal{O}_{(1+\mu)r_j}(\tilde{G}(b_j), \mathbb{R}^m)$ cover the set $G(A)$ and the sum of their volumes is less than $(1+\mu)C'$. Since $C'$ may be chosen arbitrarily close to $\mathrm{mes}_m A$, then $\mathrm{mes}_m G(A) \le (1+\mu)^m \mathrm{mes}_m A$. Applying the same arguments to the inverse map $G^{-1}$ and using (A8.4) we get:

**Lemma A8.1.** *If* $A \subset \mathbb{R}^m$ *is a Borel set and* $G = \mathrm{id} + G_1 : A \to \mathbb{R}^m$ *is a map such that* $\mathrm{Lip}\, G_1 \le \mu < 1$, *then* $(1-\mu)^m \mathrm{mes}_m A \le \mathrm{mes}_m G(A) \le (1+\mu)^m \mathrm{mes}_m A$.

# 9 Examples

## 9.1 Perturbed KdV equation

Let us consider a perturbed KdV equation under zero mean-value periodic boundary conditions:

$$\dot{u} = \frac{1}{4}u_{xxx} + \frac{3}{2}uu_x + \varepsilon\frac{\partial}{\partial x}f'_u(u, x) =: V_\varepsilon(u)(x),$$
$$u(t, x) \equiv u(t, x + 2\pi), \qquad \int_0^{2\pi} u\,dx \equiv 0, \tag{9.1}$$

where $f(u, x)$ is a $C^d$-smooth function[55] ($d \geq 1$), $\delta$-analytic in $u$. Then the non-linear part of the vector field $V_\varepsilon$ defines an analytic morphism of order 1:

$$H_0^d \longrightarrow H_0^{d-1}, \qquad u \longmapsto \tfrac{3}{2}uu_x + \varepsilon\frac{\partial}{\partial x}f'_u(u, x)$$

(see Example 1.1). The equation is Hamiltonian in the symplectic space $(H_0^d, \alpha_2)$, $\alpha_2 = (\partial/\partial x)^{-1}\,du \wedge du$, with the Hamiltonian

$$\mathcal{H}_\varepsilon = \int_0^{2\pi}\left(\frac{1}{8}u'^2 - \frac{1}{4}u^3 - \varepsilon f(u(x), x)\right) dx.$$

For $\varepsilon = 0$ this is the KdV equation and a bounded part $\mathcal{T}^{2n}$ of any finite-gap manifold $\mathcal{T}_V^{2n}$,

$$\mathcal{T}^{2n} = \bigcup\{T^n(r) \subset \mathcal{T}_V^{2n} \mid 0 < r_j < K_0\},$$

satisfies assumptions 1–5 (see Section 3.2.). The linearized KdV equation has a system of Floquet solutions which is complete non-resonant (Section 6.2). Assumption (1) of the Main Theorem now holds with $d_1 = 3$, $d_1^1 = \cdots = 0$ (see (6.17)) and assumption (2) holds since $d_H = \tilde{d} = 1$ (see Section 7.3). We get:

**Theorem 9.1.** *For any $\rho < 1$ and for sufficiently small $\varepsilon > 0$, there exists a Borel subset $R_\varepsilon^n$ of the cube $R^n = \{0 < r_j < K_0\}$ and a Lipschitz map $\Sigma^\varepsilon : R_\varepsilon^n \times \mathbb{T}^n \longrightarrow H_0^d(S^1)$,*

[55] $H^d$-smoothness is sufficient, see in Kuksin (1993).

*analytic in the second variable, such that:*

(a) $\text{mes}_n(R^n \setminus R^n_\varepsilon) \longrightarrow 0$ *as* $\varepsilon \to 0$*;*

(b) *the map* $\Sigma^\varepsilon$ *is* $\varepsilon^\rho$*-close to the map* $\Phi_0$, $\Phi_0(r, \mathfrak{z})(x) = G(Vx + \mathfrak{z}, r)$ *(see (3.16)), also in the Lipschitz norm;*

(c) *each torus* $T^n_\varepsilon(r) = \Sigma^\varepsilon(\{r\} \times \mathbb{T}^n)$, $r \in R^n_\varepsilon$, *is invariant for eqn (9.1) and is filled with its linearly stable time-quasiperiodic solutions of the form* $t \to \Sigma^\varepsilon(r, \mathfrak{z} + t\omega_\varepsilon(r))$, *where the n-vector* $\omega_\varepsilon$ *is* $C\varepsilon$*-close to* $\mathbf{W}(r)$.

To get the result we used Theorem 8.1, its Amplification, and Theorem 8.2. The last theorem applies since the linearized KdV equation is uniformly well defined due to arguments in Example 1.6 (see (1.13)).

The theorem implies that the union of all linearly stable time-quasiperiodic solutions becomes infinite-dimensional and dense in $H^d_0$, asymptotically as $\varepsilon \to 0$:

**Corollary 9.1.** *The space* $H^d_0$ *contains a subset* $Q_\varepsilon$ *filled with linearly stable time-quasiperiodic solutions of (9.1) such that its Hausdorff dimension tends to infinity when* $\varepsilon \to 0$ *and for any fixed function* $v \in H^d_0$ *we have*

$$\text{dist}_{H^d_0}(v, Q_\varepsilon) \underset{\varepsilon\to 0}{\longrightarrow} 0. \tag{9.2}$$

*In particular, the set* $\bigcup_{\varepsilon>0} Q_\varepsilon$ *is dense in* $H^d_0$.

*Proof.* We define $Q_\varepsilon$ as a union of all sets $\Sigma^\varepsilon(R^n_\varepsilon \times \mathbb{T}^n) = \widetilde{\mathcal{T}}^{2n}_\varepsilon$, corresponding to all $n$-gap manifolds $\mathcal{T}^{2n}$, $n = 1, 2, \ldots$. By Proposition 8.1, the set $\widetilde{\mathcal{T}}^{2n}_\varepsilon$ has a positive $2n$-dimensional Hausdorff measure when $\varepsilon$ is small. Thus, $\dim_{\mathcal{H}} Q_\varepsilon \longrightarrow \infty$.

To prove (9.2) we note that for any $\mu > 0$ one can find $n \geq 1$ and an $n$-gap potential $u(x)$ such that $\|u - v\|_k \leq \mu$ (this is a famous result of V.A. Marčenko; see Marčenko 1977, Theorem 3.4.3, and Garnet and Trubowitz 1987, p. 27). Accordingly, $u$ equals $\Phi_0(r, \mathfrak{z})$ with some $r \in R$ and $\mathfrak{z} \in \mathbb{T}^n$. If $\varepsilon$ is sufficiently small, then by the assertion (a) of the theorem, there exists $r_1 \in R_\varepsilon$ such that $|r - r_1| \leq \mu$. Using (b), we get that $\|u - \Sigma^\varepsilon(r_1, \mathfrak{z})\|_k \leq C\mu + \varepsilon^\rho$ and (9.2) follows since $\mu > 0$ can be chosen arbitrarily small. □

Another immediate consequence of the theorem is the observation that the Its–Matveev formula (3.15) with corrected frequency vector $\mathbf{W}$, 'almost solves' eqn (9.1) for all $t$:

**Corollary 9.2.** For any $r \in \mathbb{R}^n_+$ and any $\mathfrak{z} \in \mathbb{T}^n$ there exists an $n$-vector $\mathbf{W}_\varepsilon(r)$ and a solution $u_\varepsilon(t, x)$ of (9.1) in $H^d_0$ such that

$$\sup_t \left\| u_\varepsilon(t, \cdot) - 2\frac{\partial^2}{\partial x^2} \ln \theta(i(\mathbf{V} \cdot + \mathbf{W}_\varepsilon t + \mathfrak{z}); r)) \right\|_d \underset{\varepsilon\to 0}{\longrightarrow} 0.$$

*Proof.* Let us take any sequence $\{r_\varepsilon \in R_\varepsilon\}$ which converges to $r$ as $\varepsilon \to 0$ and take $u_\varepsilon(t) = \Sigma^\varepsilon(r_\varepsilon, \mathfrak{z} + t\omega_\varepsilon(r))$. Then

$$\|u_\varepsilon(t) - \Phi_0(r, \mathfrak{z} + \omega_\varepsilon t)\|_k \le \|u_\varepsilon(t) - \Phi_0(r_\varepsilon, \mathfrak{z} + \omega_\varepsilon t)\|_k + \|\Phi_0(r_\varepsilon, \mathfrak{z} + \omega_\varepsilon t) - \Phi_0(r, \mathfrak{z} + \omega_\varepsilon t)\|_k = o(1) \quad \text{as } \varepsilon \to 0.$$

This implies the result since

$$\Phi_0(r, \mathfrak{z} + \omega_\varepsilon t)(x) = 2\frac{\partial^2}{\partial x^2} \ln \theta(i(\boldsymbol{V}x + \boldsymbol{W}_\varepsilon t + \mathfrak{z}); r),$$

where $\boldsymbol{W}_\varepsilon = \omega_\varepsilon$ □

An easy analysis of the first step in the proof of Theorem 8.3 (see Kuksin 1989) shows that the new frequency vector $\boldsymbol{W}_\varepsilon$ has the form $\boldsymbol{W}_\varepsilon(r) = \boldsymbol{W}(r) + \varepsilon \boldsymbol{W}_1(r) + O(\varepsilon^2)$, where the components $W_1^j$ of the $n$-vector $\boldsymbol{W}_1$ are obtained by averaging along the torus $T^n(r)$[56] of the function

$$G_*\left(\frac{\partial}{\partial p_j}\right)\left(-\int_0^{2\pi} f(u, x)\, dx\right), \quad j = 1, \dots, n.$$

Here $G : (p, q, y) \mapsto u(\cdot)$ is the normal form transformation from Theorem 7.2.

Therefore the assertion of Corollary 9.2 can be viewed as an averaging theorem: for most $r$ and for all $\mathfrak{z}$ the functions

$$2\frac{\partial^2}{\partial x^2} \ln \theta(i(\boldsymbol{V}x + \boldsymbol{W}_\varepsilon t + \mathfrak{z}); r), \qquad \boldsymbol{W}_\varepsilon = \boldsymbol{W}(r) + \varepsilon \boldsymbol{W}_1(r) + O(\varepsilon^2),$$

approximate solutions of the perturbed KdV equation (9.1) for *all* $t$ and $x$, where the $n$-vector $\boldsymbol{W}_1$ is obtained by the averaging described above. Here 'for most $r$' means 'for all $r$ outside a set whose measure goes to zero with $\varepsilon$'.

Thus, the result proves a stronger version of the *Whitham averaging principle* for space-periodic solutions (classically, the Whitham principle deals with solutions which are bounded uniformly in space and locally in time; see Dubrovin and Novikov 1989).

## 9.2 Higher KdV equations

Let us consider a perturbation of the $l$th equation from the KdV hierarchy:

$$\dot{u} = \frac{\partial}{\partial x}(\nabla_u \mathcal{H}_l + \varepsilon \nabla_u H_1), \tag{9.3}$$

where

$$\mathcal{H}_l(u) = K_l \int_0^{2\pi} \left(u^{(l)^2} + \langle\text{higher-order terms with } \le l-1 \text{ derivatives}\rangle\right) dx$$

[56] With respect to the measure $(2\pi)^{-n}\, dq = (2\pi)^{-n}\, d\mathfrak{z}$.

and $H_1 = \int_0^{2\pi} f(x, u, \dots, u^{(l-1)})\, dx$. The function $f$ is assumed to be $C^d$-smooth in $x, \dots u^{(l-1)}$ and $\delta$-analytic in $u, \dots, u^{(l-1)}$. Since

$$\nabla_u H_1 = \sum_{j=0}^{l-1} (-1)^j \frac{\partial^j}{\partial x^j} f'_{u^{(j)}}(x, \dots, u^{(l-1)}),$$

then arguing as in Example 1.1, we see that $(\partial/\partial x)\nabla_u H_1$ is an analytic map of order $2l-1$: it analytically maps $H_0^d$ to $H_0^{d-2l+1}$ if $d \geq l$.

Let us take a bounded part $\mathcal{T}^{2n}$ of any $n$-gap manifold. It is invariant for the $l$th KdV equation (equal to $(9.3)_{\varepsilon=0}$) and it satisfies assumptions 1–4 (see Sections 3.6 and 6.3). The linearized equation has a complete system of Floquet solutions (see Section 6.3). Owing to (3.33) this system is non-resonant (cf. Section 6.2.1).

Now Theorem 8.1 applies to eqn (9.3) since the assumption (1) holds with $d_1 = 2l+1$, $d_1^1 = \dots = 0$ (see (6.21)) and (2) holds with $d_H = \tilde{d} = 2l+1$.

We see that *most of the n-gap solutions of the lth KdV equation persist in the perturbed equation (9.3) with sufficiently small* $\varepsilon$ in the same sense as for the KdV equation. For the solutions which persist obvious reformulations of Corollaries 9.1 and 9.2 hold.

## 9.3 Time-quasiperiodic perturbations of Lax-integrable equations

Slight modification of the Main Theorem's proof implies that most of finite-gap solutions of a Lax-integrable equation persist under a small perturbation of the equation's Hamiltonian by a time-dependent functional, provided that the functional is time-quasiperiodic and its frequency vector is 'typical' in a sense to be specified.

Below we restrict our presentation to the KdV equation, perturbed by a time-quasiperiodic forcing:

$$\begin{aligned} \dot{u} &= \frac{1}{4}u_{xxx} + \frac{3}{2}uu_x + \varepsilon\frac{\partial}{\partial x} f(t\varrho + \xi_0, x), \\ u(t,x) &\equiv u(t, x+2\pi), \qquad \int u\, dx \equiv 0. \end{aligned} \tag{9.4}$$

Here $f(\xi, x)$ is an analytic function on the torus $\mathbb{T}^M_\xi \times \mathbb{T}^1_x \sim \mathbb{T}^{M+1}$, $\varrho \in \mathbb{R}^M$ is a frequency vector, and $\xi_0 \in \mathbb{T}^M$ is a phase. The frequency vector is assumed to be a parameter of the problem. It varies in a bounded domain $\mathcal{R}$ of a positive measure:

$$\varrho \in \mathcal{R} \Subset \mathbb{R}^M, \qquad \text{mes}_M \mathcal{R} > 0.$$

Equation (9.4) is Hamiltonian and its Hamiltonian is

$$\mathcal{H}_\varepsilon(u,t) = \int \left(-\tfrac{1}{8}u'^2 + \tfrac{1}{4}u^3 + \varepsilon f(t\varrho + \xi_0, x)u(x)\right) dx.$$

Let us take any bounded part $\mathcal{T}^{2n}$ of a finite-gap manifold $\mathcal{T}_V^{2n}$ as in Section 9.1, i.e. $\mathcal{T}^{2n} = \bigcup\{T^n(r) \mid r \in \mathcal{K}\}$, where $\mathcal{K} = \{0 \leq r_j \leq K_0\}$. Subdividing in a need the cube $\mathcal{K}$ to smaller cubes and cutting out a narrow layer $\{r \in \mathcal{K} \mid 0 < r_j < \mu \text{ for some } j\}$, $0 < \mu \ll 1$, we may achieve that:[57]

(1) the KdV equation, restricted to the manifold $\mathcal{T}^{2n}$, admits there global analytic action-angle coordinates $(p, q)$, where $p \in P \Subset \mathbb{R}^n$ and $q \in \mathbb{T}^n$, and
(2) the gradient map $p \mapsto \nabla h(p)$ defines a diffeomorphism $\nabla h : P \to P' \subset \mathbb{R}^n$ (here $h$ is the KdV Hamiltonian, restricted to $\mathcal{T}^{2n}$).

Applying Theorem 7.2, we construct in the vicinity of the manifold $\mathcal{T}^{2n}$ in a space $H_0^d(S^1)$, $d \geq 3$, analytic simplectic coordinates $(p, q, y)$, where $(p, q)$ are as above and $y \in \mathcal{O}_\delta(Y_d)$. In these coordinates the Hamiltonian $\mathcal{H}_\varepsilon$ takes the form

$$\mathcal{H}_\varepsilon = \mathcal{H}^{KdV}(p, q, y) + \varepsilon h_1(p, q, y, t\varrho + \xi_0),$$

where

$$\mathcal{H}^{KdV} = h(p) + \tfrac{1}{2}\langle B(p)y, y\rangle + h_3(p, q, y)$$

and $h_1(p, q, y, \xi)$ is the functional $u(\cdot) \to \int f(\xi, x)u(x)\,dx$, written in the variables $(p, q, y)$ and depending on the parameter $\xi \in \mathbb{T}^n$. Equation (9.4) takes the form

$$\dot p = -\nabla_q \mathcal{H}_\varepsilon, \qquad \dot q = \nabla_p \mathcal{H}_\varepsilon, \qquad \dot y = J\nabla_y \mathcal{H}_\varepsilon. \tag{9.5}$$

Now we extend the phase space $P \times \mathbb{T}^n \times \mathcal{O}_\delta(Y_d) = \{(p, q, y)\}$ to the space $P \times \mathcal{O}_\delta(\mathbb{R}^M) \times \mathbb{T}^n \times \mathbb{T}^M \times \mathcal{O}_\delta(Y_d) = \{(p, I, q, \xi, y)\}$, given the symplectic form $dp \wedge dq + dI \wedge d\xi + \bar J dy \wedge dy$, and replace the non-autonomous equations (9.5) by the following autonomous system of higher dimension:

$$\begin{aligned} \dot p &= -\nabla_q \tilde{\mathcal{H}}_\varepsilon, & \dot I &= -\nabla_\xi \tilde{\mathcal{H}}_\varepsilon, \\ \dot q &= \nabla_p \tilde{\mathcal{H}}_\varepsilon, & \dot \xi &= \nabla_I \tilde{\mathcal{H}}_\varepsilon, \\ \dot y &= J\nabla_y \tilde{\mathcal{H}}_\varepsilon. \end{aligned} \tag{9.6}$$

Here $\tilde{\mathcal{H}}_\varepsilon(p, I, q, \xi, y) = \mathcal{H}^{KdV}(p, q, y) + \varrho \cdot I + \varepsilon h_1(p, q, y, \xi)$ (we note that the Hamiltonian $\tilde{\mathcal{H}}_\varepsilon$ is affine in the actions $I$). Certainly, the $(p, q, y)$-component of any solution for (9.6) such that $\xi(0) = \xi_0$ gives a solution for (9.5).

Next we perform the parameter-depending shift of the action $p$ as in Step 3 from Section 8.2:

$$p = p' + a, \qquad q = q', \ldots, y = y'; \quad a \in P.$$

Then

$$\tilde{\mathcal{H}}_\varepsilon = \text{const} + \omega \cdot p' + \varrho \cdot I' + \tfrac{1}{2}\langle B(a)y', y'\rangle + \varepsilon h_1'(p', \ldots, y'; a) + h_3'(p', \ldots, y'; a),$$

[57] Subdividing and cutting out are both unnecessary in the KdV case but they are needed for more involved equations.

where $\omega = \nabla h(a)$ and $h_3' = O(\|y'\|^3 + |p'|^2)$. Denoting $\tilde{n} = n + M$, $\tilde{P} = P \times \mathcal{O}_\delta(\mathbb{R}^M)$, and

$$(p', I') = \tilde{p} \in \tilde{P}, \qquad (q', \xi') = \tilde{q}, \qquad y' = \tilde{y}, \qquad (\omega, \varrho) = \tilde{\omega},$$

we write the Hamiltonian as

$$\tilde{\mathcal{H}}_\varepsilon(\tilde{p}, \tilde{q}, \tilde{y}; \tilde{\omega}) = \text{const} + \tilde{\omega} \cdot \tilde{p} + \tfrac{1}{2}\langle B(\omega)\tilde{y}, \tilde{y}\rangle + \varepsilon \tilde{h}_1 + \tilde{h}_3$$

(we replaced the parameter $a \in P$ by $\omega = \nabla h(a) \in P'$ using that the gradient-map is non-degenerate by assumption (2) ). The function $\tilde{h}_3$ is $O(\|\tilde{y}\|_d^3 + |\tilde{p}|^2)$ and the functions $\tilde{h}_1$ and $\tilde{h}_3$ are both $I'$-independent.

Theorem 8.3 applies to the Hamiltonian $\tilde{\mathcal{H}}_\varepsilon(\tilde{p}, \tilde{q}, \tilde{y}; \tilde{\omega})$, where the parameter $\tilde{\omega}$ belongs to the set $P' \times \mathcal{R}$. Since the functions $\tilde{h}_1$ and $\tilde{h}_3$ are $I'$-independent, then for any $m$ the functions $H_{2m}$ and $H_{3m}$ from a Hamiltonian's decomposition at the $m$th step of the KAM procedure (see Step 1 in Section 10.2 below) are $I'$-independent as well. Hence, the vectors $h^{\tilde{p}}$, $h^{1\tilde{p}}$, and $h^{0\tilde{p}}$ (see (10.16)) are such that their last $M$ components, corresponding to linear in $I'$ terms, vanish. Therefore the Hamiltonians $F$ at Step 2 also are $I'$-independent. Accordingly, the canonical transformations $S_m$ are identical in $\xi'$ and do not change the linear in $I'$ parts of the Hamiltonians $\mathcal{H}_m$: they remain equal to $\varrho \cdot I'$ (see Step 1 of the proof). Hence, the limiting map $\Lambda_\infty$ has the form $\Lambda_\infty(\omega, \varrho) = (\omega_\varepsilon(\omega, \varrho), \varrho)$.

Let us fix any $d \in \mathbb{N}$. Reformulating the theorem's assertions in terms of the original equation, we get the following result:

**Theorem 9.2.** *For any $\rho < 1$ there exist a Borel subset $Q_\varepsilon \subset \mathcal{K} \times \mathcal{R}$, a Lipschitz map $\omega_\varepsilon : Q_\varepsilon \to \mathbb{R}^n$, $C\varepsilon$-close to the map $(r, \varrho) \mapsto \mathbf{W}(r)$, and a Lipschitz map $\Sigma^\varepsilon : Q_\varepsilon \times \mathbb{T}^n \times \mathbb{T}^M \to H_0^d(S^1)$, analytic in $\mathbb{T}^n \times \mathbb{T}^M$, such that:*

*(a)* $\text{mes}_{n+M}(\mathcal{K} \times \mathcal{R} \setminus Q_\varepsilon) \to 0$ *as* $\varepsilon \to 0$;
*(b) for any $\xi \in \mathbb{T}^m$ the map $Q_\varepsilon \times \mathbb{T}^n \to H_0^d(S^1)$, $(r, \varrho, \mathfrak{z}) \mapsto \Sigma^\varepsilon(r, \varrho, \mathfrak{z}, \xi)$, is $\varepsilon^\rho$-close to the original $\rho$-independent map $\Phi_0$, also in the Lipschitz norm;*
*(c) every curve $\zeta^\varepsilon(t) = \Sigma^\varepsilon(r, \varrho, \mathfrak{z}+t\omega_\varepsilon(r, \varrho), \xi_0+t\varrho, )$, where $(r, \varrho) \in Q_\varepsilon$ and $\mathfrak{z} \in \mathbb{T}^n$, is a solution of (2.4) with a zero Lyapunov exponent.*

The solutions $\zeta^\varepsilon(t)$, constructed in the theorem, are quasiperiodic with $n + M$ frequencies. Their hulls are $(n + M)$-tori which lie in $\varepsilon^\rho$-neighbourhoods of the corresponding ('persisted') $n$-gap tori $T^n(r)$.

For most frequency vectors $\varrho$, the set $\mathcal{K}_\varrho = \{r \in \mathcal{K} \mid (r, \varrho) \in Q_\varepsilon\}$ which enumerates the finite-gap tori $T^n(r)$ which persist, approximates the whole set $\mathcal{K}$ in measure. Indeed, denoting by $\mu_n$ and $\mu_M$ the normalized Lebesgue measures on $\mathcal{K}$ and $\mathcal{R}$, respectively, we have $(\mu_n \times \mu_M)(Q_\varepsilon) = 1 - \gamma$, where $\gamma$ goes to 0 with $\varepsilon$. By the Fubini theorem,

$$\int_{\mathcal{R}} \mu_n(\mathcal{K}_\varrho)\, \mu_M(d\varrho) = 1 - \gamma.$$

In particular, for any positive $\gamma'$, the $\mu_M$-measure of the set, formed by all frequencies $\varrho \in \mathcal{R}$ such that $\mu_n(\mathcal{K}_\varrho) < 1 - \gamma'$, goes to zero with $\varepsilon$.

## 9.4 Perturbed SG equation

Now we consider a perturbed SG equation under the odd periodic boundary conditions:

$$\begin{gathered} \ddot{u} = u_{xx} - \sin u + \varepsilon f_u'(u, x), \\ u(t, x) \equiv u(t, x + 2\pi) \equiv -u(t, -x). \end{gathered} \tag{9.7}$$

Similar to the SG equation, we write (9.7) as a system of two first-order equations:

$$\dot{u} = -\sqrt{A}w, \qquad \dot{w} = \sqrt{A}\big(u + A^{-1}(\sin u - u) + \varepsilon f_u'(u, x)\big). \tag{9.8}$$

This system is Hamiltonian in the symplectic Hilbert spaces $(\{Z_s^0\}, \beta_2)$, $s \geq 0$, where $\beta_2 = \langle \bar{J}(du, dw), (du, dw)\rangle$. The corresponding Hamiltonian is $\mathcal{H}_\varepsilon = \frac{1}{2}\|(u, w)\|_0^2 + \varepsilon H_\varepsilon(u, w)$, where

$$H_\varepsilon(u, w) = \int_0^{2\pi} (\operatorname{Cos} u(x) - \varepsilon f(u, x))\, dx.$$

We recall that $\operatorname{Cos} u = -\cos u + 1 - \frac{1}{2}u^2$, that the space $Z_s^0 \subset H^{s+1}(S; \mathbb{R}^2)$ is given the $H^{s+1}$-scalar product, and that $J(u, w) = (-\sqrt{A}\, w, \sqrt{A}\, u)$ (see Sections 2.1 and 4.3).

Concerning the function $f$ we assume that:

(H1) $f(u, x)$ is a smooth function, $\delta$-analytic and even in $u$, even and $2\pi$-periodic in $x$.

If the equation is considered in a space $Z_s^o$ with small $s \geq 0$, then these assumptions may be relaxed. For example, if $s = 0$ or 1, then the following assumption suffices:

(H2) $f(u, x)$ is a $C^{s+1}$-smooth function, $\delta$-analytic in $x$, and vanishing for $u = 0, \pi$ identically in $x$.

Owing to the same calculations as in Section 2.1, $\nabla H_\varepsilon(u, w) = (A^{-1}(\sin u - u - \varepsilon f_u'(u, x), 0)$. Denoting $g(u, x) = \sin u - u - \varepsilon f_u'(u, x)$, we write $J\nabla H_\varepsilon$ as

$$J\nabla H_\varepsilon(u, w) = (0, A^{-1/2} g(u, x)). \tag{9.9}$$

Let us assume that (H1) holds. Then the map $u(x) \to g(u(x), x)$ gives rise to a zero-order analytic morphism of the Sobolev scale $H^l(S)$ for $l \geq 1$ (see Section 1.2). Therefore for any $s \geq 0$ the r.h.s. of (9.9) defines an analytic map $Z_s^o \longrightarrow H^{s+2}(S; \mathbb{R}^2)$.

Because of (H1), the function $g(u(x), x)$ is odd periodic. Hence, the range of the map (9.9) is contained in the space $Z_{s+1}^o$ and $J\nabla H_\varepsilon$ defines an analytic morphism of the scale $\{Z_s^o\}$ of order $-1$ for $s \geq 0$.

If $s = 0$ or 1 and (H2) holds, then we argue differently and view the perturbed SG equation (9.7) as an equation under Dirichlet boundary conditions

$$u(t, 0) = u(t, \pi) = 0 \tag{D}$$

(cf. the end of Section 1.2). Accordingly, we treat (9.8) as a Hamiltonian system in the symplectic space $\{Z_s, \beta_2\}$ where for $s = 1, 2$ the space $Z_s$ is

$$Z_s = \{\xi \in H^{s+1}([0, \pi]; \mathbb{R}^2) \mid \xi(0) = \xi(\pi) = 0\},$$

and for any integer $s$ the space $Z_s$ is formed by restrictions to $[0, \pi]$ of the vector-functions from $Z_s^o$.

Now (9.9) defines an analytic map $Z_s \longrightarrow H^{s+2}([0, \pi]; \mathbb{R}^2)$. If $(u, w) \in Z_s$ with $s = 0$ or 1, then the function $g(u(x), x)$ belongs to $H^{s+1}[0, \pi]$ and vanishes at $x = 0$ and $x = \pi$ (as well as $u(x)$). Hence, $(0, g) \in Z_s$ and the vector-function $A^{-1/2}(0, g) = (0, A^{-1/2}g)$ belongs to $Z_{s+1}$. Therefore, the range of the map (9.9) is contained in $Z_{s+1}$ and $J\nabla H_\varepsilon$ defines an analytic morphism of the scale $\{Z_s\}$ of order one for $0 \le s \le 1$.

For any $n$ let us take the finite-gap manifold $\mathcal{T}^{2n} = \Phi_0(R \times \mathbb{T}^n)$ as in Section 4.3. It is filled with odd periodic finite-gap solutions (4.17), where the branching points $E_1, \ldots, E_{4n}$ of the corresponding Riemann surfaces $\Gamma$ satisfy the relations (4.13)–(4.15), (4.18), and (4.21). We recall that the restrictions (4.14), (4.18), and (4.21) are imposed to guarantee that the solution (4.17) is real and odd periodic, and that assumption (4.15) is non-restrictive since there $C$ is an arbitrary number. At the same time, assumption (4.13) is superficial; see the Remark in Section 4.2 and the discussion that follows Theorem 9.4 below.

The finite-gap manifold $\mathcal{T}^{2n}$ satisfies assumptions 1–5 and the linearized SG equation has a complete non-resonant system of Floquet solutions, constructed in Section 6.4. Since $\nu(P_j) = j^* + O(j^{-1}) = j + O(j^{-1})$ (see (6.29′)), then the Main Theorem and its application apply with $d_1 = 1, d_1^1 = \cdots = 0$, and $\tilde{d} = -1$. Denoting by $\mu_n$ any finite measure on the $n$-dimensional real algebraic set $R$ which is absolutely continuous with respect to the Hausdorff measure $\text{mes}_n^{\mathcal{H}}$, we get:

**Theorem 9.3.** *Let us fix any $\rho' < 1$ and assume that the function $f(u, x)$ satisfies (H1), or (H2) if $s = 0$ or 1. Then there exists a Borel subset $R_\varepsilon \subset R$ such that $\mu_n(R \setminus R_\varepsilon) \to 0$ as $\varepsilon \to 0$ and for any $r \in R_\varepsilon$ the finite-gap torus $T^n(r) = \Phi_0(\{r\} \times \mathbb{T}^n) \subset Z_s^o$ persists as an analytic invariant $n$-torus of eqn (9.8) in $Z_s^o$ (or in $Z_s$ if $s = 0$ or 1). The invariant torus is filled with time-quasiperiodic solutions of eqn (9.8) and is $o(\varepsilon^{\rho'})$-close to $T^n(r)$.*

In difference with the KdV case, some of the time-quasiperiodic solutions which persist are not linearly stable (as well as the corresponding unperturbed finite-gap solutions).

Similar results with the same proof hold for even periodic finite-gap solutions and for finite-gap solutions with an odd number $g$ of open gaps (see Bikbaev and Kuksin 1992).

If $(u, w)$ is an odd periodic solution (4.17) that violates (4.13), then it belongs to some finite-gap manifold as in the Remark in Section 4.2. So if this solution lies in the same connected component of this manifold as the zero solution, then the Main Theorem applies to prove that most of the odd periodic finite-gap solutions (4.17), close to $(u, w)$, persist in the perturbed equation (9.8).

*Remark.* If the SG equation were considered under periodic boundary conditions (rather than under odd periodic), then its $g$-gap $x$-periodic solutions (4.17) would form $2g$-dimensional manifolds $\mathcal{T}^{2g}$ with singularities and in the vicinity of $\mathcal{T}^{2n}$ the SG equation can be put into the normal form as in Theorem 7.3 (in Section 7.3 we briefly indicated corresponding arguments, taking for granted that the system of Floquet solutions is non-resonant). The perturbed equation (9.7) has the form (8.4) and meets the assumptions of the Main Theorem with one exception: the exponents $\nu_j(r)$ are now asymptotically double and go in pairs $\nu_{j\pm}$, where both exponents $\nu_{j+}$ and $\nu_{j-}$ for $j \to \infty$ have the same asymptotic expansion as in item (1) of the theorem. Accordingly, to prove the persistence of most of the $x$-periodic finite-gap solutions of the SG equation, one needs a version of the Main Theorem that applies to equations with asymptotically double Floquet exponents. To get it one needs a corresponding version of Theorem 8.3 for perturbations of linear equations with asymptotically double frequencies $\nu_j$. Recently this result was proven by Chierchia and You (1999) (see Section 8.5). Using it one can repeat our arguments to get the KAM persistence of most of the $x$-periodic finite-gap solutions.

## 9.5 KAM persistence of lower-dimensional invariant tori of non-linear finite-dimensional systems

Let $\mathbb{R}^{2N}$ be a Euclidean space, given the usual symplectic structure, let $\mathcal{T}^{2N} = \bigcup_{r\in R} T_r^n$ be an analytic submanifold of $\mathbb{R}^{2N}$, diffeomorphic to $R\times\mathbb{T}^n$, $R \Subset \mathbb{R}^n$, and let $H_1, \ldots, H_n$ be commuting Hamiltonians, as in Proposition 5.2 (so that they are defined and analytic in the vicinity of $\mathcal{T}^{2n}$ and each torus $T_r^n$ is invariant for every Hamiltonian vector field $V_{H_j}$).

Let us take any Hamiltonian, say $H_1$. Then the vector field $V_{H_1}|_{\mathcal{T}^{2n}}$ has the form $\sum \omega_l(r)\partial/\partial\mathfrak{z}_l$ and by Proposition 5.2 the linearized equations have Floquet solutions with analytic frequencies $\nu_j(r)$.

Applying Theorem 8.1 we get that:

**Theorem 9.4.** *Let us assume that the following analytic functions do not vanish identically:*

$$l\cdot\nu(r) + s\cdot\omega(r), \quad l \in \mathbb{Z}^{N-n},\ 1 \le |l| \le 2;\ s \in \mathbb{Z}^n. \tag{9.10}$$

*Let $h$ be an analytic function, defined in the vicinity of $\mathcal{T}^{2n}$. Then most of the tori $T_r^n$ persist as invariant $n$-tori of the perturbed Hamiltonian vector field $V_{H_1+\varepsilon h}$, $0<\varepsilon\ll 1$, in the sense specified in Theorem 8.1. The invariant tori of the perturbed equation are filled with quasiperiodic solutions with zero Lyapunov exponents.*

This reduction of the Main Theorem is a much easier result than the theorem itself. Its claim remains essentially true under weaker assumptions: it suffices to check that only the functions (9.10) with $|l| = 1$ do not vanish identically; see Bourgain (1997) (we note that under this weaker assumption the claim about Lyapunov exponents is not true).

Results, similar to Theorem 9.4, were first stated in Melnikov (1965, 1968), where a scheme of a proof was given. First complete proofs of results, similar to Melnikov's theorems, arrived much later independently in the works of the author of this book, in Eliasson (1988), and next in Pöschel (1989).

# 10

# Proof of Theorem 8.3 on parameter-depending equations

As in Section 7.1 we restrict ourselves to the case when all frequencies $\nu_j(\omega)$ are real, i.e.

$$j_1 = n + 1,$$

since the general case differs from this one in more cumbersome notations only. We shall prove the theorem after some elementary transformations of the problem which we perform in Section 10.1. The proof is rather technical and a reader who is not used to the KAM techniques is advised to read the Addendum first where the classical Kolmogorov theorem is proved using the same ideas which we exploit below in a more involved situation.

## 10.1 Preliminary reductions

The proof becomes more complicated when either the frequencies $\nu_j(\omega)$ have linear growth with $j$ (i.e. in (8.10) $d_1 = 1$), or the perturbations $H_1$ and $h_3$ define Hamiltonian vector fields $J\nabla H_1$ and $J\nabla h_3$ of positive order $\tilde{d} > 0$. Since $\tilde{d} < d_1 - 1$, then these two complications cannot happen simultaneously. Equations with $\tilde{d} \leq 0$ were considered in Kuksin (1993) and Pöschel (1996*a*), and results of these works imply Theorem 8.3 and its amplification for $d_1 = 1$. Thus, it remains to prove the theorem for $d_1 > 1$. In this case it is convenient to replace assumption (1) of the theorem by the weaker assumption:

(1′) *The real functions $\nu_j(\omega)$ are Lipschitz in $\omega$ and odd in $j$, positive for positive $j$. For all $j$ and $k$ they satisfy the following inequalities*:

$$\begin{cases} K_1^{-1} j^{d_1} - K_0 \leq \nu_j(\omega) \leq K_1 j^{d_1}, & \forall\omega, \\ |\nu_j(\omega) - \nu_k(\omega)| \geq K_1^{-1} \, |j^{d_1} - k^{d_1}|, & \forall\omega, \\ \operatorname{Lip} \nu_j \leq K_1 \, j^{\tilde{d}}. \end{cases} \tag{10.1}$$

Before proving the theorem we shall have made some trivial reductions. Since $j_1 = n + 1$, then the operator $J$ is diagonal in the complex basis $\{\psi_j\}$. Therefore the operator $B(\omega)$ is diagonal in the complex basis $\{\psi_j = (\varphi_{|j|} - i \operatorname{sgn} j \, \varphi_{-|j|})/\sqrt{2}\} \mid j \in \mathbb{Z}_n\}$, as well as in the real basis $\{\varphi_j\}$, which is a symplectic basis for the form $\alpha_2$. Let us consider the linear operator $M$ which for every $j$ sends the vector $\varphi_j$ to $(\nu_j^J)^{1/2}\varphi_j$. This operator defines an isomorphism of the scale $\{Y_s\}$ of order $d_J/2$ since $J$ defines an isomorphism of order $d_J$. Since $\alpha_2 = \bar{J}\, du \wedge du$, then $M^*\alpha_2 = (M^*\bar{J}M)du \wedge du$, where

$$M^*\bar{J}M\varphi_{\pm j} = \pm\varphi_{\mp j}, \quad j \geq n+1.$$

That is, $\{\varphi_j\}$ is a Darboux basis for the form $M^*\alpha_2$. Equations (8.9) transformed by the map $\mathrm{id} \times M$ are Hamiltonian with respect to a symplectic structure defined by the form $dp \wedge dq \oplus M^*\alpha_2$. The corresponding Hamiltonian is

$$\mathcal{H}_\varepsilon \circ M = \omega \cdot p + \tfrac{1}{2}\langle \widetilde{B}(\omega)y, y\rangle + \varepsilon H_1 \circ L + h_3 \circ L,$$

where $\widetilde{B}(\omega) = M^*B(\omega)M$. So

$$\widetilde{B}(\omega)\varphi_j = |\nu_j(\omega)|\varphi_j, \quad \widetilde{B}(\omega)\psi_j = |\nu_j(\omega)|\psi_j, \quad \forall\, j \in \mathbb{Z}_n. \tag{10.2}$$

Clearly, the new Hamiltonian and the new symplectic form satisfy assumptions (1)–(3) of Theorem 8.3 with $d_J = 0$. Thus, it remains to prove the theorem with $d_J = 0$ and $\nu_j^J = \operatorname{sgn} j$. The operator $B(\omega)$ is diagonal in the bases $\{\varphi_j\}$ and $\{\psi_j\}$. Corresponding eigenvalues are $\{|\nu_j(\omega)|\}$.

Finally, we note that it suffices to prove the theorem for eqn (8.9) with $h_3 = 0$. Indeed, if we stretch the variables:

$$p = \varepsilon^{2/3}\tilde{p}, \qquad q = \tilde{q}, \qquad y = \varepsilon^{1/3}\tilde{y},$$

then in the tilde-variables we get a Hamiltonian equation with the Hamiltonian

$$\widetilde{\mathcal{H}}_\varepsilon = \omega \cdot \tilde{p} + \tfrac{1}{2}\langle B(\omega)\tilde{y}, \tilde{y}\rangle + \varepsilon^{1/3}H_1 + \varepsilon^{-2/3}h_3.$$

Denoting $\widetilde{H}_1(\tilde{p}, \tilde{q}, \tilde{y}; \omega) = (\varepsilon^{1/3}H_1 + \varepsilon^{-2/3}h_3)(\varepsilon^{2/3}\tilde{p}, \tilde{q}, \varepsilon^{1/3}\tilde{y}; \omega)$ and using (8.11) we see that both $\widetilde{H}_1$ and its gradient are $\varepsilon^{1/3}$-small. Thus, a version of Theorem 8.3 for perturbations with $h_3 = 0$ implies the general theorem for $\varepsilon$ replaced by $\varepsilon^{1/3}$ (i.e. it proves the general theorem for any $\rho < \frac{1}{9}$). Similarly with the amplification.

In Section 10.2 we prove the theorem and the amplification for $h_3 = 0$. As we have explained, these results imply the assertions we claim in Section 8.3 with a worse exponent $\rho$. To get the right exponent one should repeat the proof given below for equations with a non-zero $h_3$. All arguments and estimates remain quite similar but become longer. See Kuksin (1993) where we did this job for equations with $\tilde{d} \leq 0$.

## 10.2 Proof of the theorem

Here we prove Theorem 8.3 for $d_J = 0$ and $h_3 = 0$, i.e. for a Hamiltonian $\mathcal{H}_\varepsilon$ of the form

$$\mathcal{H}_\varepsilon = \omega \cdot p + \tfrac{1}{2}\langle B(\omega)y, y\rangle + \varepsilon H(p, q, y; \omega)$$

(we re-denoted $H_1$ as $H$). The corresponding equations are:

$$\begin{cases} \dot{p} = -\varepsilon\nabla_q\, H(\mathfrak{h}; \omega), \\ \dot{q} = \omega + \varepsilon\nabla_p\, H(\mathfrak{h}; \omega), \\ \dot{y} = J(B(\omega)y + \varepsilon\nabla_y\, H(\mathfrak{h}; \omega)), \end{cases} \tag{10.3}$$

where $(p, q, y) \in Q_\delta$, see (8.8). Below we abbreviate $(p, q, y)$ to $\mathfrak{h}$.

We shall use systematically notation for Lipschitz maps, described in Notation at the beginning of the book. In particular, if $B_1$ and $B_2$ are complex Banach spaces, $O_1$ is a domain in $B_1$, and $f$ maps $O_1 \times \Omega$ to $B_2$, then we write

$$\|f\|_{B_2}^{O_1,\Omega} = \max\Big(\sup_{b,\omega}\|f(b,\omega)\|, \sup_b \operatorname{Lip} f(b,\cdot)\Big),$$

where $\operatorname{Lip} f(b,\cdot)$ stands for a Lipschitz constant of the corresponding map from $\Omega$ to $B_2$. So our assumptions concerning the function $H(\mathfrak{h};\omega) = H_1$ (see (8.11)) mean that

$$|H|^{Q^c,\Omega} + \|\nabla_y H\|_{d-\tilde{d}}^{Q^c,\Omega} \le 1 \tag{10.4}$$

(we abbreviate $\|\cdot\|_{\mathbb{C}}^{\cdots}$ to $|\cdot|^{\cdots}$ and $\|\cdot\|_{Z_s^c}^{\cdots}$ to $\|\cdot\|_s^{\cdots}$).

We shall need some additional notation.

**Notation.** We introduce an increasing sequence $\{e(j)\}$, where $e(0) = 0$ and for $m \ge 1$

$$e(m) = (1^{-2} + \cdots + m^{-2})/K_*, \quad K_* = 2(1^{-2} + 2^{-2} + \cdots)$$

(thus $e(m) < 1/2$ for all $m$) and introduce two decreasing sequences, $\{\varepsilon_m\}$ and $\{\delta_m\}$:

$$\varepsilon_m = \varepsilon^{(1+\rho)^m}, \qquad \delta_m = \delta_0(1 - e(m)).$$

For $\delta > 0$, we denote by $U(\delta)$ the complex $\delta$-neighbourhood of the $n$-torus:

$$U(\delta) = \{q \in \mathbb{C}^n/2\pi\mathbb{Z}^n \mid |\operatorname{Im} q| < \delta\},$$

and denote by $U_m$, $m = 0, 1, \ldots,$ the complex domains $U_m = U(\delta_m)$. We also consider complex neighbourhoods $O_m$ of the torus $T_0^n = \{0\} \times \mathbb{T}^n \times \{0\}$ in $\mathcal{Y}_d^c$, where

$$O_m = \mathcal{O}_{\varepsilon_m^{2/3}}(\mathbb{C}^n) \times U_m \times \mathcal{O}_{\varepsilon_m^{1/3}}(Y_d^c) \subset \mathcal{Y}^c.$$

Moreover, we define the intermediate numbers

$$\delta_m^j = \frac{6-j}{6}\delta_m + \frac{j}{6}\delta_{m+1} = \delta_m - j/(6K_*(m+1)^2), \quad 0 \le j \le 5,$$

and the intermediate domains

$$O_m^j = \mathcal{O}_{(2^{-j}\varepsilon_m)^{2/3}}(\mathbb{C}^n) \times U(\delta_m^j) \times \mathcal{O}_{(2^{-j}\varepsilon_m)^{1/3}}(Y_d^c), \quad U_m^j = U(\delta_m^j).$$

If $\bar{\varepsilon} \ll 1$ (i.e. $\bar{\varepsilon}$ is sufficiently small), then

$$O_m \supset O_m^1 \supset \cdots \supset O_m^5 \supset O_{m+1} \supset \cdots \supset T_0^n.$$

Sometimes in the proofs of auxiliary results we use the domains $O_m^j$ with half-integer indexes $j$.

By $C$, $C_1$, etc. we denote different positive constants independent from $\varepsilon$ and $m$; by $C(m)$, $C_1(m)$, etc. we denote different functions of $m$ of the form $C(m) = C_1 m^{C_2}$; by

$C^e(m), C_1^e(m)$, etc. we denote functions of the form $\exp C(m)$, $\exp C_1(m)$. By $C_*, C_*(m)$, $C_*^e(m)$, etc. we denote fixed constants and functions. The constants $C, C_1, \ldots$ and the functions $C(m), C^e(m)$ may depend on $\gamma$. We observe that for each $C^e(m)$ and each $\sigma < 0$ the estimate $C^e(m) < \varepsilon_m^\sigma$ holds for all $m$ provided that $\bar{\varepsilon} \ll 1$. We profit from the assumption that $\varepsilon < \bar{\varepsilon}$ with sufficiently small $\bar{\varepsilon} > 0$ and use inequalities like

$$C^e(m)\varepsilon_m^\rho < 1$$

without extra remark.

**The KAM procedure.** Theorem 8.3 will be proved by the KAM procedure. That is, for $m = 0, 1, \ldots$ we shall define a subset $\Omega_m \subset \Omega$, an analytic function $\mathcal{H}_m$ on the domain $O_m$ as above, and a symplectic transformation $S_m : O_{m+1} \longrightarrow O_m$. For $m = 0$ we choose $\Omega_0 = \Omega$ and $\mathcal{H}_0 = \mathcal{H}_\varepsilon$. For every $m \geq 0$, $S_m$ transforms $\mathcal{H}_m$ to $\mathcal{H}_{m+1}$, i.e. $\mathcal{H}_m \circ S_m = \mathcal{H}_{m+1}$. We shall show that the system $V_{\mathcal{H}_m}$ on $O_m \cap \mathcal{Y}_d$ is integrable modulo a term $O(\varepsilon_m^\rho)$. So the transformation $S_0 \circ \cdots \circ S_{m-1}$ with a big $m$ 'almost integrates' the initial equations (10.3). Finally, we shall see that the limiting transformation $S_0 \circ S_1 \circ \cdots$ is well defined and integrates the equations.

We start with inductively constructing the transformation $S_m$ and the Hamiltonian $\mathcal{H}_{m+1}$ and finish with investigating the limiting transformation $S_0 \circ S_1 \circ \cdots$.

**Hamiltonians $\mathcal{H}_m$.** On a domain $O_m$ we consider a Hamiltonian $\mathcal{H}_m(\mathfrak{h}; \omega)$ of the form

$$\mathcal{H}_m = H_{0m}(p, y; \omega) + \varepsilon_m H_m(\mathfrak{h}; \omega), \tag{10.5}$$

where

$$H_{0m} = p \cdot \Lambda_m(\omega) + \tfrac{1}{2}\langle B_m(q; \omega)y, y\rangle, \tag{10.6}$$

and $\omega \in \Omega_m$, where $\Omega_m$ is a Borel subset of $\Omega$ such that

$$\operatorname{mes}(\Omega \setminus \Omega_m) \leq \gamma e(m). \tag{10.7}$$

The map $\omega \longmapsto \Lambda_m$ is Lipschitz and

$$|\Lambda_m(\omega) - \omega|^{\Omega_m, \mathrm{Lip}} \leq C\varepsilon^{1/3} e(m). \tag{10.8}$$

The operator $B_m$ is self-adjoint and is diagonal in the basis $\varphi_j^\pm$:

$$B_m \varphi_j^\pm = (\nu_j^{(m)}(\omega) + \beta_j^{(m)}(q; \omega))\,\varphi_j^\pm, \quad \forall j \in \mathbb{N}_n$$

(in particular, $B_m$ commutes with $B$). Here $\nu_j^{(m)}$ are real functions, close to the original frequencies $\nu_j$:

$$|\nu_j^{(m)} - \nu_j|^{\Omega_m, \mathrm{Lip}} \leq j^{\tilde{d}} C\varepsilon^\rho e(m). \tag{10.9}$$

The functions $\beta_j^{(m)}$ are real for real $q$ and analytically in $q$ extend to $U_m$. They are Lipschitz in $\omega \in \Omega_m$ and satisfy the following estimates:

$$\int \beta_j^{(m)} dq = 0, \quad |\beta_j^{(m)}|^{U_m,\Omega_m} \leq j^{\tilde{d}} \, C\varepsilon^{\rho} e(m).$$

In particular, $|\nabla_q \beta_j^{(m)}|^{U_m^1,\Omega_m} \leq j^{\tilde{d}} C(m)\varepsilon^{\rho}$ (the Cauchy estimate) and

$$\|\nabla_q B_m\|_{d,d_c}^{U_m^1,\Omega_m} \leq C(m)\varepsilon^{\rho}, \quad d_c := d - \tilde{d}. \tag{10.10}$$

For $-j \in -\mathbb{N}_n$ we set $\nu_{-j}^{(m)} = -\nu_j^{(m)}$ and $\beta_{-j}^{(m)} = -\beta_j^{(m)}$. Then

$$J B_m \psi_j = i(\nu_j^{(m)}(\omega) + \beta_j^{(m)}(q;\omega))\psi_j, \quad \forall j \in \mathbb{Z}_n.$$

The functional $H_m$ is assumed to be analytic in $O_m$ and to meet the following estimates:

$$|H_m|^{O_m,\Omega_m} \leq 2^m, \tag{10.11}$$

$$\|\nabla_y H_m\|_{d_c}^{O_m,\Omega_m} \leq \varepsilon_m^{-1/3} 2^m, \quad d_c = d - \tilde{d}. \tag{10.12}$$

Hamiltonian equations with the Hamiltonian $\mathcal{H}_m$ have the form

$$\dot{p} = -\tfrac{1}{2} \langle \nabla_q B_m(q;\omega)y, y\rangle - \varepsilon_m \nabla_q H_m, \quad \dot{q} = \Lambda_m(\omega) + \varepsilon_m \nabla_p H_m, \tag{10.13}$$

$$\dot{y} = J B_m(q;\omega)y + \varepsilon_m J\nabla_y H_m. \tag{10.14}$$

Clearly, the initial Hamiltonian $\mathcal{H}_\varepsilon$ has the form $\mathcal{H}_0$. (One should choose $\Lambda_0(\omega) = \omega$, $B_m = A$, $H_0 = H$, and $\Omega_m = \Omega$. The assumptions (10.7)–(10.10) with $m = 0$ become empty, while (10.11) and (10.12) follow from (10.4).)

**Transformations** $S_m$**.** Our goal is to find for every $m$ an analytic symplectomorphism $S_m : O_{m+1} \longrightarrow O_m$ which transforms the Hamiltonian $\mathcal{H}_m$ to a Hamiltonian $\mathcal{H}_{m+1} = \mathcal{H}_m \circ S_m$, where the latter has the form (10.5) with $m$ replaced by $m+1$. The transformation $S_m$ is constructed in four steps which are essentially identical to those in Kuksin (1993). The only difference comes during 'averaging' when we extract from the perturbation the whole diagonal of Hess $\varepsilon_m H_m$ and add it to the integrable part $H_{0m}$—not just the diagonal's averaging in $q$ as in Kuksin (1993).[58] Because of this, the operators $B_m$ depend on $q$ (their analogies in Kuksin (1993) are $q$-independent). Accordingly, homological equations written in terms of these operators become more involved. Their resolution is based on a theorem on first-order linear differential equations with variable coefficients, proved in Chapter 5.

We recall that *everywhere below* $\varepsilon < \bar{\varepsilon}$, *where* $\bar{\varepsilon}$ *is sufficiently small.*

---

[58] We are forced to do so since if $\tilde{d} > 0$ (and the perturbing vector field is unbounded), then to kill the diagonal part of Hess $\varepsilon_m H_m$ the transformation $S_m$ must be unbounded.

**Step 1: Averaging and splitting the perturbation.** Isolating the affine in $(p, q)$ and quadratic in $y$ parts of the Hamiltonian $H_m$, we rewrite it as

$$H_m = h^q(q;\omega) + p\cdot h^{1p}(q;\omega) + \langle y, h^y(q;\omega)\rangle + \langle h^{yy}(q;\omega)y, y\rangle + H_{3m}(\mathfrak{h};\omega), \tag{10.15}$$

where $\mathfrak{h} = (p, q, y)$ and $H_{3m} = O(|p|^2 + \|y\|_d^3 + |p|\,\|y\|_d)$. Next we change $h^q$ (and so $H_m$) by an $\omega$-dependent constant to achieve $(2\pi)^{-n}\int h^q dq = 0$ (this change is irrelevant since it does not affect the Hamiltonian equations). We denote by $h^{0p}$ averaging of the vector-function $h^{1p}$:

$$h^{0p} = (2\pi)^{-n}\int h^{1p} dq,$$

and set

$$h^p = h^{1p} - h^{0p}, \qquad \Lambda_{m+1} = \Lambda_m + \varepsilon_m h^{0p}(\omega). \tag{10.16}$$

Now we rewrite $\mathcal{H}_m = H_{0m} + \varepsilon_m H_m$ as

$$\mathcal{H}_m = H'_{0\,m+1}(p, y;\omega) + \varepsilon_m(H_{2m} + H_{3m})(\mathfrak{h};\omega),$$

where

$$H'_{0\,m+1} = p\cdot\Lambda_{m+1} + \tfrac{1}{2}\langle B_m y, y\rangle$$

and the function $H_{2m}$ equals

$$H_{2m} = h^q + p\cdot h^p + \langle y, h^y\rangle + \langle h^{yy}y, y\rangle.$$

**Lemma 10.1.** *The terms of the decomposition (10.15) are estimated as follows:*

*(a)*

$$\begin{aligned}
|h^q|^{U_m,\Omega_m} &\le 2^m,\\
|h^{1p}|^{U_m,\Omega_m} &\le 2^m\varepsilon_m^{-2/3},\\
|h^p|^{U_m,\Omega_m} &\le 2^{m+1}\varepsilon_m^{-2/3},\\
\|h^y\|_{d_c}^{U_m,\Omega_m} &\le 2^m\varepsilon_m^{-1/3},\\
\|h^{yy}\|_{d,d_c}^{U_m,\Omega_m} &\le 2^m\varepsilon_m^{-2/3}.
\end{aligned}$$

*Moreover, the operator $h^{yy}$ is symmetric and is real for real $q$.*

*(b) In the domain $O_{m+1}\subset O_m$ the term $\varepsilon_m H_{3m}$ is smaller than the admissible disparity of the next step (cf. (10.11) and (10.12)):*

$$\begin{aligned}
\varepsilon_m|H_{3m}|^{O_{m+1},\Omega_m} &\le \tfrac{2}{3}\,2^{m+1}\varepsilon_{m+1},\\
\varepsilon_m\|\nabla_y H_{3m}\|_{d_c}^{O_{m+1},\Omega_m} &\le \tfrac{2}{3}\,2^{m+1}\varepsilon_{m+1}^{2/3}.
\end{aligned}$$

*(c) The functions $H_{2m}$ and $H_{3m}$ are analytic in $\mathfrak{h}\in O_m$ and are real for real arguments.*

*Proof.* (a) The estimate for $h^q$ results from (10.11) since $h^q(q;\omega) = H_m(0,q,0;\omega)$.

To prove the estimate for $h^{1p}$ we observe that $h^{1p}(q;\omega) = \nabla_p H_m(0,q,0;\omega)$, so the estimate follows by application of the Cauchy estimate to the map $p \mapsto H_m(p,q,0;\omega)$ at $p=0$. To bound the Lipschitz constant in $\omega$ we consider the map $p \mapsto H_m(p,q,0;\omega_1) - H_m(p,q,0;\omega_2)$ and argue as above.

The estimate for $h^p$ obviously follows from the previous ones.

The estimate for $h^y$ results from (10.12) with $y=0$.

The estimate for the operator $h^{yy}$ follows by applying the Cauchy estimate to the map $\nabla_y H_m : y \mapsto \nabla_y H_m(q,0,y;\omega)$ since $h^{yy} = \frac{1}{2}\big(\nabla_y H_m(0,q,0;\omega)\big)_*$. This operator is symmetric and real (for real $q$) as a Hessian of a real function.

(b) Let $\mathfrak{h} = (p,q,y) \in O_{m+1}$ and $\nu = \varepsilon_m^{\rho/3}$. Then $\big((z/\nu)^2 p, q, (z/\nu)y\big) \in O_m$ for $z$ from the unit disc in the complex plane. Let us consider the function $z \mapsto H_m\big((z/\nu)^2 p, q, (z/\nu)y;\omega\big)$ and its Taylor series at zero:

$$H_m\left(\left(\frac{z}{\nu}\right)^2 p, q, \left(\frac{z}{\nu}\right)y;\omega\right) = h_0 + h_1 z + h_2 z^2 + \cdots .$$

By (10.11) and the Cauchy inequality, $|h_k| \le 2^m$ for all $k$. Since $H_{3m}(\mathfrak{h};\omega) = h_3\nu^3 + h_4\nu^4 + \cdots$ , then we have

$$\varepsilon_m |H_{3m}(\mathfrak{h};\omega)| = \varepsilon_m |h_3\nu^3 + h_4\nu^4 + \cdots| \le \frac{2^m \varepsilon_m^{1+\rho}}{1-\nu} \le \frac{2}{3} 2^{m+1} \varepsilon_{m+1} ,$$

if $\bar{\varepsilon}$ is sufficiently small. In a similar way one estimates the Lipschitz constant of $H_{3m}$.

To estimate $\nabla_y H_{3m}$ we consider the map

$$z \to \nabla_y H_m\left(\left(\frac{z}{\nu}\right)^2 p, q, \left(\frac{z}{\nu}\right)y;\omega\right) = h_0' + h_1' z + \cdots \in Y_{d_c}^c .$$

By (10.12), $\|h_k'\|_{d_c} \le \varepsilon_m^{-1/3} 2^m$ for all $k$. So

$$\begin{aligned}\varepsilon_m \|\nabla_y H_{3m}(\mathfrak{h};\omega)\|_{d_c} &= \varepsilon_m \|h_2'\nu^2 + h_3'\nu^3 + \cdots\|_{d_c} \\ &\le \frac{\nu^2}{1-\nu}\varepsilon_m^{2/3} 2^m \le \frac{2}{3}\varepsilon_{m+1}^{2/3} 2^{m+1} .\end{aligned}$$

A similar estimate holds for the Lipschitz constant, so assertion (b) is proved.

(c) The analyticity of the functions is evident. Their real-valuedness for real arguments results from the real-valuedness of the Hamiltonian $\mathcal{H}_m$. □

By the second estimate in item (a) of the lemma, $|h^{0p}|^{\Omega_m,\mathrm{Lip}} \le 2^m \varepsilon_m^{-2/3}$. Therefore,

$$|\Lambda_m - \Lambda_{m+1}|^{\Omega_m,\mathrm{Lip}} \le 2^m \varepsilon_m^{1/3} .$$

So the vector $\Lambda_{m+1}$ satisfies (10.8) with $m := m+1$.

**Step 2: Formal construction of the transformation $S_m$ and derivation of hological equations.** We construct the transformation $S_m$ as the time-one shift along trajectories of an auxiliary Hamiltonian vector field

$$\dot{p} = -\varepsilon_m \nabla_q F, \qquad \dot{q} = \varepsilon_m \nabla_p F, \qquad \dot{y} = \varepsilon_m J \nabla_y F, \tag{10.17}$$

where the Hamiltonian $F$ has the same structure as $H_{2m}$:

$$F = f^q(q;\omega) + p \cdot f^p(q;\omega) + \langle y, f^y(q;\omega)\rangle + \langle f^{yy}(q;\omega)y,\ y\rangle.$$

The flow $\{S^t\}$ of Hamiltonian equations (10.17) is formed by canonical transformations (see Theorem 1.7), and we set $S_m := S^t|_{t=1}$. Then formally

$$\mathcal{H}_m(S_m(\mathfrak{h};\omega);\omega) = \mathcal{H}_m(\mathfrak{h};\omega) + \varepsilon_m\{F, \mathcal{H}_m\} + O\left(\varepsilon_m^2\right),$$

where $\{\cdot,\cdot\}$ is the Poisson bracket (see Theorem 1.4 and formula (1.23)). Taking into account assertion (b) of Lemma 10.1, we get that in $O_{m+1}$ the composition $\mathcal{H}_m \circ S_m$ can be written as

$$\begin{aligned}\mathcal{H}_m \circ S_m =& H'_{0\,m+1} + \varepsilon_m\big(H_{2m} + \nabla_p F \cdot \nabla_q H'_{0\,m+1} - \nabla_q F \cdot \nabla_p H'_{0\,m+1}\\ &+ \langle J\nabla_y F, \nabla_y H'_{0\,m+1}\rangle\big) + O\left(\varepsilon_{m+1}\right).\end{aligned}$$

We observe that

$$\nabla_p H'_{0\,m+1} = \Lambda_{m+1}, \quad \nabla_q H'_{0\,m+1} = \tfrac{1}{2}\langle \nabla_q B_m\, y, y\rangle, \quad \nabla_y H'_{0\,m+1} = B_m\, y$$

and abbreviate

$$\Lambda_{m+1} = \omega', \qquad \omega' \cdot \nabla_q = \frac{\partial}{\partial \omega'}, \qquad B_m = B.$$

Now we rewrite $\mathcal{H}_m \circ S_m$ as

$$\begin{aligned}\mathcal{H}_m \circ S_m =& H'_{0\,m+1} + \varepsilon_m \Big[\tfrac{1}{2}\langle (f^p \cdot \nabla_q B)y, y\rangle - \partial f^q/\partial\omega' - p \cdot \partial f^p/\partial\omega'\\ &- \langle y, \partial f^y/\partial\omega'\rangle - \langle y, (\partial f^{yy}/\partial\omega')y\rangle + \langle By, Jf^y\rangle + 2\langle By, Jf^{yy}y\rangle\\ &+ h^q + p \cdot h^p + \langle y, h^y\rangle + \langle y, h^{yy}y\rangle\Big] + O\left(\varepsilon_{m+1}\right).\end{aligned} \tag{10.18}$$

(The term in the square brackets equals $H_{2m} + \{F, H_{0\,m+1}\}$.)

We wish to find the function $F$ in such a way that the contents of the square brackets in the r.h.s. of (10.18) vanishes up to an admissible disparity we define below. To this end $f^q$, $f^p$, $f^y$, and $f^{yy}$ should satisfy the homological equations:

$$\partial f^q/\partial\omega' = h^q(q;\omega), \qquad \partial f^p/\partial\omega' = h^p(q;\omega), \tag{10.19}$$

$$\partial f^y/\partial\omega' - BJf^y = h^y, \tag{10.20}$$

$$\partial f^{yy}/\partial\omega' + f^{yy}JB - BJf^{yy} = h^{yy} + \tfrac{1}{2}\, f^p \cdot \nabla_q B =: h^{1yy}$$

(the disparity will be introduced later). We define the functions $a_j$ as

$$a_j(q;\,\omega) = \tfrac{1}{2}\langle h^{1yy}\varphi_j^+, \varphi_j^+\rangle + \tfrac{1}{2}\,\langle h^{1yy}\varphi_j^-, \varphi_j^-\rangle, \quad \forall j \in \mathbb{N}_n,$$

and define the operator $A_m$ as

$$A_m(q;\,\omega) = \mathrm{diag}\,\{a_{n+1}, a_{n+1}, a_{n+2}, a_{n+2}, \dots\}$$

(i.e. $A_m\varphi_j^\pm = a_j\varphi_j^\pm$ for each $j$). Finally, we set

$$h^{0yy}(q;\,\omega) = h^{1yy}(q;\,\omega) - A_m(q;\,\omega).$$

We note that both operators $h^{0yy}$ and $h^{1yy}$ depend on a solution $f^p$ of the second equation in (10.19).

We observe that $JB = BJ$ and rewrite the last homological equation for $f^{yy}$ with $h^{1yy}$ replaced by $h^{0yy}$ (i.e. introducing a disparity):

$$\partial f^{yy}/\partial\omega' + [f^{yy}, JB] = h^{0yy}. \tag{10.21}$$

If $f^q, \dots, f^{yy}$ solve eqns (10.19)–(10.21), then the contents of the square brackets in (10.18) equal $\langle A_m y, y\rangle$ and

$$\{F, H'_{0\,m+1}\} = -H_{2m} + \langle A_m y, y\rangle. \tag{10.22}$$

**Step 3: Solving the homological equations.** The following result is classical for the KAM theory. For a proof see Lemmas A9.1 and A9.2 in Appendix 9 below.

**Lemma 10.2.** *Let us define the set* $\Omega^1$ *as*

$$\Omega^1 = \{\omega \in \Omega_m \,|\, |\omega'\cdot s| \le C^{-1}(m+1)^{-2}|s|^{-n} \text{ for some } s = s(\omega) \in \mathbb{Z}^n\backslash\{0\}\}.$$

*Then* mes $\Omega^1 \le \gamma(m+1)^{-2}/3K_*$[59] *if $C$ is chosen sufficiently large. For $\omega \in \Omega_m\backslash\Omega^1$, eqns (10.19) have analytic solutions, real for real arguments and such that*

$$|f^q|^{U^1_m,\Omega_m\backslash\Omega^1} \le C(m), \qquad |f^p|^{U^1_m,\Omega_m\backslash\Omega^1} \le \varepsilon_m^{-2/3}C(m).$$

Using the estimate for the solution $f^p$ as well as Lemma 10.1 (a) and (10.10), we get that

$$\|h^{1yy}\|_{d,d_c}^{U^1_m,\Omega_m\backslash\Omega^1} \le C(m)\,\varepsilon_m^{-2/3}.$$

Hence,

$$|a_j|^{U^1_m,\Omega_m\backslash\Omega^1} \le j^{\tilde d}\,C(m)\,\varepsilon_m^{-2/3}, \quad \forall\, j \ge n+1,$$

and we arrive at the following corollary.

**Corollary.** *The operator $h^{0yy}$ satisfies the estimate*

$$\|h^{0yy}\|_{d,d_c}^{U^1_m,\Omega_m\backslash\Omega^1} \le C_1(m)\,\varepsilon_m^{-2/3}.$$

---

[59] The constant $K_*$ was defined at the beginning of Section 10.2.

Equations (10.20) and (10.21) are more complicated than (10.19). We start with the more difficult equation (10.21).

**Lemma 10.3.** *There exists a Borel subset* $\Omega^2 \subset \Omega_m$ *such that* $\operatorname{mes}\Omega^2 \le \gamma(m+1)^{-2}/(3K_*)$ *and*

$$\left|\omega' \cdot s + \nu_j^{(m+1)} - \nu_k^{(m+1)}\right| \ge \frac{|j^{d_1} - k^{d_1}|}{C_{**}(m)\langle s\rangle^{c_1}},$$

*for all* $\omega \in \Omega \setminus (\Omega^2 \cup \Omega^1)$*, all* $j, k \in \mathbb{Z}_n$*, and all* $s \in \mathbb{Z}^n$*, with some constant* $C_{**}(m)$ *and some exponent* $c_1 > 0$*. Here and below for* $j \in \mathbb{Z}$ *we write* $j^{d_1} = \operatorname{sgn} j\, |j|^{d_1}$*.*

The proof follows Kuksin (1993) and will be given in Section 10.3.

We recall that the operator $JB = JB_m(q;\omega)$ is diagonal in the complex basis $\{\psi_j \mid j \in \mathbb{Z}_n\}$ and has the eigenvalues $i\tilde\nu_j$, where

$$\tilde\nu_j(q;\omega) = \nu_j^{(m)}(\omega) + \beta_j^{(m)}(q;\omega).$$

Let us denote by $\{f_{kj}(q;\omega) \mid k, j \in \mathbb{Z}_n\}$ and $\{h_{kj}(q;\omega) \mid k, j \in \mathbb{Z}_n\}$ the Hilbert matrices of the operators $f^{yy}$ and $h^{0yy}$ with respect to the complex basis $\{\psi_j\}$ of the space $Y^c$. Then $f_{kj} = \langle f^{yy}\psi_j, \psi_{-k}\rangle$ (see Appendix 2) and the operator $[f^{yy}, JB]$ has a Hilbert matrix with the entries

$$\begin{aligned}\langle (f^{yy}JB - JBf^{yy})\psi_j, \psi_{-k}\rangle &= \langle (f^{yy}JB\psi_j, \psi_{-k}\rangle + \langle f^{yy}\psi_j, BJ\psi_{-k}\rangle \\ &= i\tilde\nu_j\langle f^{yy}\psi_j, \psi_{-k}\rangle + i\tilde\nu_{-k}\langle f^{yy}\psi_j, \psi_{-k}\rangle = i(\tilde\nu_j - \tilde\nu_k) f_{kj}.\end{aligned}$$

Hence, in terms of the matrix elements $f_{kj}$ eqn (10.21) reads as

$$\frac{\partial}{\partial\omega'} f_{kj}(q;\omega) + i(\tilde\nu_j - \tilde\nu_k)(q;\omega) f_{kj} = h_{kj}(q;\omega), \tag{10.23}$$

for every $k, j \in \mathbb{Z}_n$. Owing to the definition of the operator $h^{0yy}$, its diagonal part vanishes:

$$h_{kk}(q;\omega) \equiv 0, \quad \forall k.$$

Moreover, the matrix of the operator $h^{0yy}$ as a map $Y_d^c \to Y_{d_c}^c$ is

$$\{|k|^{d_c} h_{kj} |j|^{-d} \mid k, j \in \mathbb{Z}_n\},$$

provided that the spaces $Y_d^c$ and $Y_{d_c}^c$ are given the complex Hilbert bases $\{|j|^{-d}\psi_j\}$ and $\{|j|^{-d_c}\psi_j\}$, respectively, (see (A2.2) in Appendix 2, Chapter 1). Using the corollary from Lemma 10.2, we get an estimate for the r.h.s. of (10.23):

$$|h_{kj}|^{U_m^1, \Omega_m \setminus \Omega^1} \le C(m)\,\varepsilon_m^{-2/3} |j|^d |k|^{-d_c}.$$

Let us observe that

$$\tilde\nu_j - \tilde\nu_k = (\nu_j^{(m+1)} - \nu_k^{(m+1)})(\omega) + (\beta_j^{(m+1)} - \beta_k^{(m+1)})(q;\omega)$$

is the sum of a constant which is $\ge \max(|j|, |k|)^{d_1 - 1}/C$ (due to (10.1)) and a $q$-dependent function of order

$$\varepsilon \max(|j|, |k|)^{\tilde d}.$$

Since $\tilde d$ can be positive, then (10.23) is a perturbation of a constant-coefficient equation by a variable-coefficient term which can be arbitrarily large. However, since $\tilde d < d_1 - 1$, then the 'very large' constant-coefficient part of (10.23) suppresses the 'large' variable coefficient part: Theorem 12.1, which we prove in Chapter 12, implies[60] that for $\omega \in \Omega_m \setminus (\Omega^1 \cup \Omega^2)$ eqn (10.23) has a unique analytic solution $f_{kj}$ and

$$|f_{kj}|^{U_m^2} \le C^e(m) \frac{|h_{kj}|^{U_m^1}}{|j^{d_1} - k^{d_1}|}.$$

The operator $f^{yy} : Y_d^c \longrightarrow Y_d^c$ has a Hilbert matrix $F$ with the entries $F_{kj} = |k|^d f_{kj} \, |j|^{-d}$. Using the estimate for $h_{kj}$, we get that

$$|F_{kj}(q)| \le C'^e(m)\varepsilon_m^{-2/3}|k|^{\tilde{d}}/|j^{d_1} - k^{d_1}|, \quad k \ne j,$$

for each $q \in U_m^2$. Since $F_{kk} \equiv 0$ and $d_1 > \tilde{d} + 1$, then

$$\sum_k |F_{kj}| \le \varepsilon_m^{-2/3} C_1^e(m) \left( \int_{-\infty}^{-1} + \int_1^j + \int_{j+1}^{\infty} \right) \frac{|x|^{\tilde{d}} dx}{|j^{d_1} - x^{d_1}|}$$

$$\le \varepsilon_m^{-2/3} C_2^e(m) \, |j|^{\tilde{d}+1-d_1} \log |j| \le C^e(m) \, \varepsilon_m^{-2/3}.$$

A similar estimate holds for $\ell^1$-norms of rows of the matrix $F$. Therefore a norm of the operator $f^{yy}(q) : Y_d \longrightarrow Y_d$ with any $q$ in $U_m^2$ is bounded by $C^e(m)\varepsilon_m^{-2/3}$ by the Schur criterion.

So the norm of $f^{yy}(q)$, $q \in U_m^2$, is estimated. To estimate the Lipschitz constant, we consider an increment $f_\Delta^{yy}$ of the operator $f^{yy}$, $f_\Delta^{yy} = f^{yy}(q; \omega_1) - f^{yy}(q; \omega_2)$. It satisfies the equation

$$\partial f_\Delta^{yy} / \partial \omega' + [f_\Delta^{yy}, JB] = h_\Delta^{0yy} - \nabla_q f^{yy}(q; \omega_2) \cdot (\omega_1 - \omega_2) - [f(q; \omega_2), JB_\Delta] =: H_\Delta^{yy},$$

where $h_\Delta^{0yy}$ and $B_\Delta$ stand for increments of $h^{0yy}$ and $B$. We see that for $q \in U_m^3$,

$$\|H_\Delta^{yy}(q; \omega)\|_{d,d_c} \le C_1^e(m)\varepsilon_m^{-2/3} \, |\omega_1 - \omega_2|.$$

So the arguments given above estimate the Lipschitz constant in $\omega$ for $f^{yy}$ when $q \in U_m^4$. We can use intermediate domains like $U_m^{3/2}$ to get a similar estimate for $q$ in $U_m^2$:

**Lemma 10.4.** *If $\omega \in \Omega_m \setminus (\Omega^1 \cup \Omega^2)$, then eqn (10.21) has an analytic solution $f^{yy}$ which is a symmetric in $Y^c$ operator, real for real $q$, and such that*

$$\|f^{yy}\|_{d,d}^{U_m^2, \Omega_m \setminus (\Omega^1 \cup \Omega^2)} \le C^e(m) \; \varepsilon_m^{-2/3}. \tag{10.24}$$

---

[60] Applying the theorem, one should choose $n_1 = c_1$, $n_2 = n$, $K_1 = C(m)/|j^{d_1} - k^{d_1}|$, $K_2 = Cm^2$, and $\Delta = Cm^{-2}$.

Quite similar (but simpler) arguments show the solvability of equation (10.20):

**Lemma 10.5.** *There exists a Borel subset* $\Omega^3 \subset \Omega_m$, mes $\Omega^3 \le \gamma(m+1)^{-2}/3K_*$, *such that for* $\omega \in \Omega_m \setminus (\Omega^1 \cup \Omega^3)$ *eqn (10.20) has an analytic solution* $f^y(q;\omega)$, *real for real* $q$, *and such that*

$$\|f^y\|_d^{U_m^2, \Omega_m \setminus (\Omega^1 \cup \Omega^3)} \le C^e(m)\, \varepsilon_m^{-2/3}.$$

Now we define the set $\Omega_{m+1}$ as

$$\Omega_{m+1} = \Omega_m \setminus (\Omega^1 \cup \Omega^2 \cup \Omega^3). \tag{10.25}$$

Owing to the estimates for measures of the sets $\Omega^1$, $\Omega^2$, and $\Omega^3$, obtained in Lemmas 10.2, 10.3, and 10.6, we have

$$\text{mes}\,(\Omega \backslash \Omega_{m+1}) \le \text{mes}\,(\Omega \backslash \Omega_m) + \gamma(m+1)^{-2}/K_* \le \gamma\, e(m+1).$$

So the set $\Omega_{m+1}$ satisfies (10.7) with $m := m+1$.

**Step 4: Study of the transformation $S_m$.** To carry out the arguments of this step and of the next one, we shall use the symplectic Hilbert scale $(\{Z_s = \mathbb{R}^{2n} \times Y_s\}, dp \wedge dq \oplus \alpha_2)$ and its complexification. The scalar product in $Z_0$ is denoted $\langle \cdot, \cdot \rangle$. The spaces $Z_d$ and $Z_d^c$ are covering spaces for the manifolds $\mathcal{Y} = \mathcal{Y}_d$ and $\mathcal{Y}^c = \mathcal{Y}_d^c$ with respect to the natural projections; see Section 1.3. In addition to the usual norms $\|\cdot\|_s$, we provide spaces in the scales $\{Z_s\}$ and $\{Z_s^c\}$ with the weighted norms $\|\cdot\|_{(+,s)}$ and $\|\cdot\|_{(-,s)}$, where

$$\|(p,\xi,y)\|_{(\pm,s)}^2 = |p|^2 + \varepsilon_m^{\pm\frac{4}{3}} |\xi|^2 + \varepsilon_m^{\pm\frac{2}{3}} \|y\|_s^2.$$

By $Z_s^{\pm}$ and $Z_s^{c\pm}$ we denote the spaces $Z_s$ and $Z_s^c$, given the norms we have just defined. Clearly, the spaces $Z_s^+$ and $Z_{-s}^-$ are dual with respect to the inner product $\langle \cdot, \cdot \rangle$. Therefore, for any linear operator $A : Z_a \to Z_b$ we have

$$\|A\|_{(+,a),(+,b)} = \|A^*\|_{(-,-b)(-,-a)}. \tag{10.26}$$

The weighted norms provide the manifolds $\mathcal{Y}$ and $\mathcal{Y}^c$ with distances $\text{dist}_{(\pm,d)}$. It follows from the definitions of the domains $O_m^j$ that

$$\text{dist}_{(-,d)}(O^{j+1}, \mathcal{Y}^c \setminus O^j) \ge C^{-1}(m), \quad \forall j. \tag{10.27}$$

We recall that $S_m = S^t|_{t=1}$, where $\{S^t\}$ is the flow of the system (10.17) which we now write as

$$\dot{\mathfrak{h}} = \varepsilon_m V_F(\mathfrak{h}), \qquad \mathfrak{h} = \mathfrak{h}(t) = (p,q,y)(t), \tag{10.28}$$

where $V_F(\mathfrak{h}) = V_F(\mathfrak{h};\omega) = (-\nabla_q F, \nabla_p F, J\nabla_y F)$. The estimates from Lemmas 10.2, 10.4, and 10.5 (and the Cauchy estimate) show that the vector field $V_F$ is analytic in the domain $O_m^{2.5}$ and

$$\|\varepsilon_m V_F\|_{(-,d)}^{O_m^{2.5}, \Omega_{m+1}} \le C^e(m) \varepsilon_m^{1/3}. \tag{10.29}$$

A straightforward analysis of terms forming the linearized vector field $V_{F*}$, based on the same lemmas, shows that

$$\|\varepsilon V_F(\mathfrak{h})_*\|_{\theta,\theta} \leq C^e(m)\varepsilon_m^{1/3}, \quad \forall\,|\theta| \leq d \tag{10.30}$$

and

$$\|\varepsilon V_F(\mathfrak{h})_*\|_{(-,\theta),(-,\theta)} \leq C^e(m)\varepsilon_m^{1/3}, \quad \forall\,|\theta| \leq d, \tag{10.31}$$

for every $\mathfrak{h} \in O_m^3$. The same estimates hold for the Lipschitz constants in $\omega \in \Omega_{m+1}$.

**Lemma 10.6.** *The map $S_m$ is an analytic symplectomorphism which maps $O_m^j$ to $O_m^{j-1}$ for $j = 3, 4, 5$. This map is close to the identity, namely:*

(a) $\|S_m - id\|^{O_m^3,\Omega_{m+1}} \leq C_1^e(m)\varepsilon_m^{1/3}$, *where $\|\cdot\|$ stands for the norm $\|\cdot\|_d$ or $\|\cdot\|_{(-,d)}$;*
(b) $\|S_{m*} - id\|_{\cdot,\cdot}^{O_m^4,\Omega_{m+1}} \leq C_2^e(m)\varepsilon_m^{1/3}$, *where $\|\cdot\|_{\cdot,\cdot}$ stands for the operator norm $\|\cdot\|_{\theta,\theta}$ or $\|\cdot\|_{(-,\theta),(-,\theta)}$, with any $|\theta| \leq d$;*
(c) *all the estimates, stated above for the map $S_m = S^t|_{t=1}$, remain true for the maps $S^t$ with $0 \leq t \leq 1$.*

*Proof.* Since $S^t(\mathfrak{h}) - \mathfrak{h} = \int_0^t \varepsilon_m V_F(S^\tau(\mathfrak{h}))\,d\tau$, then by (10.29) and (10.27) the map $S_m = S^1$ sends $O_m^j$ to $O_m^{j-1}$ and $\|S_m - \mathrm{id}\| \leq C^e(m)\varepsilon_m^{1/3}$. This map is an analytic symplectomorphism due to Theorem 1.3. To check its Lipschitz constant in $\omega$, we take any $\omega_1, \omega_2 \in \Omega_{m+1}$ and denote by $\mathfrak{h}_j(t)$ a solution for (10.28) with $\mathfrak{h}_j(0) = \mathfrak{h} \in O_m^j$ and $\omega = \omega_j$, $j = 1, 2$. We have to estimate the difference $\eta(t) = \mathfrak{h}_1(t) - \mathfrak{h}_2(t)$. The curve $\eta$ satisfies the equation

$$\dot\eta = \varepsilon_m V_F(\mathfrak{h}_1; \omega_1) - \varepsilon_m V_m(\mathfrak{h}_2; \omega_2).$$

Owing to (10.30) and (10.31) the map $\varepsilon_m V_F$ is Lipschitz in $\mathfrak{h}$-variable, so a norm of the r.h.s. is estimated by $C^e(m)\varepsilon_m^{1/3}(\|\eta\| + |\omega_2 - \omega_1|)$. Accordingly,

$$\frac{d}{dt}\|\eta\| \leq C^e(m)\varepsilon_m^{1/3}(\|\eta\| + |\omega_2 - \omega_1|), \quad \eta(0) = 0.$$

Using the Granwall estimate we find that

$$\|S_m(\mathfrak{h}; \omega_1) - S_m(\mathfrak{h}; \omega_2)\| = \|\eta(1)\| \leq C^e(m)\varepsilon_m^{1/3}|\omega_2 - \omega_1|.$$

So assertion (a) is proved.

To prove (b), we note that for any $\xi$ the curve $t \mapsto S^t(\mathfrak{h})_*\xi$ is a solution of the linearized equation $\dot\xi = \varepsilon_m V_F(\mathfrak{h}(t))_*\xi$. Since $S_{m*} = S^1{}_*$, then the estimates for the operator $(S_{m*} - \mathrm{id})$ follow from (10.30) and (10.31) (cf. Proposition 1.4).

The same arguments as above apply to any map $S^t$, thus proving (c). □

**Step 5: The transformed Hamiltonian.** Now we study the transformed Hamiltonian $\mathcal{H}_m \circ S_m = (H'_{0\,m+1} + \varepsilon_m(H_{2m} + H_{3m})) \circ S_m$. Since the functional $H'_{0\,m+1}$ is smooth on the space $\mathcal{Y}_d$ and the flow maps $S^t$ are $C^1$-smooth in $t$, then

$$\frac{d}{dt} H'_{0\,m+1} \circ S^t = \varepsilon_m \{F,\, H'_{0\,m+1}\} \circ S^t = -\varepsilon_m (H_{2m} - \langle A_m y,\, y\rangle) \circ S^t,$$

where the second equality follows from (10.22) and the first one follows from Theorem 1.4. Now we can calculate the second derivative:

$$\frac{d^2}{dt^2} H'_{0\,m+1} \circ S^t = -\varepsilon_m \frac{d}{dt}(H_{2m} - \langle A_m y,\, y\rangle) \circ S^t = -\varepsilon_m^2 \{F,\, H_{2m} - \langle A_m y,\, y\rangle\} \circ S^t.$$

Thus,

$$\begin{aligned} H'_{0\,m+1} \circ S_m &= H'_{0\,m+1} \circ S^t|_{t=1} \\ &= H'_{0\,m+1} + \frac{d}{dt} H'_{0\,m+1} \circ S^t|_{t=0} + \int_0^1 (1-t) \frac{d^2}{dt^2} H'_{0\,m+1} \circ S^t \, dt \\ &= H'_{0\,m+1} + \varepsilon_m \langle A_m y,\, y\rangle - \varepsilon_m H_{2m} \\ &\quad - \varepsilon_m^2 \int_0^1 (1-t)\{F,\, H_{2m} - \langle A_m y,\, y\rangle\} \circ S^t dt. \end{aligned}$$

Calculating similarly $(d/dt)(H_{2m} + H_{3m}) \circ S^t$ we find that

$$\varepsilon_m (H_{2m} + H_{3m}) \circ S_m = \varepsilon_m (H_{2m} + H_{3m}) + \varepsilon_m^2 \int_0^1 \{F,\, H_{2m} + H_{3m}\} \circ S^t dt.$$

Therefore, the transformed Hamiltonian can be written as

$$\begin{aligned} \mathcal{H}_m \circ S_m &= H_{0\,m+1} + \varepsilon_m H_{3m} \\ &\quad + \varepsilon_m^2 \int_0^1 \Big( (t-1)\{F, H_{2m} - \langle A_m y, y\rangle\} \circ S^t \Big) dt \\ &\quad + \varepsilon_m^2 \int_0^1 \{F, H_{2m} + H_{3m}\} \circ S^t dt, \end{aligned}$$

where we denoted

$$H_{0\,m+1} = H'_{0\,m+1} + \langle A_m y,\, y\rangle.$$

The Hamiltonian $H_{0\,m+1}$ has the form (10.6) with $m := m+1$ and with

$$B_{m+1} = B_m + 2\varepsilon_m A_m.$$

Since the diagonal elements $a_j$ of the operator $A_m$ are bounded by $j^{\tilde{d}} C(m)\, \varepsilon_m^{-2/3}$ (see Lemma 10.2 and its discussion), then the diagonal elements $\nu_j^{(m+1)} + \beta_j^{(m+1)}$ of the operator $B_{m+1}$ satisfy the a priori estimates (see (10.9), etc.) with $m$ replaced by $m+1$.

For $j = 1, 2, 3, 4$ we denote by $\Delta_j H$ the $j$th term in the r.h.s. of the formula for $\mathcal{H}_m \circ S_m$. To prove that the Hamiltonian $\mathcal{H}_{m+1} := \mathcal{H}_m \circ S_m$ has the form (10.5) we should check that

$$\Delta_2 H + \Delta_3 H + \Delta_4 H = \varepsilon_{m+1} H_{m+1}, \tag{10.32}$$

where $H_{m+1}$ is a function satisfying the estimates (10.11) and (10.12) in the domain $O_{m+1}$.

By Lemma 10.1, the term $\Delta_2 H$ and its $y$-gradient are smaller than two-third times the right-hand sides of (10.11) and (10.12), respectively. The estimates for $\Delta_3 H$ and $\Delta_4 H$ will follow from the following statement:

**Lemma 10.7.** *If $H$ is an analytic function such that*

$$|H|^{O^1_m, \Omega_{m+1}} \le C^e(m)\, \varepsilon_m^2, \qquad \|\nabla_y H\|_{dc}^{O^1_m, \Omega_{m+1}} \le C^e(m)\, \varepsilon_m^{5/3}, \tag{10.33}$$

*then for any $0 \le t \le 1$ we have*

$$|\{F, H\} \circ S^t|^{O^4_m, \Omega_{m+1}} \le C^e_1(m)\, \varepsilon_m^{4/3} \tag{10.34}$$

*and*

$$\|\nabla_y(\{F, H\} \circ S^t)\|_{dc}^{O^5_m, \Omega_{m+1}} \le C^e_1(m)\, \varepsilon_m. \tag{10.35}$$

Postponing the proof we complete Step 5. Application of Lemma 10.7 to the functions $H = \varepsilon_m^2(H_{2m} - \langle A_m y, y\rangle)$ and $H = \varepsilon_m^2(H_{2m} + H_{3m})$, followed by integration of the corresponding inequalities (10.34) from $t = 0$ to $t = 1$, proves that in $O_{m+1}$ the function $\Delta_3 H + \Delta_4 H$ is bounded by $2C_1^2(m)\, \varepsilon_m^{4/3} \le \frac{1}{3} 2^{m+1} \varepsilon_{m+1}$, as well as its Lipschitz constant in $\omega \in \Omega_{m+1}$. Similarly, due to (10.35) the gradient $\nabla_y(\Delta_3 H + \Delta_4 H)$ is bounded by $\frac{1}{3} 2^{m+1} \varepsilon_{m+1}^{2/3}$. Therefore the Hamiltonian $\mathcal{H}_{m+1} := \mathcal{H}_m \circ S_m$ has the required form (10.5) with $m$ replaced by $m + 1$.

*Proof of the lemma.* Owing to the first inequality in (10.33) (and, as usual, the Cauchy estimate), we have $|\nabla_p H|^{O^2_m, \Omega_{m+1}} \le C_1^e(m) \varepsilon_m^{4/3}$ and $|\nabla_q H|^{O^2_m, \Omega_{m+1}} \le C_1^e(m) \varepsilon_m^2$. Using this estimate jointly with (10.33) and (10.29) we find that

$$|\{F, H\}|^{O^3_m, \Omega_{m+1}} \le C_2^e(m) \varepsilon_m^{4/3}. \tag{10.36}$$

Since $S_m$ analytically maps $O^4_m$ to $O^3_m$ by Lemma 10.6, then we get (10.34).

To prove (10.35) we first have to bound the gradient of the Poisson bracket $\{F, H\}$. The bracket is formed by three terms, where the most difficult one is the term $\langle J\nabla_y F, \nabla_y H\rangle$. Its gradient is $\nabla H_* \Pi_y^* J \nabla F - \nabla F_* \Pi_y^* J \nabla H$ ($\Pi_y^*$ is the operator which sends a vector $y$ to $(0, 0, y)$). Using (10.29) and (10.33) we get that for $\mathfrak{h} \in O^4_m$ the

$d_c$-norm of the gradient is bounded by $C^e(m)\varepsilon_m$, as well as its Lipschitz constant in $\omega$. Analysing the two other terms in a similar way we get that

$$\|\nabla_y\{F,H\}\|_{d_c}^{O_m^4,\Omega_{m+1}} \le C^e(m)\varepsilon_m.$$

Owing to (10.36),

$$|\nabla_p\{F,H\}|^{O_m^4,\Omega_{m+1}} \le C_2^e(m)\varepsilon_m^{2/3}, \qquad |\nabla_q\{F,H\}|^{O_m^4,\Omega_{m+1}} \le C_2^e(m)\varepsilon_m^{4/3}.$$

Thus,

$$\|\nabla\{F,H\}\|_{(+,d_c)}^{O_m^4,\Omega_{m+1}} \le C_2^e(m)\varepsilon_m^{4/3}. \tag{10.37}$$

So we have

$$\|\nabla(\{F,H\}\circ S^t)\|_{(+,d_c)}^{O_m^4,\Omega_{m+1}} = \|S^{t*}(\mathfrak{h})\nabla\{F,H\}\|_{(+,d_c)}^{O_m^4,\Omega_{m+1}} \le C^e(m)\varepsilon_m^{4/3}.$$

This inequality follows from (10.37) since by (10.26) and assertions (b) and (c) of Lemma 10.6 (with $\theta = -d_c$) the map $S^{t*}(\mathfrak{h};\omega)$ defines an operator in $Z_{(+,d_c)}$, analytic in $\mathfrak{h} \in O_m^4$ and Lipschitz in $\omega \in \Omega_{m+1}$. Now (10.35) is proved since $\|\nabla_y \cdots\|_{d_c} \le \varepsilon_m^{-1/3}\|\nabla\cdots\|_{(+,d_c)}$. □

**Step 6: Transition to the limit.** Here we show that the set $(S_0 \circ S_1 \circ \dots)\,(T_0^n) \subset \mathcal{Y}_d$ is a smooth torus, invariant for eqns (10.3).

Let us denote $\Omega_\varepsilon = \bigcap \Omega_m$. Then $\Omega_\varepsilon$ is a Borel subset of $\Omega$ and by (10.7)

$$\operatorname{mes}(\Omega\backslash\Omega_\varepsilon) \le \gamma/2.$$

For $0 \le r \le N$ we denote by $\Sigma_N^r$ the map

$$\Sigma_N^r : O_N \times \Omega_N \to O_r, \qquad (\mathfrak{h},\omega) \mapsto S_r \circ \cdots \circ S_{N-1}(\mathfrak{h}),$$

where $S_j(\mathfrak{h}) = S_j(\mathfrak{h};\omega)$. As usual, $\Sigma_r^r$ stands for the projection $\Pi_{\mathcal{Y}} : O_r \times \Omega_r \longrightarrow O_r$. We claim that for all $r, m \ge 0$

$$\|\Sigma_{r+m}^r - \Pi_{\mathcal{Y}}\|_d^{O_{r+m},\Omega_\varepsilon} \le \varepsilon_r^\rho. \tag{10.38}$$

Indeed, let us rewrite the identity $\Sigma_{r+m}^r(\mathfrak{h};\omega) = S_r(\Sigma_{r+m}^{r+1}(\mathfrak{h};\omega);\omega)$ in the form

$$\Sigma_{r+m}^r - \Pi_{\mathcal{Y}} = (S_r - \Pi_{\mathcal{Y}}) \circ (\Sigma_{r+m}^{r+1} \times \Pi_\Omega) + (\Sigma_{r+m}^{r+1} - \Pi_{\mathcal{Y}}),$$

where $\Pi_\Omega(\mathfrak{h},\omega) = \omega$. By Lemma 10.6, the Lipschitz constant of the map $(S_r - \Pi_{\mathcal{Y}}) : O_{r+1} \times \Omega_r \to Z_d$ is less than $\varepsilon_r^\rho$. So, denoting the l.h.s. of (10.38) by $D_{r+m}^r$, we get that

$$D_{r+m}^r \le C^e(m)\varepsilon_r^{1/3}\,(D_{r+m}^{r+1} + 2) + D_{r+m}^{r+1}.$$

Since $D_{r+m}^{r+m} = 0$, then (10.38) follows by induction.

Let us also observe that because of Lemma 10.6, for any finite $r \le N$ and any $\mathfrak{h} \in O_N$ the tangent map $\Sigma^r_{N*}(\mathfrak{h})$ is close to the identity:

$$\|\Sigma^r_{N*}(\mathfrak{h}) - \mathrm{id}\|_{\theta,\theta} \le \varepsilon_r^\rho, \quad \forall\, |\theta| \le d \tag{10.39}$$

(abusing notation we now view $\Sigma^r_N$ as a map $O_N \to O_r$, so $\Sigma^r_{N*}$ is a map from $Z$ to $Z$).

Let us denote by $\mathcal{O}$ the set

$$\mathcal{O} = \{0\} \times U(\delta/2) \times \{0\} \subset \mathcal{Y}^c_d.$$

This is a complex neighbourhood of the torus $T^n_0 = \{0\} \times \mathbb{T}^n \times \{0\}$ in the complex cylinder $\{0\} \times (\mathbb{C}^n/2\pi\mathbb{Z}^n) \times \{0\}$, which is contained in each domain $O_m$ since $\delta_m > \delta/2$.

As a consequence of (10.38) we get that for every $m \ge 0$ and for each $\omega \in \Omega_\varepsilon$ the maps $\Sigma^m_{m+N}$ restricted to $\mathcal{O}$ converge (as $N \to \infty$) to an analytic map

$$\Sigma^m_\infty(\cdot\,;\omega) : \mathcal{O} \longrightarrow O_m \subset \mathcal{Y}^c_d,$$

and $\Sigma^m_p \circ \Sigma^p_\infty = \Sigma^m_\infty$ for all $p \ge m$.

For any $\omega \in \Omega_\varepsilon$ fixed, the linearizations $\Sigma^m_{m+N*}(\mathfrak{h})$ define analytic maps from $\mathcal{O}$ to the space of linear operators $Z^c_d \to Z^c_d$, where $Z^c_d = \mathbb{C}^{2n} \times Y^c_d$. Owing to (10.39), for any $|\theta| \le d$ the norms $\|\Sigma^m_{m+N*}(\mathfrak{h})\|_{\theta,\theta}$ are bounded uniformly in $N \ge 1$ and in $\mathfrak{h} \in \mathcal{O}$. By analyticity, the limiting map $\Sigma^m_{\infty*}$ satisfies (10.39) as well. That is,

$$\|\Sigma^m_\infty(\mathfrak{h})_* - \mathrm{id}\|_{\theta,\theta} \le \varepsilon^\rho_m, \quad \forall r,\ \forall |\theta| \le d. \tag{10.40}$$

Owing to the estimate that follows Lemma 10.1, the maps $\Lambda_m$ converge to a Lipschitz map $\Lambda_\infty : \Omega_\varepsilon \to \mathbb{R}^n$ such that $|\Lambda_\infty - \omega|^{\Omega_\varepsilon,\mathrm{Lip}} \le C\,\varepsilon^{1/3}$ and

$$|\Lambda_\infty - \Lambda_m| \le C(m)\varepsilon_m^{1/3}. \tag{10.41}$$

Now for any $\omega \in \Omega_\varepsilon$ we consider the curve

$$\mathfrak{h}_\infty(t) = (0, q_0 + t\Lambda_\infty(\varepsilon), 0) \subset T^n_0$$

and the curves $\mathfrak{h}_m(t) = \Sigma^m_\infty(\mathfrak{h}_\infty(t)) \subset O_m$. We are going to show that $\mathfrak{h}_0(t)$ is a solution of eqn (10.3). To do this, we first use (10.40) to get that

$$\dot{\mathfrak{h}}_m = \Sigma^m_{\infty*}(\mathfrak{h}_\infty)\dot{\mathfrak{h}}_\infty = (0, \Lambda_\infty, 0) + O\,(\varepsilon^\rho_m) \in Z_d.$$

Next, abbreviating eqns (10.13) and (10.14) to

$$\dot{\mathfrak{h}} = V_{\mathcal{H}_m}(\mathfrak{h}), \qquad \mathfrak{h} \in O_m, \tag{10.42}$$

and using the estimates (10.11), (10.12), and (10.41) we see that

$$V_{\mathcal{H}_m}(\mathfrak{h}_m) = (0, \Lambda_m, 0) + O(\varepsilon_m^\rho) = (0, \Lambda_\infty, 0) + O(\varepsilon_m^\rho)$$

in the space $Z_{d-d_1}$. Therefore,

$$\dot{\mathfrak{h}}_m - V_{\mathcal{H}_m}(\mathfrak{h}_m) = O(\varepsilon_m^\rho)$$

in $Z_{d-d_1}$. Since $\Sigma_{m*}^0(\mathfrak{h}_m)\big(V_{\mathcal{H}_m}(\mathfrak{h}_m)\big) = V_{\mathcal{H}_0}(\mathfrak{h}_0)$ and $\Sigma_{m*}^0(\mathfrak{h}_m)\dot{\mathfrak{h}}_m = \dot{\mathfrak{h}}_0$, then, applying the operator $\Sigma_{m*}^0(\mathfrak{h}_m)$ to the last equality and using (10.39) with $\theta = d - d_1$, we get that

$$\dot{\mathfrak{h}}_0 - V_{\mathcal{H}_0}\mathfrak{h}_0 = O(\varepsilon_m^\rho)$$

in $Z_{d-d_1}$. Since $m$ can be taken arbitrarily large, then the l.h.s. is zero and $\mathfrak{h}_0(t)$ is a solution of the system (10.3) (which coincides with (10.12)–(10.14) when $m = 0$).

Now the assertions of Theorem 8.3 follow if we choose $\Sigma_\varepsilon(q, \omega) = \Sigma_\infty^0(0, q, 0; \omega)$ and $\omega' = \Lambda_\infty(\omega)$.

To get the results stated in the amplification, the first few steps of the KAM procedure should be done 'by hand', using the fact that the original Hamiltonian $H_0$ has the radius of analyticity of order one rather than $\varepsilon_0^{2/3}$, see Kuksin (1993, p. 51).

## 10.3 Proof of Lemma 10.3 (estimation of the small divisors)

We denote $\Lambda_{jk}(\omega) = \nu_j^{(m+1)}(\omega) - \nu_k^{(m+1)}(\omega)$ and rewrite the assertion of the lemma as

$$|\omega' \cdot s + \Lambda_{jk}(\omega)| \geq \kappa := \frac{|j^{d_1} - k^{d_1}|}{C_{**}(m)\langle s\rangle^{c_1}}, \tag{10.43}$$

for all $j, k \in \mathbb{Z}\backslash\{0\}$ and all $\omega$ in $\widetilde{\Omega}\backslash\Omega^2$, where $\widetilde{\Omega} = \Omega \setminus \Omega^1$. Here the constants $C_{**}$, $c_1$, and the Borel subset $\Omega^2 \subset \Omega$ such that mes $\Omega^2 \leq \gamma(m+1)^{-2}/(3K_*)$, are to be found.

If $|s| \leq M_1$ and $j \leq j_2$, then (10.8), (10.9), and assumption (8.12) of Theorem 8.3 jointly imply (10.43) (provided that $\bar{\varepsilon}$ is sufficiently small), so henceforth we may suppose that

$$|s| \geq M_1 \quad \text{or} \quad j \geq j_2, \tag{10.44}$$

where $M_1$ and $j_2$ will be chosen later.

Let us denote for a moment $D(j, k, s) = \omega' \cdot s + \Lambda_{jk}(s)$. Then

$$D(j, k, s) = D(-k, -j, s) = -D(-j, -k, -s),$$

so to prove (10.43) it is sufficient to consider $j$ and $k$ such that

$$j \geq 1, \qquad j \geq |k|, \qquad j \neq k \tag{10.45}$$

(for $j = k$ the estimate (10.43) is trivial). For later usage we note that $j$ and $k$ as above satisfy the elementary inequality:[61]

$$|j^{d_1} - k^{d_1}| \ge d_1(\tfrac{1}{2})^{d-1} j^{d_1-1}. \tag{10.46}$$

Now we observe that

$$|\Lambda_{jk}| \ge C_0^{-1}\ |j^{d_1} - k^{d_1}|. \tag{10.47}$$

Indeed, if $j > j_2$, then the estimate (10.47) with $C_0 = 2K_1$ follows from (10.1) and (10.9), (10.46), while for $j \le j_2$ the estimate with $C_0^{-1} = K_3/2$ results from the assumption (8.12) with $s = 0$ and from (10.9).

By virtue of (10.47), the estimate (10.43) holds trivially if $|s| \le C^{-1}\,|\,j^{d_1} - k^{d_1}|$, where $C$ is any constant bigger than $2C_0|\omega'|$, say $C = 2C_0(K+1)$ (see assumption (3) of the theorem). So we can assume below that

$$|s| \ge C^{-1}\,|\,j^{d_1} - k^{d_1}|. \tag{10.48}$$

In particular, $s \ne 0$.

Let us denote by $\mathcal{L}$ the set of all triples $(k, j, s)$ as in (10.44), (10.45), and (10.48). For any $(k, j, s) \in \mathcal{L}$ we define $\Omega(k, j, s) \subset \widetilde{\Omega}$ as a set of all $\omega \in \widetilde{\Omega}$ violating (10.43) for the chosen triple $(k, j, s)$. Let us take for $\Omega^2$ the union

$$\Omega^2 = \bigcup \{\Omega(k, j, s) \mid (k, j, s) \in \mathcal{L}\}.$$

Clearly, (10.43) holds for $\omega$ outside $\Omega^2$. So it remains to estimate the measure of $\Omega^2$. Here the key is the following result:

**Lemma 10.8.** *For any triple* $(k, j, s) \in \mathcal{L}$ *we have*

$$\operatorname{mes} \Omega(k, j, s) \le C\kappa,$$

*provided that* $j_2$ *and* $M_1$ *are sufficiently large in terms of the quantities listed in Theorem 8.3 (*$\kappa$ *was defined in (10.43)).*

*Proof.* By (10.8), the map

$$\widetilde{\Omega} \ni \omega \longmapsto \omega' = \Lambda_{m+1}(\omega) \in \mathbb{R}^{n+1}$$

is Lipschitz-close to the identity. So it is a Lipschitz homeomorphism which changes the diameters of sets and their Lebesgue measure no more than twice (see Lemma A8.1 in Appendix 8). Therefore to estimate the measure of the set $\Omega(k, j, s)$ is equivalent to estimating the measure of its image $\Omega'$:

$$\Omega' = \Lambda_{m+1}(\Omega(k, j, s)).$$

To do this we express $\nu_k$, $\nu_j$, and $\Lambda_{kj}$ as functions of $\omega'$ and write the set $\Omega'$ as

$$\Omega' = \{\omega' \in \Lambda_{m+1}(\widetilde{\Omega}) \bigm| |\omega' \cdot s - \Lambda_{kj}(\omega')| \le \kappa\}.$$

The set $\Omega'$ is bounded since it is contained in the bounded set $\omega'(\Omega)$. So by the Fubini theorem to majorize a measure of this set it suffices to majorize the one-dimensional

[61] For $j = 1$ the inequality is obvious. For $j \ge 2$ it holds since the l.h.s. is $\ge j^{d_1} - (j-1)^{d_1} > d_1(j-1)^{d_1-1} \ge d_1(j/2)^{d_1-1}$.

measure of the intersection of $\Omega'$ with any line in $\mathbb{R}^n$, parallel to the vector $S = s/|s|$; that is, with any line $L_\eta = \{\eta + tS \mid t \in \mathbb{R}\}$, $\eta \in \mathbb{R}^2$. The intersection of $\Omega'$ with $L_\eta$ corresponds to $t$ from the set

$$\{t \mid |\Gamma(t)| \leq \kappa\}, \tag{10.49}$$

where $\Gamma$ is the function

$$\Gamma(t) := \big(\omega'(t) \cdot s + \Lambda_{kj}(\omega'(t))\big), \quad \omega'(t) = \eta + tS.$$

Let us observe that $(\partial/\partial t)\omega' \cdot s = |s|$ and that $\operatorname{Lip} \Lambda_{jk} \leq C j^{\tilde{d}}$, where $\operatorname{Lip} \Lambda_{jk}$ stands for a Lipschitz constant of the map $\omega' \to \Lambda_{jk}$ (we use (8.10) and (10.9)). Then for any $t_1 \geq t_2$ we have

$$\begin{aligned}\Gamma(t_1) - \Gamma(t_2) &\geq |s|(t_1 - t_2) - (t_1 - t_2)\operatorname{Lip} \Lambda_{kj}\\ &\geq (t_1 - t_2)\,(|s| - C\,j^{\tilde{d}}) \geq C^{-1}(t_1 - t_2)\,(j^{d_1} - k^{d_1} - C_1\,j^{\tilde{d}})\\ &\geq C_2^{-1}(t_1 - t_2)\,(j^{d_1 - 1} - C_3\,j^{\tilde{d}})\end{aligned} \tag{10.50}$$

(we use (10.48) in the third inequality and (10.46) in the fourth one). So if $j > j_2$ and $j_2$ is sufficiently large, then

$$\Gamma(t_1) - \Gamma(t_2) \geq t_1 - t_2.$$

If $j \leq j_2$, then by (10.44) $|s| \geq M_1$. Using the second estimate in (10.50) we get that

$$\Gamma(t_1) - \Gamma(t_2) \geq (t_1 - t_2)\,(M_1 - C j_2^{\tilde{d}}) \geq t_1 - t_2,$$

if we choose $M_1 \geq C j_2^{\tilde{d}} + 1$.

Thus, the measure of the set (10.49) is less than $2\kappa$. Since $\operatorname{diam} \omega'(\Omega) \leq 2 \operatorname{diam} \Omega \leq 2K_2$, then by Fubini $\operatorname{mes} \Omega' \leq 2\kappa c_{n-1} K_2^{n-1}$, where $c_{n-1}$ is a volume of the 1-ball in $\mathbb{R}^{n-1}$. Since $\operatorname{mes} \Omega(k, j, s) \leq 2 \operatorname{mes} \Omega'$, then the lemma is proved. □

Now an estimate of the measure of $\Omega^2$ is straightforward:

$$\operatorname{mes} \Omega^2 \leq \sum_{\mathcal{L}} \operatorname{mes} \Omega(k, j, s) \leq \frac{C}{C_{**}(m)} \sum_s \langle s \rangle^{-c_1} \sum_{\substack{j,k\\(j,k,s)\in\mathcal{L}}} (j^{d_1} - k^{d_1}).$$

By (10.46) and (10.48), $j \leq C|s|^{d_0}$, where $d_0 = 1/(d_1 - 1)$. Since $|k| \leq j$, then the cardinality of the set $\{(j, k, s) \in \mathcal{L} \mid s \text{ is fixed}\}$ is less than $2C|s|^{2d_0}$. Using (10.48) we see that the inner sum in the r.h.s. is estimated as follows:

$$\sum_{\substack{j,k\\(j,k,s)\in\mathcal{L}}} (j^{d_1} - k^{d_1}) \leq C \sum_{\substack{j,k\\(j,k,s)\in\mathcal{L}}} |s| \leq C_1 \langle s \rangle^{2d_0+1}.$$

Therefore,

$$\operatorname{mes} \Omega^2 \leq \frac{CC_1}{C_{**}(m)} \sum_s \langle s \rangle^{2d_0+1-c_1} \leq \frac{\gamma}{3(m+1)^2 K_*},$$

if $c_1 > 2\,d_0 + n + 1$ and $C_{**}(m)$ is sufficiently large.

Lemma 10.3 is proved.

## Appendix 9. Some inequalities for Fourier series

Let $B^c$ be a complex Banach space and $f : U(\delta) \longrightarrow B^c$ be a complex-analytic map such that $\|f\|_B \leq 1$. We can write $f$ as a Fourier series,

$$f(q) = \sum_{s \in \mathbb{Z}^n} f_s \, e^{is \cdot q}, \quad f_s = \int_{\mathbb{T}^n} f(q) \, e^{-iq \cdot s} dq/(2\pi)^n \in B^c.$$

Let us replace the integration over $\mathbb{T}^n$ by integration over the shifted torus $\mathbb{T}^n - i(\delta - \varepsilon)(s/|s|) \subset U(\delta)$. Since after the shift we have $|e^{iq \cdot s}| \leq e^{-|s|(\delta-\varepsilon)}$, then $\|f_s\|_B \leq e^{-|s|(\delta-\varepsilon)}$ for every positive $\varepsilon$. Thus, we have

$$\|f_s\|_B \leq e^{-|s|\delta}, \quad \forall s \in \mathbb{Z}^n. \tag{A9.1}$$

Conversely, if for some $d \geq 0$ we have $\|f_s\|_B \leq \langle s \rangle^d \, e^{-|s|\delta}$ for every $s$, then

$$\|f\|_B^{U(\delta-\varrho)} \leq \sum_{s \in \mathbb{Z}^n} \langle s \rangle^d \, e^{-\varrho|s|} \leq C \int_{\mathbb{R}^n} |x|^d \, e^{-\varrho|x|} dx$$

$$= C\rho^{-n-d} \int_{\mathbb{R}^n} |y|^d \, e^{-|y|} dy = C_d \varrho^{-n-d}. \tag{A9.2}$$

As a consequence of the estimates (A9.1) and (A9.2) we get:

**Lemma A9.1.** *If $f : \mathbb{T}^n \to B^c$ is a zero-mean-value map, analytically extendable to $U(\delta)$, and $\omega$ is a Diophantine $n$-vector, namely*

$$|\omega \cdot s| \geq |s|^{-d}/C_* \quad \forall s \in \mathbb{Z}^n \setminus 0, \tag{A9.3}$$

*with some positive $d$ and $C_*$, then the equation*

$$\frac{\partial u}{\partial \omega}(q) = f(q), \quad \frac{\partial u}{\partial \omega} := \sum \omega_j \frac{\partial u}{\partial q_j}, \tag{A9.4}$$

*has a unique zero-mean-value analytic solution $u(q)$ and*

$$\|u\|_B^{U(\delta-\rho)} \leq C_* C_d \rho^{-n-d} \|f\|_B^{U(\delta)} \tag{A9.5}$$

*for any $0 < \rho < \delta$. If $f = f(q; a)$ is a Lipschitz function of an additional parameter $a \in \mathfrak{A}$, then*

$$\|u\|_B^{U(\delta-\rho),\mathfrak{A}} \leq C_* C_d \rho^{-n-d} \|f\|_B^{U(\delta),\mathfrak{A}}. \tag{A9.6}$$

**Lemma A9.1′.** *If a map $f : \mathbb{T}^n \to B^c$ analytically extends to $U(\delta)$ and an $n$-vector $\omega$ is incommensurable with a real constant $E$, namely*

$$|\omega \cdot s + E| \geq (|s|+1)^{-d}/C_*, \quad \forall s \in \mathbb{Z}^n, \tag{A9.7}$$

*then the equation*

$$\frac{\partial u}{\partial \omega}(q) + iEu = f(q) \tag{A9.8}$$

*has a unique analytic solution $u(q)$. This solution satisfies (A9.5). If $f = f(q; a)$ is Lipschitz in $a \in \mathfrak{A}$, then $u = u(q; a)$ satisfies (A9.6).*

To prove (A9.5) we expand $u(q)$ and $f(q)$ to Fourier series denoting by $u_s$ and $f_s$ the corresponding Fourier coefficients. Then $f_0 = u_0 = 0$ and $u_s = f_s/(i\omega\cdot s)$ if $s \neq 0$. Thus,

$$\|u_s\|_B \leq C_*|s|^d \|f_s\|_B \leq C_*|s|^d \mathrm{e}^{-|s|\delta} \|f\|_B^{U(\delta)}$$

by (A9.1), and the estimate (A9.5) follows by (A9.2).

To get (A9.6) it is sufficient to apply (A9.5) to an increment $u(q; a_1) - u(q; a_2)$ of the solution $u$.

The proof of Lemma A9.1′ is quite similar.

**Remark 1.** If $B^c$ is a complexification of a real Banach space $B$ and the map $f$ is real, i.e. $f(q) \in B$ for $q \in \mathbb{T}^n$, then the solution $u(q)$ of eqn (A9.4) is real since $\overline{u(\bar{q})}$ is an analytic map which also solves (A9.4); so it must be equal to $u(q)$. If $u(q)$ solves (A9.8) with real $f(q)$, then $v = \overline{u(\bar{q})}$ is the unique analytic solution of the adjoint equation $\partial v/\partial\omega - iE = f$.

If $d > n$ and $\Omega$ is a bounded subset of $\mathbb{R}^n$, then the set $\Omega_{C_*}$ formed by all $\omega \in \Omega$ which violate the Diophantine assumption (A9.3) has a measure $O(C_*^{-1})$:

**Lemma A9.2.** *If* $d > n - 1$, *then* $\mathrm{mes}_n \Omega_{C_*} \leq C(d, \Omega)/C_*$.

*Proof.* The set $\Omega_{C_*}$ is a union of subsets $\Omega_s \subset \Omega$,

$$\Omega_s = \{\omega \in \Omega \mid |\omega\cdot s| \leq |s|^{-d}/C_*\}, \quad s \in \mathbb{Z}^n \setminus 0.$$

Each set $\Omega_s$ is an intersection of $\Omega$ with the set $\{|\omega\cdot s| \leq |s|^{-d}/C_*\}$ which is a strip of width $|s|^{-d-1}/C_*$ in $\mathbb{R}^n$. Thus, $\mathrm{mes}_n \Omega_s \leq C(\Omega)|s|^{-d-1}/C_*$ and

$$\mathrm{mes}_n \Omega_{C_*} \leq \sum_{s\neq 0} \mathrm{mes}_n \Omega_s \leq \frac{C(\Omega)}{C_*} \sum_{s\neq 0} |s|^{-d-1} = \frac{C(d,\Omega)}{C_*}.$$

□

A similar result with the same proof holds for relation (A9.7):

**Lemma A9.3.** *If* $d > n$ *and* $|E| \geq C_*^{-1}$, *then the subset of all* $\omega \in \Omega$ *which violate (A9.7) is a measurable set of measure* $\leq C(d, \Omega, E)/C_*$.

If for any analytic function $f(q)$ such that $\|f\|_B^{U(\delta)} \leq 1$, we cut off its low-frequency part, namely for any $R > 1$ define $f^R$ as

$$f^R(q) = \sum_{|s|\geq R} f_s \mathrm{e}^{is\cdot q},$$

then by (A9.1) for any positive $\rho < \delta$ we have

$$\|f^R\|_B^{U(\delta-\varrho)} \leq \sum_{|s|\geq R} \mathrm{e}^{-|s|\varrho} \leq C \int_R^\infty \mathrm{e}^{-t\varrho} t^{n-1} dt$$

$$= C \sum_{m=1}^{n} \varrho^{-m} \frac{(n-1)!}{(n-m)!} \mathrm{e}^{-\varrho R} R^{n-m}. \qquad \text{(A9.9)}$$

Take any $k > 0$. Then by (A9.9),

$$\|f^R\|_B^{U(\delta-\varrho)} \leq CR^{-k}\rho^{-n-k}\sum_{m=1}^{n}\frac{(n-1)!}{(n-m)!}\mathrm{e}^{-\rho R}(\rho R)^{n-m+k}$$
$$\leq C_{n,k}\,R^{-k}\varrho^{-n-k}, \tag{A9.10}$$

since $\mathrm{e}^{-x}x^{n-m+k} \leq C_k$ for every $x \geq 0$ and any $k \geq 0,\ m \leq n$.

## Appendix 10. On the Craig–Wayne–Bourgain KAM scheme

There is an alternative KAM approach to prove that for small $\varepsilon$ and for most parameters $\omega$ eqn (10.3) has an invariant torus, close to the torus $T_0^n = \{0\} \times \mathbb{T}^n \times \{0\}$. This approach is due to Craig–Wayne–Bourgain (Craig and Wayne 1992; Bourgain 1998).

In this appendix we describe the corresponding scheme in comparison with the one used in Chapter 10 (and in the Addendum). Our description is very vague. In particular, we do not specify which function norms for functions of $q \in \mathbb{T}^n$ have to be used.

**Domains and Hamiltonians.** We use a suitable family of domains $\mathcal{Y} \supset Q_0 \supset Q_1 \supset \cdots \supset T_0^n$, $\cap Q_j = T_0^n$, and of Hamiltonians $\mathcal{H}_m$, defined on these domains. Every Hamiltonian $\mathcal{H}_m$ has the form

$$\mathcal{H}_m = p \cdot \Lambda_m(\omega) + \tfrac{1}{2}\langle B_m(q;\omega)y, y\rangle + \epsilon_m H_m(\mathfrak{h};\omega), \tag{A10.1}$$

where $\omega \in \Omega_m$ and $\Omega_m$ is a 'large' Borel subset of $\Omega$ (e.g. it satisfies (10.7)). The self-adjoint operator $B_m$ *is not assumed to commute with* $B$, but it is close to this operator:

$$\|B_m(q;\omega) - B(\omega)\| \leq Ce(m)\varepsilon. \tag{A10.2}$$

The sequence $0 = e(0) < e(1) < \cdots < 1/2$ is defined as above in Section 10.2; the sequence $\{\epsilon_m\}$ decays to zero 'sufficiently fast' (but $\epsilon_{m+1} > \epsilon_m^2$) and $\epsilon_0 = \varepsilon$. In particular,

$$\epsilon_m < C_c(m)c^m, \quad \forall m \geq 1$$

for any positive $c$. The corresponding Hamiltonian equations are:

$$\dot p = -\tfrac{1}{2}\langle \nabla_q B_m(q;\omega)y, y\rangle - \epsilon_m \nabla_q H_m, \qquad \dot q = \Lambda_m(\omega) + \epsilon_m \nabla_p H_m,$$
$$\dot y = JB_m(q;\omega)y + \epsilon_m J\nabla_y H_m. \tag{A10.3}$$

For $m = 0$ we have $\mathcal{H}_0 = \mathcal{H}_\varepsilon$, so $(\text{A10.3})_{m=0} = (10.3)$.

We note that the torus $T_0^n$ is invariant for eqn (A10.3) up to terms of order $\epsilon_m$. As in Chapter 10, we wish to construct symplectic transformations $S_m : Q_{m+1} \to Q_m$ such that $\mathcal{H}_m \circ S_m = \mathcal{H}_{m+1}$. Then the limiting transformation $\boldsymbol{S} = S_0 \circ S_1 \ldots$ (if it is well defined) provides us with an invariant torus $\boldsymbol{S}(T_0^n)$ of eqn (10.3), filled with quasiperiodic solutions $\boldsymbol{S}(0, q + t\Lambda_\infty, 0)$.

To construct the transformation $S_m$ given a Hamiltonian $\mathcal{H}_m$, we first isolate an affine in $p$, quadratic in $y$ part of $H_m$ and write $H_m$ in the form (10.15):

$$H_m = h^q(q;\omega) + p \cdot h^{1p}(q;\omega) + \langle y, h^y(q;\omega)\rangle + \langle h^{yy}(q;\omega)y, y\rangle + H_{3m}(\mathfrak{h};\omega),$$

where $\mathfrak{h} = (p, q, y)$. Subtracting from $H_m$ a $\mathfrak{h}$-independent constant, we achieve $\langle h^q\rangle = 0$, where $\langle \ldots \rangle$ stands for the averaging $(2\pi)^{-n}\int \ldots dq$. As at Step 1 (see Section 10.2 or the Addendum), we denote $h^{0p} = \langle h^{1p}\rangle$, $h^p = h^{1p} - h^{0p}$. In crucial difference with the proof of Theorem 8.3, we do not average the quadratic part $h^{yy}$, but add the whole of it to the integrable part. Accordingly, we write $\mathcal{H}_m$ as

$$\mathcal{H}_m = p \cdot \underbrace{\left(\Lambda_m + \epsilon_m h^{0p}\right)}_{\Lambda_{m+1}} + \tfrac{1}{2}\Big\langle \underbrace{\left(B_m + 2\epsilon_m h^{yy}\right)}_{B'_m} y, y\Big\rangle$$
$$+\, \epsilon_m \underbrace{\left(h^q + p \cdot h^p + \langle y, h^y\rangle\right)}_{H_{1m}} + \epsilon_m H_{3m}.$$

As in Section 10.2, we assume that the domains $Q_m$ shrink to $T_0^n$ sufficiently fast so that $|\epsilon H_{3m}| \le \frac{1}{2}\epsilon_{m+1}$ in $Q_{m+1}$. Accordingly, $\epsilon_m H_{3m}$ is an admissible part of the term $\epsilon_{m+1}H_{m+1}$ and it remains to kill the term $\epsilon_m H_{1m}$. To do this we use a transformation $S_m$ which is a time-one shift along trajectories of a Hamiltonian vector field $V_{\epsilon_m F}$, where the Hamiltonian $F$ has the same structure as $H_{1m}$, i.e.

$$F = f^q(q;\omega) + p \cdot f^p(q;\omega) + \langle y, f^y(q;\omega)\rangle.$$

Abbreviating $\Lambda_{m+1} = \omega'$, $\omega' \cdot \nabla_q = \partial/\partial\omega'$ and arguing as at Step 2, we get that

$$\begin{aligned}\mathcal{H}_m \circ S_m &= \mathcal{H}_m + \epsilon_m\{F, \mathcal{H}_m\} + O(\epsilon_m^2)\\ &= p \cdot \Lambda_{m+1} + \tfrac{1}{2}\langle B'_m y, y\rangle + \epsilon_m \Big[ -\partial f^q/\partial\omega' - p \cdot \partial f^p/\partial\omega'\\ &\quad - \langle y, \partial f^y/\partial\omega'\rangle - \langle y, (\partial f^{yy}/\partial\omega')y\rangle + \langle B'_m y, Jf^y\rangle + H_{1m}\Big]\\ &\quad + \frac{\epsilon_m}{2}\langle (f^p \cdot \nabla_q B'_m)y, y\rangle + O(\epsilon_{m+1}).\end{aligned}$$

Therefore, if the functions $f^p$, $f^q$, and $f^y$ satisfy the homological equations

$$\partial f^q/\partial\omega' = h^q(q;\omega), \qquad \partial f^p/\partial\omega' = h^p(q;\omega), \tag{A10.4}$$
$$\partial f^y/\partial\omega' - B'_m J f^y = h^y + O(\epsilon_{m+1}/\epsilon_m), \tag{A10.5}$$

then the transformed Hamiltonian $\mathcal{H}_m \circ S_m$ takes the form (A10.1) with $m = m + 1$, where $B_{m+1} = B'_m + \epsilon_m(f^p \cdot \nabla_q)B'_m$.

As at the Step 3, the eqns (A10.4) are classical and can be solved easily if $\omega \in \Omega_{m+1}$ with an appropriate set $\Omega_{m+1}$. At the same time the eqn (A10.5) is much more difficult than eqn (10.21), obtained at Step 3, since the operators $B'_m(q;\omega)J$, $q \in \mathbb{T}^n$, do not commute. All known ways to solve 'non-commutative' equations (A10.5) are perturbative. They use assumption (A10.2) as well as additional properties of the perturbation $(B'_m - B)J$ and of the spectrum $\{\pm i\lambda_j\}$ of the operator $BJ$. The first results on equations of the type (A10.5), (A10.2) were obtained by Fröhlich and Spencer in their works on the Andersen localisation (see Fröhlich and Spencer 1983). There they called an operator that resolves the equation the *Green function*. Since then Green functions have been studied in a number of papers (the best results at the time of this writing are due to Bourgain 1998), but the 'right' conditions that would imply solvability of (A10.5) and (A10.2) are still missing. So every time when an equation of this kind arrives, one has to solve it anew. See Craig and Wayne (1992), Bourgain (1998), and Kriecherbauer (1998) and references therein.

After eqn (A10.5) has been resolved, one constructs the transformation $S_m$ and obtains the new Hamiltonian $\mathcal{H}_{m+1} = \mathcal{H}_m \circ S_m$. The limiting transformation $S = S_0 \circ S_1 \circ \cdots$ provides the invariant torus $S(T_0^n)$, filled with the quasiperiodic solutions of eqn (10.3).

# 11
# Linearized equations

In this chapter we consider linearization of eqns (10.3) about any solution $\mathfrak{h}_0(t)$, constructed in Theorem 8.3, and prove in Theorem 8.4. We abbreviate (8.3) as

$$\dot{\mathfrak{h}} = V_{\mathcal{H}_\varepsilon}(\mathfrak{h}(t)), \qquad \mathfrak{h} = (p, q, y),$$

and write the linearized equations as

$$\dot{\eta} = V_{\mathcal{H}_\varepsilon}(\mathfrak{h}_0(t))_* \eta. \tag{11.1}$$

The analysis of eqn (11.1) given below uses the symplectic transformations $S_l$ and their compositions $\Sigma_N^r = S_r \circ \cdots \circ S_{N-1}$, defined at Step 6 of the proof of Theorem 8.3.

To study (11.1) we consider the linearization of any transformed equation (10.42) about the transformed solution $\mathfrak{h}_m = \left(\Sigma_m^0\right)^{-1} \mathfrak{h}_0 = \Sigma_\infty^m \mathfrak{h}_\infty$:

$$\dot{\eta}_m = V_{\mathcal{H}_m}(\mathfrak{h}_m(t))_* \eta_m. \tag{11.1$_m$}$$

This equation coincides with (11.1) if $m = 0$, and the linear transformation

$$\mathcal{L}_m(t) := \Sigma_m^0(\mathfrak{h}_m(t))_*$$

sends solutions of (11.1$_m$) to solutions of (11.1). By (10.39) limiting linear maps $\mathcal{L}_\infty(t)$, $t \in \mathbb{R}$, exist and define zero-order automorphisms of the scale $\{Z_s = \mathbb{R}^{2n} \times Y_s\}$ for $|s| \le d$. Moreover, each map $\mathcal{L}_m(t)$ is symplectic since the maps $S_l$ are symplectomorphisms. The limiting maps $\mathcal{L}_\infty(t)$ are symplectic as well.

For any $0 \le m \le \infty$ and any $t$ the map $\mathcal{L}_m$ satisfies the estimates

$$\|\mathcal{L}_m(t)\|_{\theta,\theta} + \|\mathcal{L}_m^{-1}(t)\|_{\theta,\theta} \le 3, \quad |\theta| \le d. \tag{11.2}$$

Since the linearized equation (11.1) is uniformly well defined by the assumptions of Theorem 8.4, then due to (11.2) eqns (11.1$_m$) also are uniformly well defined: for any $m$, the flow maps $(S_{(m)\tau}^{\tau+t})_{**}(\mathfrak{h}_m(\tau))$ of (11.1$_m$) are such that

$$\|(S_{(m)\tau}^{\tau+t})_{**}(\mathfrak{h}_m(\tau))\|_{\theta,\theta} \le C e^{C_2 t}, \quad \text{for any } t \text{ and any } |\theta| \le d.$$

Because of (4.2) with $m = \infty$, to estimate the solutions $\eta_0(t)$ of (11.1$_m$) with $m = 0$ is equivalent to estimating their transformations $\eta_\infty(t) = (\mathcal{L}_\infty(t))^{-1}\eta_0(t)$. We cannot directly go to the limit in (11.1$_m$) to write a limiting equation (11.1$_\infty$) for $\eta_\infty(t)$. Instead, we shall obtain estimates for the limiting curve $\eta_\infty$ by examining the $p$-, $q$-, and $y$-components of the solutions $\eta_m$ with large $m$.

For any $0 \le r \le m \le \infty$ we define the linear transformations $\mathcal{L}_m^r$ as $\mathcal{L}_m^r(t) = \Sigma_m^r(\mathfrak{h}_m(t))_*$. Clearly, $\mathcal{L}_m^r = (\mathcal{L}_r)^{-1} \circ \mathcal{L}_m$. Using (10.39) once again we find that

$$\|\mathcal{L}_m^r - \mathrm{id}\|_{\theta,\theta} \le C\varepsilon_r^{\rho}. \tag{11.3}$$

Now we write $(11.1_m)$ as a system of equations for $\eta_m = (\eta_p, \eta_q, \eta_y)$, omitting the dependence on $m$ (and on the parameter $\omega$ which is now irrelevant):

$$\begin{cases} \dot\eta_p &= -\varepsilon_m \nabla_{q,p} H_m \eta_p - \varepsilon_m \nabla_{q,q} H_m \eta_q - \varepsilon_m \nabla_{q,y} H_m \eta_y, \\ \dot\eta_q &= \varepsilon_m \nabla_{p,p} H_m \eta_p + \varepsilon_m \nabla_{p,q} H_m \eta_q + \varepsilon_m \nabla_{p,y} H_m \eta_y, \\ \dot\eta_y &= J A_m(q_m(t))\eta_y + \varepsilon_m J \nabla_{y,p} H_m \eta_p \\ & \quad + \varepsilon_m J \nabla_{y,q} H_m \eta_q + \varepsilon_m J \nabla_{y,y} H_m \eta_y. \end{cases} \tag{$11.1'_m$}$$

Here $\nabla_{q,p} H_m$ is a linearization in $p$ of the gradient map $\nabla_q H_m$, i.e. a linear map $\mathbb{R}^n \to \mathbb{R}^n$, etc.

We need a refinement of the estimates (10.11) and (10.12):

**Lemma 11.1.** *The Hamiltonian $\varepsilon_m H_m$ meets the following estimates:*

$$\left\| \frac{\partial}{\partial p_j} \nabla_y (\varepsilon_m H_m)(\mathfrak{h}) \right\|_{d_c} \le C(1 + e(m)), \quad j = 1, \ldots, n, \quad \mathfrak{h} \in O_m \tag{11.4}$$

*(the numbers $e(m)$ were defined in Section 10.2, $C$ is an $m$-independent constant), and*

$$\left| \frac{\partial}{\partial p_j} \frac{\partial}{\partial p_k} (\varepsilon_m H_m)(\mathfrak{h}) \right| \le C^e(m), \quad j, k = 1, \ldots, n.$$

*Proof.* For $m = 0$ the estimate (11.4) follows from (10.4) and the Cauchy estimate since the domain of analyticity $Q^c$ of the function $H_0 = H$ is $\varepsilon$-independent. Now we suppose that the estimate is proved for $m = m$ and show that it holds for $m = m + 1$. Since $(\partial/\partial p_j)\nabla_y H_{2m} = 0$ (see Step 1 in Section 10.2) and $H_m = H_{2m} + H_{3m}$, then $\varepsilon_m H_{3m}$ also meets (11.4). By our constructions, the next-step perturbation $\varepsilon_{m+1} H_{m+1}$ is $\varepsilon_{m+1} H_{m+1} = \varepsilon_m H_{3m} + \Delta_3 H + \Delta_4 H$; see (10.32). By Lemma 10.7 the gradients in $y$ of the terms $\Delta_3 H$ and $\Delta_4 H$ are majorized in the domain $O_m^5$ by $C_1^e(m)\varepsilon_m$. So for any $\mathfrak{h} \in O_{m+1}$ and any $j = 1, \ldots, n$ we have

$$\left\| \frac{\partial}{\partial p_j} \nabla_y \Delta_l H(\mathfrak{h}) \right\|_{d_c} \le C^e(m)\varepsilon_m^{1/3} \le \frac{1}{2K_*(m+1)^2}, \qquad l = 3, 4$$

(the Cauchy estimate). Since the term $\varepsilon_m H_{3m}$ satisfies (11.4) for $\mathfrak{h} \in O_m$, then (11.4) with $m = m + 1$ follows.

The proof of the second estimate is analogous. □

By (10.11), (10.12), and Lemma 11.1, the system $(11.1'_m)$ can be abbreviated as

$$\begin{cases} \dot\eta_p = O_{p,\eta}(\varepsilon_m^{\rho})\eta, \\ \dot\eta_q = O_{q,p}(C^e(m))\eta_p + O_{q,q}(\varepsilon_m^{\rho})\eta_q + O_{q,y}(1)\eta_y, \\ \dot\eta_y = J A_m(q_m(t))\eta_y + O_{y,p}(1)\eta_p + O_{y,q}(\varepsilon_m^{\rho})\eta_q, \end{cases} \tag{11.5}$$

where $O_{p,\eta}(\varepsilon_m^\rho)$ stands for a time-dependent linear operator $Z_d \to \mathbb{R}^n$, $\eta \mapsto p$, of the norm $O(\varepsilon_m^\rho)$, and similarly with $O_{q,p}(C^e(m)), \dots, O_{q,q}(\varepsilon_m^\rho)$. The linear operators $O_{y,p}(1)$ and $O_{y,\eta}(\varepsilon_m^\rho)$ are bounded as operators valued in $Y_{d_c}$.

For $j = n+1, n+2, \dots$ let us denote by $\xi_{j0} \in Z_d$ any unit vector of the form

$$\xi_{j0} = (0, 0, y_{j0}), \qquad y_{j0} = y^j \psi_j + \bar{y^j} \psi_{-j}, \qquad \|y_{j0}\|_d = 1 \tag{11.6}$$

(the complex basis $\psi_j$ of the space $Y_d^c$ was defined above) and denote

$$\xi_{j0}^{(m)} := \mathcal{L}_\infty^m(0)\, \xi_{j0} = \xi_{j0} + O(\varepsilon_m^\rho),$$

where the second equality follows from (11.3). Let $\xi_j^{(m)}(t)$ be a solution of (11.1$_m$) such that $\xi_j^{(m)}(0) = \xi_{j0}^{(m)}$. For $m = 0, 1, \dots$ the map $\mathcal{L}_m$ sends $\xi_j^{(m)}(t)$ to $\xi_j^{(0)}(t)$.

A diagonal element $\nu_j^{(m)} + \beta_j^{(m)}$ of the operator $B_m(q;\omega)$ (defined in Section 10.2) equals

$$\nu_j(\omega) + 2\varepsilon_1 a_j^{(1)}(q;\omega) + \cdots + 2\varepsilon_m a_j^{(m)}(q;\omega),$$

where $2\varepsilon_l a_j^{(l)}(q;\omega)$ is a diagonal element of the quadratic part of the perturbation $\varepsilon_l H_l$ at the $l$th step of the KAM procedure. Since any function $a_j^{(l)}(\cdot;\omega)$ is analytic in $U_l^1$ and is bounded there by $j^{\tilde{d}} C(l) \varepsilon_l^{-2/3}$ (see the discussion that follows Lemma 10.2), then for any $\omega$ in $\Omega_\varepsilon$ we have the following convergences:

$$\beta_j^{(m)}(q;\omega) \longrightarrow \beta_j^\infty(q;\omega) \quad \text{and} \quad \nu_j^{(m)}(\omega) \longrightarrow \nu_j^\infty(\omega) \quad \text{as } m \to \infty,$$

where the functions $q \mapsto \beta_j^\infty$ are analytic with zero mean value. Letting $m \to \infty$ in the estimates for the functions $\beta_j^{(m)}$ and $\nu_j^{(m)}$, we get:

$$|\beta_j^\infty|^{U(\delta/2),\Omega_\varepsilon} + |\nu_j^\infty - \nu_j|^{\Omega_\varepsilon,\mathrm{Lip}} \le C\varepsilon_0^\rho\, j^{\tilde{d}}.$$

Denoting by $B_\infty$ the limiting operator $B_\infty(q;\omega) = \operatorname{diag}\{\nu_j^\infty + \beta_j^\infty \mid j \ge n+1\}$, we consider the corresponding non-autonomous linear equation in the space $Y_d$:

$$\dot{y}(t) = J B_\infty(q_0 + \omega' t;\omega) y(t), \qquad \omega' = \Lambda_\infty(\omega). \tag{11.7}$$

Let us consider a solution $y(t) = y_j(t)$ of (11.7) such that

$$y_j(0) = y_{j0} \quad \text{as in (11.6)}. \tag{11.8}$$

It has the form $y_j(t) = y^j(t)\psi_j + y^{-j}(t)\psi_{-j}$, where $y^j$ and $y^{-j}$ are complex-conjugated functions and $y^j$ satisfies the equation

$$\dot{y}^j(t) = i(\nu_j^\infty + \beta_j^\infty(q_0 + \omega' t)) y^j(t).$$

Since $\beta_j^\infty$ and $\nu_j^\infty$ are real functions, then $|y^j(t)| = \text{const}$. That is, $\|y_j(t)\|_d \equiv 1$.

Now let us consider the curve $\eta_j^{(m)}(t)$ in $Z_d$,

$$\eta_j^{(m)}(t) = \left(0, \int_0^t O_{q,y}(1) y_j(\tau) d\tau, \ y_j(\tau)\right),$$

where the curve $y_j$ is as above and $O_{q,y}(1)$ is the linear operator from the second equation in (11.5)= (11.1$'_m$). Clearly, its $Z_d$-norm is bounded by $Ct+1$. Analysis of eqns (11.5) shows that since $y_j$ satisfies (11.7), then $\eta_j^m$ solves eqn (11.5) with a disparity formed by the terms $O_{p,\eta}(\varepsilon_m^\rho)\eta$, $O_{q,q}(\varepsilon_m^\rho)\eta_q$, and $O_{y,q}(\varepsilon_m^\rho)\eta_q$ with $\eta = (\eta_p, \eta_q, \eta_y) = \eta_j^{(m)}(t)$. This disparity majorizes by $C'(t+1)\varepsilon_m^\rho$. Since $\eta_j^{(m)}(0) = (0, 0, y_j(0)) = \xi_{j0}^{(m)}$ and the linearized equation (11.1$_m$) is well defined, then we get the estimate for divergence of $\eta_j^{(m)}(t)$ from the exact solution $\xi_j^{(m)}(t)$:

$$\|\xi_j^{(m)}(t) - \eta_j^{(m)}(t)\|_{d_c} \le C\varepsilon_m^\rho e^{C_1 t} \tag{11.9}$$

with some $C$ and $C_1$.

The operator $\mathcal{L}_m^r$ sends $\xi_j^{(m)}$ to $\xi_j^{(r)}$ and satisfies (11.3). Therefore by (11.9) $\eta_j^{(m)}(t)$ converges (as $m$ grows) to

$$\xi_j^{(\infty)}(t) = (\mathcal{L}_\infty)^{-1}\xi_j^{(0)}(t)$$

uniformly for bounded $t$'s. Denoting by $\Pi_p$, $\Pi_q$, and $\Pi_y$ the natural projectors which send $Z_d$ to $\mathbb{R}_p^n$, $\mathbb{R}_q^n$, and $Y_d$, respectively, we get from this convergence that

$$\Pi_p \xi_j^{(\infty)} \equiv 0, \qquad \|\Pi_y \xi_j^{(\infty)}(t)\|_d \equiv 1. \tag{11.10}$$

For $\tau_1 \le \tau_2$ let $S_{\tau_1 **}^{\tau_2} = S_{\tau_1 **}^{\tau_2}(\mathfrak{h}_0(\tau_1))$ be the flow maps of eqn (11.1) and $\tilde{S}_{\tau_1 **}^{\tau_2}$ be the conjugated maps

$$\tilde{S}_{\tau_1 **}^{\tau_2} = \mathcal{L}_\infty(\tau_2)^{-1} \circ S_{\tau_1 **}^{\tau_2} \circ \mathcal{L}_\infty(\tau_1)$$

(the linear operator $\tilde{S}_{\tau_1 **}^{\tau_2}$ sends $\xi_j^\infty(\tau_1)$ to $\xi_j^\infty(\tau_2)$). We write $Z_d$ as $\mathbb{R}_p^n \times \mathbb{R}_q^n \times Y_d$ and accordingly write $\tilde{S}_{\tau_1 **}^{\tau_2}$ in the block form:

$$\tilde{S}_{\tau_1 **}^{\tau_2} = \begin{pmatrix} s_{pp} & s_{pq} & s_{py} \\ s_{qp} & s_{qq} & s_{qy} \\ s_{yp} & s_{yq} & s_{yy} \end{pmatrix}.$$

Since $\xi_j^{(0)} = \mathcal{L}_\infty(0)\xi_j$, then $\tilde{S}_{0**}^t(\xi_j) = \xi_j^{(\infty)}(t)$ and we get from (11.10) that

$$s_{py} = 0, \qquad \|s_{yy}\|_{d,d} \equiv 1. \tag{11.11}$$

For each $q \in \mathbb{T}^n$, the map $\Sigma_\omega$ sends the curve $q + \omega' t \in \mathbb{T}^n$ to a solution of the initial equation (10.3). So $\Sigma_\omega$ conjugates the translation of $\mathbb{T}^n$ in the direction $\omega'$ with the flow of (10.3) and its linearization $\Sigma_{\omega *} = \mathcal{L}_\infty\big|_{\{0\}\times\mathbb{R}^n_q\times\{0\}}$ conjugates the linearization of the translation with the corresponding operator $\tilde{S}$. This means that

$$s_{pq} = 0, \qquad s_{qq} = \mathrm{id}, \qquad s_{yq} = 0. \tag{11.12}$$

Each map $\tilde{S}^{\tau_2}_{\tau_1 **}$ is symplectic as a composition of symplectic maps. Hence,

$$\alpha_2[\tilde{S}^{\tau_2}_{\tau_1 **}(\delta p_1, 0, 0),\ \tilde{S}^{\tau_2}_{\tau_1 **}(0, \delta q_2, 0)] = \langle \delta p_1, \delta q_2\rangle_{\mathbb{R}^n}, \quad \forall\, \delta p_1, \delta q_2 \in \mathbb{R}^n.$$

Because of (11.11) and (11.12) this implies that $\langle s_{pp}\delta p_1, \delta q_2\rangle_{\mathbb{R}^n} \equiv \langle \delta p_1, \delta q_2\rangle_{\mathbb{R}^n}$. Hence,

$$s_{pp} = \mathrm{id}. \tag{11.13}$$

Since the flow maps $S^{\tau_2}_{\tau_1 **}$ are uniformly well defined, then

$$\|\tilde{S}^{\tau_2}_{\tau_1 **}\|_{d,d} \le C\|S^{\tau_2}_{\tau_1 **}\|_{d,d} \le Ce^{C_1|\tau_1 - \tau_2|}. \tag{11.14}$$

Now we can estimate the norm of the operator $\tilde{S}^T_{0**}$ with large $T$. To do this let us write $Z_d$ as

$$Z_d = \mathbb{R}^n_p \times E, \quad E = \mathbb{R}^n_q \times Y_d = \{\mu = (q, y)\}.$$

Enlarging the blocs of $\tilde{S}^{\tau_2}_{\tau_1 **}$ accordingly, we write this operator as

$$\tilde{S}^{\tau_2}_{\tau_1 **} = \begin{pmatrix} \mathfrak{s}_{pp} & \mathfrak{s}_{p\mu} \\ \mathfrak{s}_{\mu p} & \mathfrak{s}_{\mu\mu} \end{pmatrix}.$$

By (11.12), (11.13), and (11.14) we have

$$\mathfrak{s}_{p\mu} = 0, \qquad \mathfrak{s}_{pp} = \mathrm{id}, \qquad \|\mathfrak{s}_{\mu\mu}\| = 1, \qquad \|\mathfrak{s}_{\mu p}\| \le Ce^{C_1|\tau_1-\tau_2|}. \tag{11.15}$$

For any $(p_0, \mu_0) \in Z_d$ and $T \in \mathbb{N}$ we can write $\tilde{S}^T_{0**}(p_0, \mu_0)$ as

$$\tilde{S}^T_{0**}(p_0, \mu_0) = \tilde{S}^T_{T-1} \circ \cdots \circ \tilde{S}^1_{0**}(p_0, \mu_0).$$

Denoting $(p_j, \mu_j) = \tilde{S}^j_{j-1**} \circ \cdots \circ \tilde{S}^1_{0**}(p_0, \mu_0)$ and using (11.15) we see that

$$|p_j| = |p_{j-1}|, \qquad \|\mu_j\|_d \le \|\mu_{j-1}\|_d + C_2|p_{j-1}|,$$

where $C_2 = Ce^{C_1}$. Therefore we get the following component-wise inequality:

$$\begin{pmatrix} |p_T| \\ \|\mu_T\|_d \end{pmatrix} \le \begin{pmatrix} \mathrm{id} & 0 \\ C_2 & \mathrm{id} \end{pmatrix}^T \begin{pmatrix} |p_0| \\ \|\mu_0\|_d \end{pmatrix} = \begin{pmatrix} |p_0| \\ \|\mu_0\| + (C_2 + T)|p_0| \end{pmatrix}.$$

We have seen that any solution $\eta(t)$ of (11.1) meets the estimate

$$\|\eta(t)\|_d \le 3\|\eta_\infty(t)\|_d \le (C_1 + tC_2)\|\eta(0)\|_d,$$

where $\eta_\infty(t) = (\mathcal{L}^0_\infty)^{-1}(t)\eta(t)$. So Theorem 11.4 is proved.

# 12

# First-order linear differential equations on the $n$-torus

It is well known (see Lemma A9.1′ in Appendix 9) that the first-order constant coefficient differential equation

$$-i\frac{\partial x}{\partial \omega} + Ex = b(q), \quad q \in \mathbb{T}^n, \tag{12.1}$$

where $E$ is a non-zero real constant and $\partial x/\partial \omega = \nabla_q x(q) \cdot \omega$ with a fixed real $n$-vector $\omega$, has a unique analytic solution $x(q)$ if the function $b(q)$ is analytic and the vector $\omega$ is incommensurable with $E$. Namely,

$$|\omega \cdot s + E| \geq \left(|s| + 1\right)^{-n_1}/K_1, \qquad \text{for all } s \in \mathbb{Z}^n, \tag{12.2}$$

for some $n_1 \geq 0$ and $K_1 \geq E^{-1}$. If $b(q)$ is analytic in $U(\delta)$ (we recall that $U(\delta)$ stands for the complex $\delta$-neighbourhood of the real $n$-torus) and

$$|b|^{U(\delta)} \equiv \sup_{q \in U(\delta)} |b(q)| \leq 1,$$

then the solution $x$ also is analytic in $U(\delta)$ and

$$|x|^{U(\delta-\Delta)} \leq CK_1\Delta^{-n-n_1}, \qquad \text{for } 0 < \Delta < \delta. \tag{12.3}$$

If we replace (12.1) by an equation with variable coefficients

$$-i\frac{\partial x}{\partial \omega} + Ex + Bh(q)x = b(q), \tag{12.4}$$

where $B$ is a real parameter and $h$ is an analytic in $U(\delta)$ function such that

$$|h|^{U(\delta)} \leq 1, \qquad \int_{\mathbb{T}^n} h(q)\, dq = 0,$$

then we can find an analytic function $H(q)$ such that $\partial H/\partial \omega = h$, provided that the vector $\omega$ is Diophantine. Namely,

$$|\omega \cdot s| \geq |s|^{-n_2}/K_2, \qquad \text{for all } s \in \mathbb{Z}^n \setminus 0, \tag{12.5}$$

with some $n_2 > 0$ and $K_2 \geq 1$. Moreover, $|H|^{U(\delta-\Delta)} \leq CK_2\Delta^{-n-n_2}$ (see Lemma A9.1). The substitution $x = \mathrm{e}^{-iBH}y$ reduces (12.4) to the equation with constant coefficients

$$-i\frac{\partial y}{\partial \omega} + Ey = \mathrm{e}^{iBH}b =: \beta(q).$$

According to the above, this equation has a unique analytic solution $y(q)$ and $|y|^{U(\delta-2\Delta)} \leq C_1K_1\Delta^{-n-n_1}\exp(CK_2B\Delta^{-n-n_2})$. Thus (12.4) has a unique analytic solution $x(q)$ and

$$|x|^{U(\delta-\Delta)} \leq CK_1\Delta^{-n-n_1}\exp\left(C_2K_2B\Delta^{-n-n_2}\right).$$

The last estimate becomes void if we have no upper bound for $B$. Our goal in this chapter is to majorize the solution $x$ by a $B$-independent constant, provided that $E \gg B$, and more specifically, provided that

$$E \geq C_1 > 0 \quad \text{and} \quad E^\theta \geq CB, \tag{12.6}$$

where $C, C_1 > 0$ and $\theta \in (0, 1)$ are fixed constants.

The 'right' estimate for the solution $x$ turns out to be independent of $B$ and $E$. This is stated by the following theorem.

**Theorem 12.1.** *Under assumptions (12.2), (12.5), and (12.6) eqn (12.4) with* $|h|^{U(\delta)}$, $|b|^{U(\delta)} \leq 1$ $(0 < \delta \leq 1)$ *has a unique analytical solution* $x(q)$. *For any* $0 < \Delta < \delta$ *this solution satisfies the estimate*

$$|x|^{U(\delta-\Delta)} \leq CK_1\Delta^{-n-n_1}\exp\left(C_1K_2^{\frac{1}{1-\theta}}\Delta^{-n-n_2-d_1}\right), \tag{12.7}$$

*where*

$$d_1 = (n + n_2 + 2)\frac{\theta}{1-\theta}.$$

In the theorem and in its proof $C, C_1, \ldots$ are different positive constants, independent of $\omega, \Delta, \delta, \theta, E, K_1$, and $K_2$.

The estimate (12.7) is crucial to prove Lemmas 10.4 and 10.5 (with exponents $n_1, n_2$, and constants $K_1$ and $K_2$ specified in Chapter 10).

*Proof of the theorem.* Let us denote

$$C_* = C_{*0}K_2^{1/(n+n_2+2)}$$

with $C_{*0} \geq 1$ to be chosen later. We may assume that

$$B \geq (C_*/\Delta)^{d_1}, \tag{12.8}$$

since otherwise we would write $Bh$ as $BK'(K'^{-1}h)$, where $K'$ is a sufficiently large constant, and replace $B$ by $BK'$ and $h$ by $h/K'$.

To prove (12.7) under the assumption (12.8) we shall approximate the Diophantine vector $\omega$ in (12.4) by the vectors $\widetilde{\omega} = \widetilde{\omega}_\ell$ with rationally dependent coefficients ($\ell = 2, 3, \ldots$) and find an integral representation for an approximate solution for eqn (12.4) with $\omega$ replaced by $\widetilde{\omega}$. We show that the approximate solutions satisfy (12.7). Next we send $\ell$ to infinity to get the estimate (12.7) for the unique exact solution of (12.4).

All constants $C, C_1, \ldots$ below are $\ell$-independent. □

**Step 1. Approximations for the frequency vector.** For an integer $\ell \geq 2$ we consider the vector $\ell\omega \in \mathbb{R}^n$ and define $N_\ell \in \mathbb{Z}^n$ as an integer vector which is the closest to $\ell\omega$. Then

$$|\omega - \ell^{-1} N_\ell| \leq \frac{\sqrt{n}}{2\ell}. \tag{12.9}$$

For any vector $s \in \mathbb{Z}^n$ we denote $\langle s \rangle = |s| + 1$.

**Lemma 12.1.** *There exist constants $r \in (1-\ell^{-1}, 1+\ell^{-1})$ and $\tilde{C} \geq 2$ such that $\ell E \notin r\mathbb{Z}$ and the vector $\widetilde{\omega}$, defined as*

$$\widetilde{\omega} = \widetilde{\omega}_{\ell,r} := \frac{N_\ell}{\tilde{\ell}}, \quad \tilde{\ell} = \frac{\ell}{r},$$

*is incommensurable with $E$. Namely,*

$$|s \cdot \widetilde{\omega} + E| \geq \frac{\langle s \rangle^{-n-n_1-1}}{\tilde{C} K_1}, \qquad \forall s \in \mathbb{Z}^n. \tag{12.10}$$

It is clear from (12.9) that the vector $\tilde{\omega}$, constructed in this lemma, is such that

$$|\tilde{\omega}| \leq 2\left(|\omega| + \frac{\sqrt{n}}{2\ell}\right) \quad \text{and} \quad |\omega - \tilde{\omega}| \leq \frac{1}{\ell}\left(\frac{\sqrt{n}}{2} + |\omega| + \frac{\sqrt{n}}{2\ell}\right) = \frac{C}{\ell}.$$

*Proof.* By (12.2) and (12.9) any vector $\tilde{\omega}$ as above satisfies the estimate

$$|\widetilde{\omega} \cdot s + E| \geq \langle s \rangle^{-n_1}/K_1 - C\,|s|/\ell \geq \tfrac{1}{2}\,\langle s \rangle^{-n_1}/K_1$$

if $\ell^{-1} \leq \langle s \rangle^{-n_1-1}/(2CK_1)$ or, equivalently, if

$$|s| \leq \left(\frac{\ell}{2CK_1}\right)^{1/(n_1+1)} =: N_0.$$

So below we shall consider $|s| > N_0$ only.

Take any $s_0 \in \mathbb{Z}^n$ that violates (12.10) for some choice of $r \in S := (1-\ell^{-1}, 1+\ell^{-1})$. Then $|s_0 \cdot \widetilde{\omega}| \geq E/2$, since $K_1 \geq E^{-1}$ and $\tilde{C} \geq 2$. Therefore the set

$$A_{s_0} = \left\{ r \in S \;\middle|\; |s_0 \cdot \widetilde{\omega}_{\ell,r} + E| \leq \frac{\langle s \rangle^{-n-n_1-1}}{\tilde{C} K_1} \right\}$$

is a segment of length $\leq 4\langle s\rangle^{-n-n_1-1}/\tilde{C}K_1$. So

$$\operatorname{mes} \bigcup_{|s|\geq N_0} A_s \leq \frac{C}{\tilde{C}K_1} N_0^{-n_1-1} = \frac{C}{\tilde{C}K_1}\frac{2CK_1}{\ell},$$

which is less than $\ell^{-1}$ if $\tilde{C}$ is chosen sufficiently large.

Therefore, there exists a point $r \in S$ which lies outside all the sets $A_s$ with $|s| \geq N_0$. The corresponding vector $\widetilde{\omega} = \widetilde{\omega}_{\ell,r}$ satisfies all estimates (12.10). We can choose $r$ to be different from the numbers $\ell E/j$, $j = \pm 1, \pm 2, \ldots$, and the lemma is proved. □

Since $|\omega - \widetilde{\omega}| \leq C/\ell$ and the vector $\omega$ is Diophantine (see (12.5)), then

$$|s \cdot \widetilde{\omega}| \geq (2K_2\,|s|^{n_2})^{-1} \quad \text{if} \quad 0 < |s| \leq (\ell/2CK_2)^{1/(n_2+1)} =: L. \tag{12.11}$$

Let us denote by $h_s$ the Fourier coefficients of $h(q)$. Then $|h_s| \leq \mathrm{e}^{-\delta|s|}$ by estimate (A9.1) in Appendix 9. Moreover, $h_0 = 0$ since the mean value of $h$ vanishes. Now we define the resonant and the regular parts of $h$ as

$$h^{res}(q) = \sum_{\substack{s\neq 0\\ s\cdot\widetilde{\omega}=0}} h_s \mathrm{e}^{i\,s\cdot q}, \qquad h_{reg}(q) = \sum_{\substack{s\\ s\cdot\widetilde{\omega}\neq 0}} h_s \mathrm{e}^{i\,s\cdot q},$$

so $h = h^{res} + h_{reg}$.

For $j = 1, 2, 3$ we denote

$$U^j = U(\delta - j\Delta/4).$$

**Lemma 12.2.** *The functions $h^{res}$ and $h_{reg}$ are analytic in $U^1$ and*

$$|h^{res}|^{U^1} \leq C\Delta^{-n-1}(\ell/K_2)^{-1/(n_2+1)}, \qquad |h_{reg}|^{U^1} \leq C\Delta^{-n}.$$

*Proof.* The estimate for $h_{reg}$ is obvious (see (A9.1) and (A9.2) in Appendix 9). In order to estimate $h^{res}$ we observe that if $s \cdot \widetilde{\omega} = 0$, then by (12.11) $|s| \geq L$ and for $q$ in $U^1$ we have

$$|h^{res}| \leq \sum_{|s|\geq L} \mathrm{e}^{-|s|\Delta/4} \leq C\Delta^{-n-1}L^{-1}$$

(see estimate (A.9.10) with $R = L$ and $k = 1$). Thus, the estimate for $h^{res}$ also is proved. □

**Lemma 12.3.** *There exists a function $\widetilde{H}$, analytic in $U^1$, such that $\partial\widetilde{H}/\partial\widetilde{\omega} = h_{reg}$ and $|\widetilde{H}|^{U^1} \leq CK_2\Delta^{-n-n_2}$.*

*Proof.* Let us define $\widetilde{H}$ as a Fourier series with coefficients $\widetilde{H}_s$, where

$$\widetilde{H}_s = \begin{cases} 0, & \text{if } s \cdot \widetilde{\omega} = 0, \\ h_s \,/\, (s \cdot \widetilde{\omega}), & \text{otherwise.} \end{cases}$$

Since the modulus of any non-zero denominator is bigger than $1/\tilde{\ell} \geq 1/(2\ell)$, then by (12.11), for any $q$ in $U^1$, we have

$$|\widetilde{H}(q)| \leq 2 \sum_{|s| \leq L} |s|^{n_2} K_2 \mathrm{e}^{-|s|\,\Delta/4} + 2\ell \sum_{|s|>L} \mathrm{e}^{-|s|\,\Delta/4}.$$

Now the assertion follows. The first sum is obviously bounded by

$$2K_2 \sum_{s \in \mathbb{Z}^n} |s|^{n_2}\, \mathrm{e}^{-|s|\Delta/4} \leq C K_2 \int_{\mathbb{R}^n} |x|^{n_2}\, \mathrm{e}^{-|x|\Delta/4}\, dx$$

$$= C' K_2 \Delta^{-n-n_2} \int_{\mathbb{R}^n} |y|^{n_2}\, \mathrm{e}^{-|y|\Delta/4}\, dy = C_1 K_2 \Delta^{-n-n_2},$$

and the second one is bounded by $C_2 K_2 \Delta^{-n-n_2-1}$ due to the estimate (A9.10) with $k = n_2 + 1$. □

**Step 2. Approximating equations.** Let us approximate eqn (12.4) by replacing the vector $\omega$ by $\widetilde{\omega} = \widetilde{\omega}_{\ell,r}$ and replacing $h(q)$ by its regular part $h_{reg}$. This gives the equation

$$-i\, \frac{\partial x}{\partial \widetilde{\omega}} + Ex + Bh_{reg}x = b(q). \tag{12.12}$$

The substitution $x = \mathrm{e}^{-iB\widetilde{H}} y$ with $\widetilde{H}$ as in Lemma 12.3 reduces (12.12) to

$$-i\, \frac{\partial y}{\partial \widetilde{\omega}} + Ey = \mathrm{e}^{iB\widetilde{H}} b =: \beta(q). \tag{12.13}$$

By Lemma 12.1 this equation meets the condition (12.2) with $n_1 := n + n_1 + 1$, so for any analytic $\beta$ it has a unique analytic solution $y$. The estimate (12.3) for $|y|$ is insufficient for our purposes and we shall get a better one using an integral representation for $y$. To this end, we consider the equation

$$-i\mu\, \frac{\partial z}{\partial t} + Ez = f(t), \quad t \in S^1 = \mathbb{R}/2\pi\mathbb{Z}. \tag{12.14}$$

If $E \notin \mu\mathbb{Z}$, then the unique periodic solution of (12.14) can be written as

$$z(t) = \frac{\mathcal{K}_{E/\mu}}{\mu} \int_0^{2\pi} \mathrm{e}^{-i(E/\mu)\tau} f(t-\tau)\, d\tau,$$

where $\mathcal{K}_r = i/(1 - \mathrm{e}^{-i2\pi r})$. Indeed, for $f = \mathrm{e}^{ikt}$ we have $z = \mathrm{e}^{ikt}/(E + k\mu)$, which is the periodic solution of (12.14). An arbitrary periodic $f$ can be expanded in Fourier series, and the assertion follows.

Next, we take any $R \in \mathbb{T}^n$ and consider the solenoid through $R$:

$$t \longmapsto R + t\,\tilde{\ell}\,\widetilde{\omega} \in \mathbb{T}^n = \mathbb{R}^n/2\pi\,\mathbb{Z}^n. \tag{12.15}$$

Since $\tilde{\ell}\,\widetilde{\omega} = N_\ell$ is an integer vector, then the solenoid is a $2\pi$-periodic loop in $\mathbb{T}^n$. On the other hand, for a function on $\mathbb{T}^n$ and for its restriction to the solenoid one has $\partial/\partial t = \tilde{\ell}\;\partial/\partial\widetilde{\omega}$. Then eqn (12.13), restricted to the loop (12.15), takes the form (12.14) with

$$\mu = \tilde{\ell}^{-1}, \qquad f(t) = \beta(R + \tilde{\ell}\widetilde{\omega}t).$$

The assumption $E \notin \mu\mathbb{Z}$ is satisfied since $\ell E \notin r\mathbb{Z}$ by Lemma 12.2. Therefore,

$$y(R) = \mathcal{K}_{E\tilde{\ell}}\,\tilde{\ell}\int_0^{2\pi} \mathrm{e}^{-iE\tilde{\ell}\tau}\beta(R - \tilde{\ell}\widetilde{\omega}\tau)\,d\tau.$$

Finally, we denote $\nu = \widetilde{\omega}/|\widetilde{\omega}|^2$, $z = \tilde{\ell}\tau$ (so $E\tilde{\ell}\tau = E\nu\cdot\widetilde{\omega}z$), and obtain the integral representation for the (unique) solution $x$ of (12.12):

$$x(q) = \mathcal{K}_{E\tilde{\ell}}\int_0^{2\pi\tilde{\ell}} \mathrm{e}^{-iE(\nu\cdot Q + (B/E)(\widetilde{H}(q) - \widetilde{H}(q-Q))}\,b(q-Q)\big|_{Q=\widetilde{\omega}z}dz. \tag{12.16}$$

Here we treat $Q$ as a point in $\mathbb{R}^n$ and $H$ and $b$ as analytic $2\pi$-periodic functions.

The constant $E$ is an unbounded real parameter, so we have represented $x(q)$ as a rapidly oscillating Fourier integral. Its phase function is complex whenever $q$ is complex.

**Step 3. Study of the oscillating integral (12.16).** Denoting $\varrho = B/E$ and $\Psi(q, Q) = \widetilde{H}(q) - \widetilde{H}(q-Q)$ we observe that

(1) $\varrho \le C^{-1/\theta}B^{1-1/\theta} \le C^{-1/\theta}(\Delta/C_*)^{d_1(1/\theta-1)} = C^{-1/\theta}(\Delta/C_*)^{n+n_2+2}$ (see (12.6) and (12.8));

(2) $\Psi(q, 0) \equiv 0$;

(3) for $q$ in $U^2$ the function $\Psi$ is analytic in $Q$ and

$$|\nabla_Q\Psi(q,\cdot)|^{U(\Delta/2)} + |\Psi(q,\cdot)|^{U(\Delta/2)} \le CK_2\Delta^{-n-n_2-1}$$

(by Lemma 12.3 and the Cauchy estimate);

(4) the phase function of the Fourier integral (12.16) can be written as $-iE(\nu\cdot Q + \rho\Psi)$.

Let us consider the substitution

$$Q = R + f(R)\widetilde{\omega} \equiv \Phi(R),$$

where $R \in \mathbb{T}^n$ and $f$ is a complex function. Then

$$\nu\cdot Q + \varrho\Psi(q, Q)\big|_{Q=\Phi(R)} = \nu\cdot R + f(R) + \varrho\Psi(q, R + f(R)\,\widetilde{\omega}).$$

In order to simplify the phase function we wish to make the sum of the last two terms in the r.h.s. vanish. To achieve this aim the function $f$ has to satisfy the following equation:

$$f(R) + \varrho\Psi(q, R + f(R)\,\widetilde{\omega}) = 0.$$

If $C_{*0}$ is sufficiently large, then by (1) and (2) the function $\Psi$ satisfies the following estimates:

$$|\varrho\Psi| + |\varrho\nabla_Q\Psi| \le (\Delta/C_*)^{n+n_2+2} C K_2 \Delta^{-n-n_2-1} = C C_{*0}^{-n-n_2-2}\Delta$$

for $q \in U^2$, $R \in U(\Delta/2)$, and $|f| \le \Delta/C_\sharp$, where $C_\sharp = (|\omega|+1)$. Since the r.h.s. of the last inequality is smaller than $\Delta/C_\sharp$ provided that $C_{*0}$ is sufficiently large, then by the Implicit Function Theorem the equation has a unique solution $f(R) = f(q, R)$ which is a complex-analytic function of the argument $R \in U(\Delta/2)$. This solution satisfies the estimate

$$|f|^{U(\Delta/2)} \le \Delta/C_{*1},$$

where $C_{*1}$ goes to infinity with $C_{*0}$. On the other hand, due to (2), one has $f(0, q) \equiv 0$.

With this choice of the function $f$ the map $R \mapsto \Phi(R)$ analytically extends to $U(\Delta/2)$ and is there close to the identity.

Now let us view (12.16) as an integral of a holomorphic function along the segment $S = [0, 2\pi\tilde{\ell}] \cdot \widetilde{\omega}$ in the complex plane $\mathbb{C}^1 = \mathbb{C}\widetilde{\omega} \subset \mathbb{C}^n$, namely

$$x(q) = \mathcal{K}_{E\tilde{\ell}} \int_S \mathrm{e}^{-iE(\nu\cdot R + \varrho\Psi(q,R))}\, b(q - R)\, dR/|\widetilde{\omega}|\,.$$

In this integral we can replace the contour $S = \{R\}$ by $\Phi(S) = \{Q\} \subset \mathbb{C}^1$ since both contours lie in the domain of analyticity and their end-points coincide. Since $f(R) + \varrho\Psi(q, \Phi(R)) \equiv 0$, then

$$\begin{aligned}
x(q) &= \mathcal{K}_{E\tilde{\ell}} \int_{\Phi(S)} \mathrm{e}^{-iE(\nu\cdot Q + \varrho\Psi(q,Q))}\, b(q - Q)\,\frac{dQ}{|\widetilde{\omega}|} \\
&= \mathcal{K}_{E\tilde{\ell}} \int_S \mathrm{e}^{-iE\nu\cdot R}\, b(q - Q(R))(1 + |\widetilde{\omega}|\, f'(R))\frac{dR}{|\widetilde{\omega}|} \\
&= \mathcal{K}_{E\tilde{\ell}} \int_S \mathrm{e}^{-iE\nu\cdot R} g(R)\,\frac{dR}{|\widetilde{\omega}|}\,,
\end{aligned} \tag{12.17}$$

where we use the same notation $f$ for the function $f$ restricted to $\mathbb{C}^1$ and denote

$$g(R) = b(q - Q(R))\,(1 + |\widetilde{\omega}|\, f'(R)), \quad R \in \mathbb{C}^1.$$

This function is analytic in $U(\Delta/4)$ and is bounded there by some constant $C_1$.

In order to estimate the r.h.s. of (12.17) we expand $g$ in Fourier series:

$$g = \sum g_s \, \mathrm{e}^{is\cdot R}, \qquad |g_s| \le C_1 \, \mathrm{e}^{-|s|\Delta/4}, \tag{12.18}$$

(see (A9.1)). Now we have

$$x(q) = \mathcal{K}_{E\tilde{\ell}}, \sum_s g_s \int_0^{2\pi\tilde{\ell}} \mathrm{e}^{-i(E-\widetilde{\omega}\cdot s)t} \, dt = \mathcal{K}_{E\tilde{\ell}} \sum_s \frac{i g_s}{E - \widetilde{\omega}\cdot s} (\mathrm{e}^{-iE2\pi\tilde{\ell}} - 1),$$

since $\widetilde{\omega}\cdot\tilde{\ell}$ is an integer. Therefore $x(q) = \sum g_s/(E - \widetilde{\omega}\cdot s)$. By (12.10), (12.18), and (A9.2) for $q \in U^2$ the solution $x$ is estimated as follows:

$$|x(q)| \le \frac{CK_1}{E} \sum \langle s\rangle^{n+n_1+1} \mathrm{e}^{-|s|\Delta/4} \le C_1 K_1 \Delta^{-n-n_1-1}. \tag{12.19}$$

We stress that this estimate is independent of $\ell$.

**Step 4. Transition to the limit.** Changing the notation, we denote by $x_\ell(q)$ the solution of (12.12) that we have constructed, and rewrite (12.12) as

$$-i\widetilde{\omega}_\ell \cdot \nabla x_\ell + E\, x_\ell + Bh(q)x_\ell = b(q) + z_\ell(q),$$

where $z_\ell = B\, h^{res} x_\ell$. By (12.19) and Lemma 12.2, $|z_\ell|^{U^2} \le M\, \ell^{-1/(n_2+1)}$ with some $M$ independent of $\ell$. Moreover, still by (12.19), the sequence $\{x_\ell\}$ contains a subsequence such that both $\{x_\ell\}$ and $\{\nabla x_\ell\}$ converge uniformly in $U^3 \supset U(\delta-\Delta)$. Namely, $x_\ell \longrightarrow x$ and $\nabla x_\ell \longrightarrow \nabla x$, where

$$|x(q)|^{U(\delta-\Delta)} \le C_1 K_1 \Delta^{-n-n_1-1}. \tag{12.20}$$

Since $z_\ell \longrightarrow 0$ and $\widetilde{\omega}_\ell \longrightarrow \omega$, then $x(q)$ is a solution of (12.4).

Since (12.20) implies (12.7), then Theorem 12.1 is proved. □

# Addendum. The theorem of A.N. Kolmogorov

## A.1 Introduction

The celebrated theorem of Kolmogorov states that most (in the sense of measure) quasi-periodic solutions of an integrable analytic Hamiltonian equation persist under analytic perturbations of the Hamiltonian, provided that the Hessian of the Hamiltonian does not vanish identically. Kolmogorov stated this result and sketched its proof in Kolmogorov (1985). The proof was written later in full detail by Arnold and Moser, who used similar ideas to tackle other problems, thus originating the KAM theory (see, for example, Arnold 1963 and Moser 1973).

During more than 40 years of its history the theorem has been sharpened and new important related results have been proved. Many of them can be found in Arnold *et al.* (1989), Broer *et al.* (1996), Herman (1983, 1986), Lazutkin (1993), Moser (1973), Siegel and Moser (1971), and Treschev (1998).

Despite the improvements and developments, the Kolmogorov result still remains 'the KAM theorem', both because of its beauty and its huge interdisciplinary importance (this result is quoted and discussed in the majority of scientific works devoted to chaotic and regular dynamics).

Below we present a proof of the theorem based on the techniques and ideas developed to prove the abstract KAM theorem of this book. Some of these techniques are due to the author,[62] some were developed by other mathematicians.[63] We follow closely the proof of Theorem 8.3. Namely, we keep its notation, and some fragments of arguments below are identical to the corresponding fragments of the proof of Theorem 8.3.

## A.2 Theorems A and B

Let $P$ be a connected bounded domain in $\mathbb{R}^n$. In the symplectic space $(P \times \mathbb{T}^n,\ dp \wedge dq)$ we consider an integrable Hamiltonian system with the analytic Hamiltonian $h(p)$:

$$\dot{p} = 0, \qquad \dot{q} = \nabla_p h(p), \tag{A.1}$$

and its perturbation:

$$\dot{p} = -\nabla_q \mathcal{H}_\epsilon(p, q), \qquad \dot{q} = \nabla_p \mathcal{H}_\epsilon(p, \epsilon). \tag{A.2}$$

---

[62] In particular, the idea to treat Hamiltonians as Lipschitz (rather than analytic) functions of the frequency vector.

[63] In particular, the idea to pass from Theorem A below to Theorem B is due to J. Moser. It has been systematically used by J. Pöschel.

Here $0 \le \varepsilon \le 1$ and $\mathcal{H}_\epsilon = h(p) + \epsilon H_1(p, q)$ with some analytic function $H_1$. The phase-space $P \times \mathbb{T}^n$ is filled with Lagrangian tori $T_p^n = \{p\} \times \mathbb{T}^n$, which are invariant for the integrable equation (A.1). The theorem of Kolmogorov states that most of them persist as analytic invariant tori of the perturbed equation (A.2), provided that $\epsilon$ is sufficiently small and

$$\operatorname{Hess} h(p) \not\equiv 0. \tag{A.3}$$

More specifically, the following result holds for any $\rho_0 \in (0, 1/9)$:

**Theorem A.** *Let (A.3) hold. Then there exist a Borel subset $P_\epsilon \subset P$ and a Lipschitz embedding $\Sigma_\epsilon : P_\epsilon \times \mathbb{T}^n \to P \times \mathbb{T}^n$, analytic in the second variable, such that:*

*(a)* $\operatorname{mes}_n(P \setminus P_\epsilon) \to 0$ *as* $\epsilon \to 0$;

*(b) the map $\Sigma_\epsilon$ is $C\epsilon^{\rho_0}$-close to the identity map, both in the uniform and in the Lipschitz norms;*

*(c) each torus $T^n_{p,\epsilon} = \Sigma_\epsilon(T_p^n)$, $p \in P_\epsilon$, is invariant for eqn (A.2) and is filled with its time-quasiperiodic solutions $\mathfrak{h}_\epsilon(t)$ of the form $\mathfrak{h}_\epsilon(t) = \mathfrak{h}_\epsilon(t; p, q) = \Sigma_\epsilon(p, q + t\omega_\epsilon(p))$ $(p \in P_\epsilon,\ q \in \mathbb{T}^n)$, where $\omega_\epsilon = \omega_\epsilon(p)$ and $|\omega_\epsilon - \nabla h(p)| \le C\epsilon^{\rho_0}$.*

Since the function $h$ is analytic, then due to (A.3) the set $\{p \mid \operatorname{Hess} h(p) = 0\}$ is a closed zero-measure set. Hence, for any $\gamma > 0$ we can find a finite system of open connected subsets $P_j \subset P$ such that $\operatorname{mes}(P \setminus \cup P_j) < \gamma$ and $\nabla h$ defines the diffeomorphisms $\nabla h : \overline{P_j} \to \mathbb{R}^n$. Accordingly, it is sufficient to prove the theorem with (A.3) replaced by the stronger assumption:

$$\text{the map } \nabla h : \overline{P} \longrightarrow \Omega \Subset \mathbb{R}^n \text{ is a diffeomorphism.} \tag{A.4}$$

(To get Theorem A from this new result it suffices to apply it to the sets $P_j$ and then send $\gamma$ to zero.)

To prove the theorem we have to check that a 'typical' torus $T_a^n$, $a \in P$, persists under the perturbation. After our goal is formulated in this way, it is natural to scale the equation near the torus $T_a^n$:

$$p = a + \epsilon^{2/3}\tilde{p}, \qquad q = \tilde{q}. \tag{A.5}$$

Since $d\tilde{p} \wedge d\tilde{q} = \epsilon^{-2/3} dp \wedge dq$, then in the tilde-variables the Hamiltonian takes the form[64]

$$\begin{aligned}\mathcal{H}_\epsilon(\tilde{p}, \tilde{q}; a) &= \epsilon^{-2/3}(h(a + \epsilon^{2/3}\tilde{p}) + \epsilon H_1(a + \epsilon^{2/3}\tilde{p}, \tilde{q})) \\ &= \epsilon^{-2/3} h(a) + \nabla h(a) \cdot \tilde{p} + \epsilon^{1/3}(H_1 + \epsilon^{-1} h_2).\end{aligned}$$

Here $h_2 = h(a + \epsilon^{2/3}\tilde{p}) - \epsilon^{2/3}\nabla h(a) \cdot \tilde{p}$, so $\epsilon^{-1} h_2 = \epsilon^{1/3} O(|\tilde{p}|^2)$. Accordingly, $H_1 + \epsilon^{-1} h_2$ is an analytic function such that

$$|H_1 + \epsilon^{-1} h_2| \le C, \quad \text{for } \tilde{p} \in \mathcal{O}_\delta(\mathbb{C}^n), \qquad a \in P + \frac{\delta}{2} \subset \mathbb{C}^n, \qquad |\operatorname{Im} \tilde{q}| < \frac{\delta}{2}, \tag{A.6}$$

---

64 This is one of the basic properties of Hamiltonian equations (see Arnold 1974; cf. the Corollary to Theorem 1.12). It can be trivially checked by substituting (A.5) into eqns (A.2).

uniformly in $0 \le \epsilon \le 1$. Owing to (A.4), we can replace the parameter $a \in P$ of the substitution (A.5) by the parameter $\omega$:

$$\omega = \nabla h(a) \in \Omega = \nabla h(P).$$

Let us denote

$$H(\tilde{p}, \tilde{q}; \omega, \varepsilon) = H_1 + \epsilon^{-1} h_2 \mid_{a=(\nabla h)^{-1}(\omega)} .$$

Ignoring the irrelevant constant $\epsilon^{-2/3} h(a)$, we write the Hamiltonian $\mathcal{H}_\epsilon$ as

$$\mathcal{H}_\epsilon(\tilde{p}, \tilde{q}; \omega, \varepsilon) = \omega \cdot \tilde{p} + \epsilon^{1/3} H(\tilde{p}, \tilde{q}; \omega, \varepsilon).$$

Owing to estimates (A.6), the function $H$ is Lipschitz in $\omega \in \Omega$, analytic in $\tilde{p}$ and $\tilde{q}$, and

$$|H|^{\mathcal{O}_\delta(\mathbb{C}^n) \times U(\delta/2), \Omega} \le C,$$

uniformly in $\epsilon$. Here for any $\delta' > 0$ we denote

$$U(\delta') = \{q \in \mathbb{C}^n / 2\pi \mathbb{Z}^n \mid |\operatorname{Im} q| < \delta'\}.$$

Concerning the norm $|\cdot|^{\mathcal{O}_\delta(\mathbb{C}^n) \times U(\delta/2), \Omega}$, see Notation at the beginning of the book.

Now Theorem A follows from its sibling (which is another appearance of Kolmogorov's theorem):

On the domain $(\mathcal{O}_\delta \times \mathbb{T}^n, dp \wedge dq)$, where $\mathcal{O}_\delta$ is short for $\mathcal{O}_\delta(\mathbb{R}^n)$, let us consider the linear Hamiltonian $H_0 = \omega \cdot p$, depending on the parameter $\omega \in \Omega \Subset \mathbb{R}^n$, and its analytic perturbation $H_\varepsilon$:

$$H_\varepsilon = \omega \cdot p + \varepsilon H(p, q; \omega, \varepsilon).$$

The corresponding perturbed Hamiltonian equations are

$$\dot{p} = -\varepsilon \nabla_q H, \qquad \dot{q} = \omega + \varepsilon \nabla_p H. \tag{A.7}$$

Choosing any $\rho \in (0, 1/3)$ and denoting by $\Psi_0$ the map $\mathbb{T}^n \times \Omega \to \mathcal{O}_\delta \times \mathbb{T}^n$ which sends a point $(q, \omega)$ to $(0, q)$, we have:

**Theorem B.** *Let $H$ be an analytic function of the $(p, q)$-variables such that $|H|^{\mathcal{O}_\delta(\mathbb{C}^n) \times U(\delta), \Omega} \le 1$ with some $\delta > 0$, uniformly in $0 \le \varepsilon \le 1$. Then there exist a Borel subset $\Omega_\varepsilon \subset \Omega$ and a Lipschitz map $\Psi_\varepsilon : \mathbb{T}^n \times \Omega_\varepsilon \to \mathcal{O}_\delta \times \mathbb{T}^n$, analytic in the first variable, such that:*

*(a)* $\operatorname{mes}_n(\Omega \setminus \Omega_\varepsilon) \to 0$ *as* $\varepsilon \to 0$,

*(b)* $|\Psi_\varepsilon - \Psi_0|^{\mathbb{T}^n \times \Omega_\varepsilon, Lip} \le C\varepsilon^\rho$,

*(c) each torus $\Psi_\varepsilon(\mathbb{T}^n \times \{\omega\})$, $\omega \in \Omega_\varepsilon$, is invariant for the flow of eqn (A.7) and is filled with its quasiperiodic solutions $\Psi_\varepsilon(q_0 + \omega' t, \omega)$, where $q_0 \in \mathbb{T}^n$, $\omega' = \omega'(\omega)$ and $|\omega' - \omega| \le C\varepsilon^\rho$.*

To show how Theorem B with $\delta$ replaced by $\delta/2$ and $\varepsilon$ equal to $C\epsilon^{1/3}$ implies Theorem A, we choose $P_\varepsilon = (\nabla h)^{-1}\Omega_\varepsilon$ and define the map $\Sigma_\varepsilon : P_\varepsilon \times \mathbb{T}^n \to P \times \mathbb{T}^n$ as follows:

$$\Sigma_\varepsilon(p,q) = \left(p + \varepsilon^{2/3}\Pi_p\Psi_\varepsilon\left(q, \nabla h(p)\right), \Pi_q\Psi_\varepsilon\left(q, \nabla h(p)\right)\right)$$

($\Pi_p$ and $\Pi_q$ stand for the natural projectors on $\mathbb{R}^n$ and $\mathbb{T}^n$, respectively). Since the substitution (A.5) transforms a solution $(\tilde{p}, \tilde{q})$ of (A.7) to the solution of (A.2), then the curves $\Sigma_\varepsilon(p, q + t\omega_\varepsilon(p))$, where $\omega_\varepsilon(p) = \omega'(\nabla h(p))$, satisfy eqn (A.2). Clearly, the maps $\Sigma_\varepsilon$ and $\omega_\varepsilon$ meet the estimates in assertions (b) and (c) of Theorem A, so the theorem follows.

The restriction $\rho_0 < 1/9$, imposed in Theorem A, looks unnatural and indeed it is superficial: the theorem remains true for any $\rho_0 < 1$. To get this result, the first few steps of the KAM procedure which prove the theorem should be done 'by hand', see Kuksin (1993, Refinement 2, p. 51).

## A.3 Sketch of the proof

The proof of Theorem B, presented below, uses a version of the KAM procedure. We start with a brief description.

Let us introduce the sequence of real numbers $\{\varepsilon_m\}$ which converge to zero 'very fast':

$$\varepsilon_m = \varepsilon^{(1+\rho)^m}, \qquad m \geq 0,$$

and a decreasing sequence of complex neighbourhoods $O_m$ of the torus $\{0\} \times \mathbb{T}^n$:

$$O_m = \mathcal{O}_{\varepsilon_m^{2/3}}(\mathbb{C}^n) \times U(\delta_m).$$

Here $\{\delta_m\}$ is the decreasing sequence $\delta = \delta_0 > \delta_1 > \delta_2 \cdots > \delta/2$ defined below in Section A.5. By $O_m^r$ we denote a real part of the complex domain $O_m$.

The KAM procedure we use is given by the following construction. For $m = 0, 1, \ldots$ we find:

(1) an analytic function $\mathcal{H}_m$ on the domain $O_m$ which is $\varepsilon_m^{1/3}$-close to an appropriate linear function $p \cdot \Lambda_m$ (for $m = 0$, the function $\mathcal{H}_0$ equals $H_\varepsilon$); this function is treated as a Hamiltonian of the corresponding Hamiltonian system;

(2) a Borel set $\Omega_m \subset \Omega$ such that $\Omega_m \subset \Omega_{m-1}$ and $\Omega_0 = \Omega$;

(3) a symplectic transformation $S_m(\cdot;\omega) : O_{m+1}^r \to O_m^r$, defined for $\omega$ in $\Omega_{m+1}$, which analytically extends to $O_{m+1}$ and transforms the function $\mathcal{H}_m$ to $\mathcal{H}_{m+1}$.

When the objects above are obtained, we note that the transformation $S_0 \circ \cdots \circ S_{m-1}$ with a large $m$ 'almost integrates' eqn (A.7). Indeed, since $\mathcal{H}_m$ 'almost equals' $p \cdot \Lambda_m$, then the curves $t \mapsto (0, q + \Lambda_m t)$ 'almost satisfy' an equation with the hamiltonian $\mathcal{H}_m$ and the curves $t \mapsto (S_0 \circ \cdots \circ S_{m-1})(0, q + \Lambda_m t)$ 'almost satisfy' the original one, provided that $\omega \in \Omega_m$. The limiting transformation $S_0 \circ S_1 \circ \cdots$ is defined on the torus $\{0\} \times \mathbb{T}^n$ if $\omega \in \Omega_\varepsilon := \cap\Omega_m$ and sends the limiting curves $(0, q + \Lambda_\infty t)$ to exact solutions.

## A.4 Reformulation of the theorem's assertion

We note that Theorem B, which we are going to prove, is equivalent to the following result: for any $\gamma > 0$, there exists a Borel subset $\Omega_\gamma^\varepsilon \subset \Omega$ such that $\text{mes}_n(\Omega \setminus \Omega_\gamma^\varepsilon) < \gamma$ and assertions (b) and (c) of the theorem hold as soon as $\varepsilon < \bar{\varepsilon}(\gamma)$, where $\bar{\varepsilon}(\gamma) > 0$ is continuous in $\gamma$ and goes to zero with $\gamma$. This function may be assumed to be monotonic in $\gamma$.[65] So the inverse function $\gamma(\varepsilon)$,

$$\gamma(\varepsilon) = \min\{\gamma \mid \bar{\varepsilon}(\gamma) = \varepsilon\},$$

is positive for $\varepsilon > 0$, goes to zero with $\varepsilon$, and the set $\Omega_\varepsilon := \Omega^\varepsilon_{\gamma(\varepsilon)}$ satisfies all claims of Theorem B.

## A.5 Proof of Theorem B

We introduce an increasing sequence $\{e(j)\}$ as in Section 10.2. That is, $e(0) = 0$ and

$$e(m) = (1^{-2} + \cdots + m^{-2})/K_*, \qquad K_* = 2(1^{-2} + 2^{-2} + \cdots), \tag{A.8}$$

so $e(m) < 1/2$ for all $m$. Now we define a 'radius of analyticity $\delta_m$ at the $m$th step' as

$$\delta_m = \delta_0(1 - e(m)).$$

We shall use the sequence $\{\varepsilon_m\}$ and the domains $O_m$, defined earlier. Moreover, we define the intermediate numbers $\delta_m^j$:

$$\delta_m = \delta_m^0 > \delta_m^1 > \cdots > \delta_m^6 = \delta_{m+1}, \qquad \delta_m^j = \frac{6-j}{6}\delta_m + \frac{j}{6}\delta_{m+1},$$

and the intermediate domains $O_m^j$ and $U_m^j$:

$$O_m = O_m^0 \supset O_m^1 \supset \cdots \supset O_m^6 \supset O_{m+1}, \qquad O_m^j = \mathcal{O}_{(2^{-j}\varepsilon_m)^{2/3}} \times U(\delta_m^j),$$
$$U_m = U_m^0 \supset U_m^1 \supset \cdots \supset U_m^6 = U_{m+1}, \qquad U_m^j = U(\delta_m^j)$$

(the inclusion $O_m^6 \supset O_{m+1}$ holds provided that $\varepsilon$ is sufficiently small).

Below (as well as in the proofs of Part II) $C$, $C_1$, etc. stand for different positive constants, independent of $m$ and $\varepsilon$; $C(m)$, $C_1(m)$, etc. stand for different functions of the form $C(m) = C_1 m^{C_2}$. The constants $C_j$ may depend on $\gamma$.

All arguments will be done under the assumption that $\varepsilon$ is sufficiently small, i.e. $\varepsilon < \bar{\varepsilon}$ for some positive $\bar{\varepsilon}(\gamma)$. Since the sequence $\varepsilon_m$ decays with $m$ faster than any exponent, then choosing $\bar{\varepsilon}$ sufficiently small we may achieve that

$$C(m)\varepsilon_m^\nu < 1, \qquad \forall m \geq 0,$$

for any fixed $C(m)$ and $\nu > 0$. We shall use this estimate *without further remarks*, decreasing $\bar{\varepsilon}$ *finitely* many times, if necessary.

---

[65] Since if $\bar{\varepsilon}$ is not monotonic, then we can replace it by the bigger (i.e. 'better') function $\tilde{\varepsilon}(\gamma) = \max\{\bar{\varepsilon}(\tau) \mid 0 \leq \tau \leq \gamma\}$, modifying the sets $\Omega_\gamma^\varepsilon$ accordingly.

**Hamiltonians** $\mathcal{H}_m$. For any $m \geq 0$ we consider an analytic Hamiltonian $\mathcal{H}_m(p, q; \omega)$ on the domain $O_m$, depending on the parameter $\omega \in \Omega_m \subset \Omega$. For $m = 0$ this Hamiltonian equals $H_\varepsilon$. For any $m \geq 0$ it has the form

$$\mathcal{H}_m = H_{0m}(p; \omega) + \varepsilon_m H_m(p, q; \omega). \tag{A.9}$$

The term $H_{0m}$ is a linear function

$$H_{0m} = p \cdot \Lambda_m(\omega);$$

this is an 'essential part' of the Hamiltonian. The term $\varepsilon_m H_m$ is viewed as a perturbation. The set $\Omega_m$ is a Borel subset of $\Omega$ such that

$$\text{mes}\,(\Omega \setminus \Omega_m) \leq \gamma e(m). \tag{A.10}$$

The map $\omega \mapsto \Lambda_m$ is Lipschitz and is close to the identity

$$|\Lambda_m(\omega) - \omega|^{\Omega_m, \text{Lip}} \leq 2K_* \varepsilon^{1/3} e(m) \tag{A.11}$$

($|\cdot|^{\Omega_m,\text{Lip}}$ stands for the Lipschitz norm, see Notation, and $K_*$ is defined in (A.8)). The function $H_m$ is assumed to be analytic in $O_m$ and to satisfy the following estimate:

$$|H_m|^{O_m, \Omega_m} \leq 2^m. \tag{A.12}$$

The corresponding Hamiltonian equations take the form

$$\dot{p} = -\varepsilon_m \nabla_q H_m, \qquad \dot{q} = \Lambda_m + \varepsilon_m \nabla_p H_m. \tag{A.13}$$

The original equations (A.7) are eqns (A.13)$|_{m=0}$. The Hamiltonian $H_0 = H$ and the frequency vector $\Lambda_0 = \omega$ clearly satisfy (A.12) and (A.11) with $m = 0$.

Now our goal is to construct the chain of symplectic transformations $S_0, S_1, \ldots$ which successively transform the Hamiltonian $\mathcal{H}_0 = H_\varepsilon$ to $\mathcal{H}_1$, $\mathcal{H}_2$, etc. as was indicated above.

**Step 1: Averaging.** Isolating an affine in $p$ part of the Hamiltonian $H_m$, we write it as

$$H_m = h^q(q; \omega) + p \cdot h^{1p}(q, \omega) + H_{2m}(p, q; \omega),$$

where $H_{2m} = O(|p|^2)$. Subtracting from $H_m$ the irrelevant constant, equal to its mean value in $q$, we achieve that $(2\pi)^{-n} \int h^q dq = 0$. The $q$-component of eqn (A.13) for $p = 0$ is

$$\dot{q} = \Lambda_m + \varepsilon_m h^{1p}(q; \omega).$$

Following the general ideology of averaging (see Arnold *et al.* 1989), we calculate the averaged frequency $\Lambda_{m+1}(\omega)$,

$$\Lambda_{m+1} = \Lambda_m + \varepsilon_m h^{0p}, \qquad h^{0p} = (2\pi)^{-n} \int h^{1p} dq\,,$$

and modify accordingly the essential part $H_{0m}$ of the Hamiltonian. Namely, denoting $h^p = h^{1p} - h^{0p}$ we rewrite $\mathcal{H}_m$ as

$$\mathcal{H}_m = \underbrace{p \cdot \Lambda_{m+1}}_{H_{0m+1}} + \varepsilon_m \underbrace{(h^q + p \cdot h^p)}_{H_{1m}} + \varepsilon_m H_{2m}. \tag{A.14}$$

Clearly,

$$H_{1m} + H_{2m} = H_m - p \cdot h^{0p}(\omega).$$

**Lemma A.1.** *The terms of the decomposition (A.14) are estimated as follows:*

*(a)*

$$|h^q|^{U_m, \Omega_m} \le 2^m,$$
$$|h^{0p}|^{\Omega_m, Lip} \le 2 \cdot 2^m \varepsilon_m^{-2/3},$$
$$|h^p|^{U_m, \Omega_m} \le 2^{m+1} \varepsilon_m^{-2/3}.$$

*(b) In the domain $O_{m+1} \subset O_m$ the term $\varepsilon_m H_{2m}$ is twice as small as the bound for a perturbation $\varepsilon_{m+1} H_{m+1}$ of the next step:*

$$\varepsilon_m |H_{2m}|^{O_{m+1}, \Omega_m} \le 2^m \varepsilon_{m+1}.$$

*(c) The functions $H_{1m}$ and $H_{2m}$ are analytic in $O_m$ and are real for real arguments.*

*Proof.* (a) The estimates for $h^q$ and its Lipschitz constant follow from (A.12) since $h^q(q;\omega) = H_m(0, q; \omega)$.

Since $h^{1p}(q;\omega) = \nabla_p H_m(0, q; \omega)$, then (A.12) and the Cauchy estimate imply that

$$|h^{1p}| \le 2^m \varepsilon_m^{-2/3}.$$

Since $h^{0p}$ is an average of $h^{1p}$, then its norm is bounded by $2^m \varepsilon_m^{-2/3}$ and the norm of $h^p = h^{1p} - h^{0p}$ is bounded by $2 \cdot 2^m \varepsilon_m^{-2/3}$. So to prove (a) it remains to estimate the Lipschitz constants in $\omega$. To bound a Lipschitz constant of $h^{1p}$ we consider the vector-function $(\nabla_p H_m(0, q; \omega_1) - \nabla_p H_m(0, q; \omega_2))/|\omega_1 - \omega_2|$ and argue as above. This bound implies the claimed estimates for the Lipschitz constants of the functions $h^{0p}$ and $h^p$.

(b) Let $(p, q) \in O_{m+1}$ and $\nu = \varepsilon_m^{2\rho/3}$. Then for any $z$ from the unit complex disc we have $((z/\nu)p, q) \in O_m$. On this disc let us consider the function $z \mapsto H_m((z/\nu)\rho, q; \omega)$ and its Taylor series at zero:

$$H_m\left(\frac{z}{\nu} p, q; \omega\right) = h_0 + h_1 z + h_2 z^2 + \cdots,$$

where $h_k = h_k(q;\omega)$. By the Cauchy inequality and (A.12), $|h_k| \le 2^m$ for every $k$. Therefore,

$$|\varepsilon_m H_{2m}(p, q)| = \varepsilon_m |h_2 \nu^2 + h_3 \nu^3 + \cdots| \le \varepsilon_m \nu^2 2^m |1 + \nu + \nu^2 + \cdots|$$
$$\le \varepsilon_m^{1+4\rho/3} \frac{1}{1-\nu} 2^m \le \varepsilon_{m+1} 2^m,$$

if $\bar{\varepsilon}$ is sufficiently small. A similar estimate holds for the Lipschitz constant, so the assertion is proven.

(c) The analyticity is obvious; the functions are real for real arguments since the Hamiltonian $\mathcal{H}_m$ is. □

Owing to item (a) of the lemma and (A.11),

$$|\Lambda_{m+1} - \omega|^{\Omega_m,\mathrm{Lip}} \le 2K_*\varepsilon^{1/3}e(m) + 2^{m+1}\varepsilon_m^{1/3} \le 2K_*\varepsilon^{1/3}e(m+1) \qquad \text{(A.15)}$$

since $2^{m+1}\varepsilon_m^{1/3} \le 2\varepsilon^{1/3}(m+1)^{-2}$ for every $m \ge 0$ if $\bar{\varepsilon}$ is sufficiently small. Hence, $\Lambda_{m+1}$ satisfies (A.11) with $m := m+1$.

**Step 2: Formal construction of the transformation $S_m$ and derivation of the homological equations.** We construct the transformation $S_m$ as the time-one shift along the trajectories of an auxiliary autonomous Hamiltonian vector field

$$\dot{p} = -\varepsilon_m \nabla_q F, \qquad \dot{q} = \varepsilon_m \nabla_p F. \qquad \text{(A.16)}$$

The transformation $S_m$ has to kill an 'essential part' of the perturbation in Hamiltonian (A.14), where the 'perturbation' is given by the terms of order $\varepsilon_m$. Owing to item (b) of Lemma A.1, the term $\varepsilon_m H_{2m}$ is irrelevant, so the essential one is $\varepsilon_m H_{1m}$. The informal rule to kill a term is that the auxiliary Hamiltonian has to be similar to a term to be killed. Accordingly, we take the Hamiltonian $F$ of the same form as $H_{1m}$:

$$F = f^q(q;\omega) + p \cdot f^p(q;\omega).$$

The flow of eqn (A.16) is formed by canonical transformations $S^t$ and

$$\frac{d}{dt}\mathcal{H}_m \cdot S^t \,|_{t=0} = \varepsilon_m\{F, \mathcal{H}_m\} + O(\varepsilon_m^2),$$

where $\{F, \mathcal{H}_m\} = \nabla_p F \cdot \nabla_q \mathcal{H}_m - \nabla_q F \cdot \nabla_p \mathcal{H}_m$ (cf. Theorem 1.7). Since $\mathcal{H}_m = H_{0\,m+1} + \varepsilon_m H_{1m} + \varepsilon_m H_{2m}$ and $\varepsilon_m H_{2m} = O(\varepsilon_{m+1})$ in the domain $O_{m+1}$ by Lemma A.1, then for $(p,q) \in O_{m+1}$ the transformed Hamiltonian $\mathcal{H}_m \circ S_m = \mathcal{H}_m \circ S^t \,|_{t=1}$ equals

$$\mathcal{H}_m(S_m(p,q;\omega);\omega) = H_{0m+1} + \varepsilon_m(H_{1m} + \{F, \mathcal{H}_m\}) + O(\varepsilon_{m+1}).$$

Noting that $\nabla_p H_{0m+1} = \Lambda_{m+1}$, $\nabla_q H_{0m+1} = 0$, and abbreviating

$$\Lambda_{m+1} = \omega', \qquad \omega' \cdot \nabla_q = \frac{\partial}{\partial \omega'},$$

we have $\{F, H_{0\,m+1}\} = -\partial F/\partial\omega'$. Since formally

$$\varepsilon_m(H_{1m} + \{F, \mathcal{H}_m\}) = \varepsilon_m(H_{1m} + \{F, H_{0\,m+1}\}) + O(\varepsilon_m^2),$$

then

$$\mathcal{H}_m \circ S_m = H_{0\,m+1} + \varepsilon_m\left(h^q + p\cdot f^q - \frac{\partial f^q}{\partial\omega'} - p\cdot\frac{\partial f^p}{\partial\omega'}\right) + O(\varepsilon_{m+1}).$$

Therefore we shall have

$$H_{1m} + \{F, H_{0\,m+1}\} = 0 \qquad \text{(A.17)}$$

and the transformed Hamiltonian $\mathcal{H} \circ S_m$ will (formally) take the desired form $p \cdot \Lambda_{m+1} + O(\varepsilon_{m+1})$ in the domain $O_{m+1}$ (cf. (A.9) with $m := m+1$), provided that the functions $f^q$ and $f^p$ satisfy the following homological equations:

$$\frac{\partial f^q}{\partial \omega'} = h^q(q;\omega),$$
$$\frac{\partial f^p}{\partial \omega'} = h^p(q;\omega).$$

**Step 3: Solving the homological equations.** This step is described by the following lemma:

**Lemma A.2.** *Let us define the set $\Omega_{m+1}$ as $\Omega_m \setminus \Omega'$, where*

$$\Omega' = \Big\{\omega \in \Omega_m \mid |\omega' \cdot s| \le C^{-1}(m+1)^{-2}|s|^{-n} \quad \text{for some } s = s(\omega) \in \mathbb{Z}^n \setminus \{0\}\Big\},$$

*and $C = C(\gamma)$ is sufficiently large. Then*

*(a)* $\mathrm{mes}_n \Omega' \le \gamma(m+1)^{-2}/K_*$ *(for the constant $K_*$ see (A.8));*

*(b) for any $\omega \in \Omega_{m+1}$ the homological equations have unique zero-mean-value analytic solutions $f^q$ and $f^p$, real for real arguments, and such that*

$$|f^q|^{U^1_m,\Omega_{m+1}} \le C(m), \qquad |f^p|^{U^1_m,\Omega_{m+1}} \le C(m)\varepsilon_m^{-2/3}.$$

*Proof.* As $\omega' = \Lambda_{m+1}$ satisfies (A.15), then the map $\Omega_m \ni \omega \mapsto \omega'$ is Lipschitz-close to the identity. So it changes the $n$-dimensional Lebesgue measure no more than twice (see Lemma A8.1 in Appendix 8). Therefore, $\mathrm{mes}_n \Omega' \le 2\,\mathrm{mes}_n \tilde{\Omega}$, where

$$\tilde{\Omega} = \Big\{\omega \in \Omega + 1 \mid |\omega' \cdot s| \le C^{-1}(m+1)^{-2}|s|^{-n} \text{ for some } s \ne 0\Big\}$$

(here $\Omega + 1$ is the 1-neighbourhood of $\Omega$ in $\mathbb{R}^n$. This set clearly contains the range of the map $\omega \to \omega'$).

By Lemma A9.2 from Appendix 9, $\mathrm{mes}_n \tilde{\Omega} \le C(\Omega)(m+1)^{-2}/C$. So (a) follows, if we choose $C$ sufficiently large.

Assertion (b) results from Lemma A9.1 in the same Appendix with $C_* = C(m+1)^2$ and $\rho = \delta_m - \delta^1_m = \delta_0/6K_*(m+1)^2$ since analytic norms of the functions $h^p$ and $h^q$ are bounded in Lemma A.1. □

**Step 4: Study of the transformation $S_m$.** The transformation $S_m$ is a time-one shift along the trajectories of the Hamiltonian equations (A.16), which we now write as

$$\frac{d}{dt}(p,q) = \varepsilon_m\Big(-\nabla_q F(p,q;\omega),\ f^p(q,\omega)\Big) =: \varepsilon_m V(p,q;\omega). \tag{A.18}$$

We abbreviate $(p,q) = \mathfrak{h}$, so these equations abbreviate to

$$\dot{\mathfrak{h}} = \varepsilon_m V(\mathfrak{h};\omega).$$

We shall study eqns (A.18) in the domains $O_m^j$, $j \geq 2$, supplied with new distance $\mathrm{dist}_-$. The distance corresponds to the weighted norm $|\cdot|_-$ in the space $\mathbb{C}^n \times \mathbb{C}^n = \mathbb{C}^{2n}$, where

$$|(p,\xi)|_- = |p|^2 + \varepsilon_m^{-4/3}|\xi|^2.$$

The space $\mathbb{C}^{2n}$, given this norm, is denoted by $\mathbb{C}^{2n}_-$. It follows from Lemma A.2 and the Cauchy estimate that

$$|\varepsilon_m V|_-^{O_m^2,\Omega_{m+1}} \leq C(m)\varepsilon_m^{1/3}. \tag{A.19}$$

Identifying the tangent spaces $T_\mathfrak{h} O_m^2$ with $\mathbb{C}^n \times \mathbb{C}^n$, we write the linearized vector field $\varepsilon_m V_*$ as the block matrix

$$\varepsilon_m \begin{pmatrix} -\frac{\partial f^p}{\partial q} & -\frac{\partial^2 F}{\partial q^2} \\ 0 & \frac{\partial f^p}{\partial q} \end{pmatrix}.$$

A straightforward analysis of the blocks (again based on Lemma A.2 and the Cauchy estimate) shows that

$$\|\varepsilon_m V_*(\mathfrak{h})\|^{O_m^2,\Omega_{m+1}} \leq C(m)\varepsilon_m^{1/3}, \tag{A.20}$$

where $\|\cdot\|$ stands for the operator norm $\mathbb{C}^{2n} \to \mathbb{C}^{2n}$ or $\mathbb{C}^{2n}_- \to \mathbb{C}^{2n}_-$.

**Lemma A.3.** *The map $S_m$ is an analytic symplectomorphism which maps $O_m^j$ to $O_m^{j-1}$ for $j = 3, 4, 5$. It is close to the identity, namely:*

*(a)* $|S_m - \mathrm{id}|_-^{O_m^3,\Omega_{m+1}} \leq C_1(m)\varepsilon_m^{1/3}$.

*(b)* $\|S_{m*} - \mathrm{id}\|^{O_m^4,\Omega_{m+1}} \leq C_2(m)\varepsilon_m^{1/3}$, *where $\|\cdot\|$ stands for the operator norm $\mathbb{C}^{2n} \to \mathbb{C}^{2n}$ or $\mathbb{C}^{2n}_- \to \mathbb{C}^{2n}_-$.*

*(c) All the results, stated above for the map $S_m = S^1$, remain true for any map $S^\theta$, $0 \leq \theta \leq 1$.*

*Proof.* Since

$$\mathrm{dist}_-(O_m^{j+1}, O_m \setminus O_m^j) \geq C^{-1}(m), \qquad \forall j \geq 1,$$

then by virtue of estimate (A.19) the map $S_m$ is an analytic symplectomorphism which maps each domain $O_m^j$, $j \geq 3$, to $O_m^{j-1}$.

Since

$$S_m(\mathfrak{h};\omega) - \mathfrak{h} = \varepsilon_m \int_0^1 V(S^t(\mathfrak{h};\omega);\omega)dt,$$

then (A.19) implies the estimate for $S_m - \mathrm{id}$, claimed in (a). To bound the Lipschitz constant in $\omega$, we denote $\eta(t) = S^t(\mathfrak{h};\omega_1) - S^t(\mathfrak{h};\omega_2)$ and note that this curve satisfies the equation

$$\dot\eta = \varepsilon_m V(\mathfrak{h}_1;\omega_1) - \varepsilon_m V(\mathfrak{h}_2;\omega_2).$$

Owing to (A.20), the Lipschitz constant of the map $\varepsilon_m V$ in $\mathfrak{h}$, calculated both in the weighted and non-weighted norms, is bounded by $C(m)\varepsilon_m^{1/3}$. Accordingly,

$$\frac{d}{dt}|\eta|_- \leq C(m)\varepsilon_m^{1/3}(|\eta|_- + |\omega_2 - \omega_1|), \qquad \eta(0) = 0.$$

So $|\eta(1)|_- \le C_1(m)\varepsilon_m^{1/3}|\omega_2 - \omega_1|$ by the Granwall lemma and assertion (a) is proved completely.

To prove (b) we note that for any $\xi$ the curves $t \mapsto S^t(\mathfrak{h})_*\xi$ satisfy the linearized equation $\dot{\xi} = \varepsilon_m V_*(\mathfrak{h}(t)\xi$, so the estimates for the map $S_{m*} - \mathrm{id}$ follow from (A.19) and (A.20).

The same arguments as above apply to any map $S^\theta, 0 \le \theta \le 1$, thus proving (c). □

**Step 5: The transformed Hamiltonian.** In this step we study the transformed Hamiltonian

$$\mathcal{H}_m \circ S_m = H_{0\,m+1} \circ S_m + \varepsilon_m(H_{1m} + H_{2m}) \circ S_m. \tag{A.21}$$

Since

$$f(1) = f(0) + f_t(0) + \int_0^1 (1-t) f_{tt}(t) dt$$

for any $C^2$-smooth function $f(t)$, then

$$H_{0\,m+1} \circ S_m = H_{0\,m+1} \circ S^1 = H_{0\,m+1} + \frac{d}{dt} H_{0\,m+1} \circ S^t \mid_{t=0}$$
$$+ \int_0^1 (1-t) \frac{d^2}{dt^2} H_{0\,m+1} \circ S^t dt.$$

Using (A.17) we get:

$$\frac{d}{dt} H_{0\,m+1} \circ S^t = \varepsilon_m \{F,\ H_{0\,m+1}\} \circ S^t = -\varepsilon_m H_{1m} \circ S^t$$

and

$$\frac{d^2}{dt^2} H_{0\,m+1} \circ S^t = -\varepsilon_m \frac{d}{dt} H_{1m} \circ S^t = -\varepsilon_m^2 \{F,\ H_{1m}\} \circ S^t.$$

Therefore,

$$H_{0\,m+1} \circ S_m = H_{0\,m+1} - \varepsilon_m H_{1m} - \varepsilon_m^2 \int_0^1 (1-t)\{F,\ H_{1m}\} \circ S^t dt.$$

Similarly, since

$$\frac{d}{dt}(H_{1m} + H_{2m}) \circ S^t = \varepsilon_m \{F,\ H_{1m} + H_{2m}\} \circ S^t,$$

then

$$\varepsilon_m(H_{1m} + H_{2m}) \circ S_m = \varepsilon_m(H_{1m} + H_{2m}) + \varepsilon_m^2 \int_0^1 \{F,\ H_{1m} + H_{2m}\} \circ S^t dt.$$

Substituting the obtained relation into (A.21) we find that

$$\mathcal{H}_m \circ S_m = H_{0\,m+1} - \varepsilon_m H_{1m} + \varepsilon_m(H_{1m} + H_{2m})$$
$$- \varepsilon_m^2 \int_0^1 (1-t)\{F,\ H_{1m}\} \circ S^t dt + \varepsilon_m^2 \int_0^1 \{F,\ H_{1m} + H_{2m}\} \circ S^t dt.$$

That is, $\mathcal{H}_m \circ S_m = H_{0\,m+1} + \varepsilon_{m+1} H_{m+1}$, where

$$\varepsilon_{m+1} H_{m+1} = \varepsilon_m H_{2m} + \varepsilon_m^2 \int_0^1 \big((t-1)\{F, H_{1m}\} + \{F, H_m - p \cdot h^{0\,p}\}\big) \circ S^t \, dt\,. \tag{A.22}$$

We checked at the end of Step 1 that the frequency map $\Lambda_{m+1}$ satisfies (A.11). Now we claim that also the domain $\Omega_{m+1}$ and the Hamiltonian $\mathcal{H}_{m+1} := \mathcal{H}_m \circ S_m$ satisfy the corresponding estimates (A.10) and (A.12) (with $m$ replaced by $m+1$). Indeed, since $\Omega_{m+1} = \Omega_m \setminus \Omega'$, then using Lemma A.2 we get:

$$\begin{aligned} \operatorname{mes}(\Omega \setminus \Omega_{m+1}) &\le \operatorname{mes}(\Omega \setminus \Omega_m) + \operatorname{mes}\Omega' \\ &\le \gamma e(m) + \gamma (m+1)^{-2}/K^* = \gamma e(m+1), \end{aligned}$$

so $\Omega_{m+1}$ satisfies (A.10).

It remains to check that the term $\varepsilon_{m+1} H_{m+1}$, defined by (A.22), satisfies (A.12) with $m := m+1$. The term $\varepsilon_m H_{2m}$ was treated in Lemma A.1. To estimate the integral terms we note that by (A.12), Lemmas A.1, A.2, and the Cauchy estimate, everywhere in $O_m^2$ we have:

$$\|\nabla_p K\|^{O_m^2,\, \Omega_{m+1}} \le C(m) \varepsilon_m^{-2/3}, \qquad \|\nabla_q K\|^{O_m^2,\, \Omega_{m+1}} \le C(m),$$

where $K = F$ or $K = H_{1m}$ or $K = H_m - p \cdot h^{0\,p}$. Therefore all the Poisson brackets that enter (A.22), for all $t$, are bounded by $C(m)\varepsilon_m^{-2/3}$ everywhere in $O_m^2$, as well as their Lipschitz constants. Owing to Lemma A.3, the transformations $S^t$ with $0 \le t \le 1$ map $O_m^3$ to $O_m^2$ and they are Lipschitz-close to the identity. Hence, the integral in the r.h.s. of (A.22) and its Lipschitz constant in $\omega \in \Omega_{m+1}$ are bounded by $C(m)\varepsilon_m^{4/3}$.

**Step 6: Transition to the limit.** Here we show that the set $(S_0 \circ S_1 \circ \cdots)(\{0\} \times \mathbb{T}^n) \subset \mathbb{R}^n \times \mathbb{T}^n$ is an analytic torus, invariant for eqn (A.7). By $\mathfrak{h}$ we denote the points $(p, q) \in \mathbb{R}^n \times \mathbb{T}^n$; by $\Pi_{\mathfrak{h}}$ and $\Pi_\omega$ we denote the projectors $(\mathfrak{h}; \omega) \mapsto \mathfrak{h}$ and $(\mathfrak{h}; \omega) \mapsto \omega$, respectively. Moreover, we set

$$\Omega_\varepsilon = \cap \Omega_m$$

and

$$\mathcal{O} = \{0\} \times U(\delta/2) \subset \mathbb{C}^n \times (\mathbb{C}^n / 2\pi \mathbb{Z}^n)\,.$$

Then $\Omega_\varepsilon$ is a Borel subset of $\Omega$ and $\operatorname{mes}(\Omega \setminus \Omega_\varepsilon) \le \gamma/2$ due to (A.10). The set $\mathcal{O}$ is a neighbourhood of the torus $\{0\} \times \mathbb{T}^n$ in the complex cylinder $\{0\} \times (\mathbb{C}^n/2\pi\mathbb{Z}^n)$, which is contained in every domain $O_m$ since $\delta_m > \delta/2$.

For $0 \le r \le N$ we consider the maps

$$\Sigma_N^r : O_N \times \Omega_N \to O_r, \qquad (\mathfrak{h}; \omega) \mapsto S_r \circ \cdots \circ S_{N-1}(\mathfrak{h}),$$

where $S_j(\mathfrak{h}) = S_j(\mathfrak{h}; \omega)$ (by definition, $\Sigma_r^r$ is the projection $\Pi_{\mathfrak{h}}$). We note that the domain of definition of every map $\Sigma_N^r$ contains the set $\mathcal{O} \times \Omega_\varepsilon$.

We claim that

$$|\Sigma^r_{r+M} - \Pi_{\mathfrak{h}}|^{O_{r+M},\,\Omega_\varepsilon} \le \varepsilon_r^\rho, \tag{A.23}$$

uniformly in $M \ge 0$. The estimate follows by indication in $M$. Indeed, for $M = 0$ it is obvious. If $M \ge 1$, then

$$\Sigma^r_{r+M} - \Pi_{\mathfrak{h}} = (S_r - \Pi_{\mathfrak{h}}) \circ (\Sigma^{r+1}_{r+M} \times \Pi_\omega) + (\Sigma^{r+1}_{r+M} - \Pi_{\mathfrak{h}}).$$

Let us denote the l.h.s. of (A.23) as $D^r_{r+M}$. Using the last identity, Lemma A.3, and the base of induction we find that

$$D^r_{r+M} \le C(r)\varepsilon_r^{1/3}(D^{r+1}_{r+M} + 2) + D^{r+1}_{r+M} \le 3C(r)\varepsilon_r^{1/3} + \varepsilon^\rho_{r+1} \le \varepsilon^\rho_r,$$

so (A.23) follows.

Similar to (A.23),

$$\left\|\frac{\partial}{\partial \mathfrak{h}}\Sigma^r_N(\mathfrak{h};\omega) - \mathrm{id}\right\| \le \varepsilon^\rho_r, \tag{A.24}$$

for any $r \le N$ and any $\mathfrak{h} \in O_N$ and $\omega \in \Omega_N$. To prove the estimate it suffices to write $\partial\Sigma^r_N/\partial\mathfrak{h}$ using the chain rule, and next apply Lemma A.3.

Owing to (A.23) for every $m \ge 0$ and for each $\omega \in \Omega_\varepsilon$, the maps $\Sigma^m_{m+N}(\cdot;\omega)$, restricted to $\mathcal{O}$, uniformly converge as $N \to \infty$ to a limiting analytic map

$$\Sigma^m_\infty(\cdot;\omega) : \mathcal{O} \to O_m,$$

and $\Sigma^m_p \circ \Sigma^p_\infty = \Sigma^m_\infty$ for all $p \ge m$. By analyticity, the derivatives $\partial\Sigma^m_{m+N}/\partial\mathfrak{h}$ converge to a derivative of the limiting map. Using (A.24) we get that the latter satisfies the estimate

$$\left\|\frac{\partial}{\partial \mathfrak{h}}\Sigma^m_\infty(\mathfrak{h};\omega) - \mathrm{id}\right\| \le \varepsilon^\rho_m, \quad \forall(\mathfrak{h},\omega) \in \mathcal{O} \times \Omega_\varepsilon. \tag{A.25}$$

Now we discuss the frequency vectors $\Lambda_m$. Because of the recurrent definition of $\Lambda_{m+1}$ in terms of $\Lambda_m$ and item (a) of Lemma A.1, $|\Lambda^{m+1} - \Lambda^m|^{\Omega^{m+1},\mathrm{Lip}} \le 2^{m+1}\varepsilon_m^{1/3}$. So the maps $\Lambda_m : \Omega_m \to \mathbb{R}^n$, restricted to $\Omega_\varepsilon$, converge to a limiting Lipschitz transformation $\Lambda_\infty : \Omega_\varepsilon \to \mathbb{R}^n$ such that $|\Lambda_\infty - \mathrm{id}|^{\Omega_\varepsilon,\mathrm{Lip}} \le C\varepsilon^{1/3}$ and

$$|\Lambda_\infty - \Lambda_m| \le 2^{m+2}\varepsilon_m^{1/3}.$$

Let us fix any $\omega \in \Omega_\varepsilon$ and $q_0 \in \mathbb{T}^n$. We consider the curve

$$\mathfrak{h}_\infty(t) = (0, q_0 + t\Lambda_\infty(\omega), 0) \subset \{0\} \times \mathbb{T}^n$$

and its images under the maps $\Sigma^m_\infty$, i.e. the curves $\mathfrak{h}_m(t) = \Sigma^m_\infty\mathfrak{h}_\infty(t) \subset O_m$. We shall show that $\mathfrak{h}_0(t)$ is a solution for (A.7). To do this we first use (A.25) to get that

$$\dot{\mathfrak{h}}_m = \Sigma^m_{\infty *}(\mathfrak{h}_\infty)\dot{\mathfrak{h}}_\infty = (0, \Lambda_\infty) + O(\varepsilon^\rho_m) \subset \mathbb{R}^{2n}.$$

Let us denote by $V_m$ a Hamiltonian vector field with the Hamiltonian $\mathcal{H}_m$. By (A.12), $V_m(\mathfrak{h}_m) = (0, \Lambda_m) + O(\varepsilon_m^\rho)$. Since $\Lambda_m = \Lambda_\infty + O(\varepsilon_m^\rho)$, then $V_m(\mathfrak{h}_m) = (0, \Lambda_\infty) + O(\varepsilon_m^\rho)$ and we get that

$$\dot{\mathfrak{h}}_m = V_m(\mathfrak{h}_m) + O(\varepsilon_m^\rho). \tag{A.26}$$

The linear map $\Sigma^0_{m*}(\mathfrak{h}_m)$ sends $\dot{\mathfrak{h}}_m$ to $\dot{\mathfrak{h}}_0$, sends $V_m(\mathfrak{h}_m)$ to $V_0(\mathfrak{h}_0)$, and its norm is bounded by two due to (A.24). Applying this map to (A.26) we get that

$$\dot{\mathfrak{h}}_0 = V_0(\mathfrak{h}_0) + O(\varepsilon_m^\rho)$$

for every $m$. Hence, $\dot{\mathfrak{h}}_0 = V_0(\mathfrak{h}_0)$. That is, the curve $\Sigma^0_\infty(0, q_0 + t\Lambda_\infty(\omega))$ is a solution of eqn (A.7) for any $q_0 \in \mathbb{T}^n$, if $\omega \in \Omega_\varepsilon$.

This proves Theorem B if we choose $\Sigma_\varepsilon(q, \omega) = \Sigma^0_\infty(0, q; \omega)$ and $\omega' = \Lambda_\infty(\omega)$.

# References


Adams, R.S. (1975). *Sobolev spaces*. Academic Press, New York–London.

Arnold, V.I. (1963). Proof of a theorem of A.N. Kolmogorov on the conservation of quasiperiodic motions under a small change of the Hamiltonian function. *Usp. Mat. Nauk*, **18**(5), 13–40. [English translation in *Russ. Math. Surv.*, **18**(5), 9–36 (1963).]

Arnold, V.I. (1974). *Mathematical methods in classical mechanics* (3rd edn). Nauka, Moscow. [English translation, 2nd edn, Springer-Verlag, Berlin (1989).]

Arnold, V.I. and Givental, A.B. (1989). Symplectic geometry. In *Encyclopaedia of mathematical sciences*, Vol. 4. Springer-Verlag, Berlin.

Arnold, V.I., Kozlov, V.V., and Neistadt, A.I. (1989). Mathematical aspects of classical and celestial mechanics. In *Encyclopaedia of mathematical sciences*, Vol. 3. Springer-Verlag, Berlin.

Babin, A.V. and Vishik, M.I. (1989). *Attractors of evolution equations*. Nauka, Moscow. [English translation, North-Holland, Amsterdam, (1991).]

Baker, H.F. (1897). *Abelian functions*. Cambridge University Press.

Bättig, D., Kappeler, T., and Mityagin, B. (1996). On the KdV equation: convergent Birkhoff normal form. *J. Funct. Anal.*, **140**, 335–58.

Belokolos, E.D., Bobenko, A.I., Enolskii, V.Z., Its, A.R., and Matveev, V.B. (1994). *Algebro-geometric approach to nonlinear integrable equations*. Springer-Verlag, Berlin.

Bikbaev, R.F. and Kuksin, S.B. (1992). A periodic boundary value problem for the Sine–Gordon equation, small Hamiltonian perturbations of it, and KAM-deformations of finite-gap tori. *Algebra i Analiz*, **4**(3), 42–78. [English translation in *St.-Petersburg Math. J.*, **4**, 439–68 (1993).]

Bikbaev, R.F. and Kuksin, S.B. (1993). On the parametrization of finite-gap solutions by frequency vector and wave-number vector and a theorem of I. Krichever. *Lett. Math. Phys.*, **28**, 115–22.

Bobenko, A.I. and Kuksin, S.B. (1991). Finite-gap periodic solutions of the KdV equation are nondegenerate. *Phys. Lett.* A **161**(3), 274–6.

Bobenko, A.I. and Kuksin, S.B. (1995*a*). The nonlinear Klein–Gordon equation on an interval as a perturbed Sine–Gordon equation. *Comment. Math. Helv.*, **70**, 63–112.

Bobenko, A.I. and Kuksin, S.B. (1995*b*). Small-amplitude solutions of the Sine–Gordon equation on an interval under Dirichlet or Neumann boundary conditions. *J. Nonlinear Sci.*, **5**, 207–32.

Bochner, S. and Martin, W.T. (1948). *Several complex variables*. Princeton University Press.

Bourgain, J. (1997). On Melnikov's persistency problem. *Math. Res. Lett.*, **4**, 445–58.

Bourgain, J. (1998). Quasi-periodic solutions of Hamiltonian perturbations of 2D linear Schrödinger equation. *Ann. Math.*, **148**, 363–439.

Broer, H.W., Huitema, G.B., and Sevryuk, M.B. (1996). *Quasi-periodic motions in families of dynamical systems*, Lecture Notes Math. 1645, Springer-Verlag, Berlin.

Cartan, H. (1967*a*). *Calculus différerentiel*. Hermann, Paris.

Cartan, H. (1967*b*). *Formes différerentielles*. Hermann, Paris.

Chierchia, L. and You, J. (2000). *KAM tori for 1D nonlinear wave equations with periodic boundary conditions*. *Commun. Math. Phys.*, in press.

Craig, W. and Wayne, C.E. (1992). Newton's method and periodic solutions of nonlinear wave equation, *Commun. Pure Appl. Math.*, **46**, 1409–98.

Dubrovin, B.A. (1981). Theta functions and nonlinear equations. *Usp. Mat. Nauk*, **36**(2), 11–80. [English translation in *Russ. Math. Surv.*, **36** (1981).]

Dubrovin, B.A. and Natanzon, S.M. (1982). Real two-gap solutions of the Sine–Gordon equation. *Funkt. Anal. Prilozh.*, **16**(1), 27–43. [English translation in *Funct. Anal. Appl.*, **16**, 21–33 (1982).]

Dubrovin, B.A. and Novikov, S.P. (1989). Hydrodynamics of weakly deformed soliton lattices. Differential geometry and Hamiltonian theory, *Usp. Mat. Nauk*, **44**(6), 29–98. [English translation in *Russ. Math. Surv.*, **44**, 35–124 (1989).]

Dubrovin, B.A., Matveev, V.F., and Novikov, S.P. (1976). Nonlinear equations of Korteweg-de Vries type, finite zone linear operators, and Abelian varieties. *Usp. Mat. Nauk*, **31**(1), 55–136. [English translation in *Russ. Math. Surv.*, **31**(1) (1976).]

Eliasson, L.H. (1988). Perturbations of stable invariant tori. *Ann. Scuola. Super. Pisa, Cl. Sci.*, IV Ser., **15**, 115–47.

Ercolani, N., Forest, M.G., and McLaughlin, D. (1987). *Geometry of the modulational instability*, Part 1. Preprint, University of Arizona.

Ercolani, N., Forest, M.G., and McLaughlin, D. (1990). Geometry of the modulational instability, Part 3. *Physica*, **43** D, 349–84.

Faddeev, L.D. and Takhtajan, L.A. (1986). *Hamiltonian methods in the theory of solitons*. Nauka, Moscow. [English translation, Springer-Verlag, Berlin, (1987).]

Falconer, K. (1990). *Fractal geometry*. John Wiley & Sons.

Fay, J.D. (1973). *Theta-functions on Riemann surfaces*. Lecture Notes Math. 352, Springer-Verlag, Berlin.

Federer, H. (1969). *Geometric measure theory*. Springer-Verlag, Berlin.

Flaschka, H. and Mclaughlin, D.W. (1976). Canonically conjugate variables for the KdV equation and the Toda lattice with periodic boundary conditions. *Progr. Theoret. Phys.*, **55**, 438–56.

Fröhlich, I. and Spencer, T. (1983). Localization for a class of one-dimensional quasi-periodic Schrödinger operators. *Commun. Math. Phys.* **88**, 151–84.

Garnet, J. and Trubowitz, E. (1987). Gaps and bands on one dimensional Schrödinger operator II. *Comment. Math. Helv.*, **62**, 18–37.

Godbillon, C. (1969). *Géométrie différentielle et mècanique analytique*. Hermann, Paris.

Guillemin, V. and Sternberg, H. (1977). *Geometric asymptotics*, Amer. Math. Soc., Providence.

Gunning, R.C. and Rossi, H. (1965). *Analytic functions of several complex variables*. Prentice-Hall, Inc., Englewood Cliffs, N.J.

Hartman, Ph. (1964). *Ordinary differential equations*. John Wiley & Sons, New York–London–Sydney.

Herman, M. (1983, 1986). Sur les courbes invariantes par les difféomorphismes de l'anneau, Vol. 1,2. *Astérisque*, **103–104** and **144**.

Herman, M. (1989). Inégalités a priori pour des tores Lagrangiens invariants par des difféomorphismes symplectiques. *Publ. Math. IHES*, **70**.

Hofer, H. and Zehnder, E. (1994). *Symplectic invariants and Hamiltonian dynamics*. Birkhaser, Basel.

Ito, H. (1989). Convergence of Birkhoff normal forms for integrable systems. *Comment. Math. Helv.*, **64**, 412–61.

Kappeler, T. (1991). Fibration of the phase space for the Korteweg–de Vries equation. *Ann. Inst. Fourier*, **41**, 539–75.

Kato, T. (1966). *Perturbation theory for linear operators*. Springer-Verlag, Berlin.

Kato, T. (1975). *Quasi-linear equations of evolutions, with applications to partial differential equations*, Lecture Notes Math. **448**, Springer-Verlag, Berlin, 25–70.

Klainerman, S. (1983). Long-time behaviour of solutions to nonlinear wave equations. In *Proceedings of International Congress of Mathematics* (Warsaw, 1983), pp. 1209–15.

Kolmogorov, A.N. (1985). On the conservation of conditionally periodic motions for a small change in Hamilton's function. *Selected works of A.N. Kolmogorov*, Vol. 1. Nauka, Moscow. [English translation, Kluwer, (1991).]

Kozel, V.A. and Kotlyarov, V.P. (1976). Quasiperiodic solutions of the equation $U_{tt} - U_{xx} + \sin U = 0$, *Dokl. Akad. Nauk Ukr. SSR*, **10A**, 878–81 (Russian).

Krichever, I.M. (1989). Spectral theory of two dimensional periodic operators and its applications. *Usp. Mat. Nauk*, **44**(2), 121–84. [English translation in *Russ. Math. Surv.*, **44**(2), 145–225 (1989).]

Krichever, I.M. (1991). Perturbation theory in periodic problems for two-dimensional integrable systems, *Sov. Sci. Rev. C. Math. Phys.*, **9**, 1–101.

Kriecherbauer, T. (1998). Estimates on Green's functions of quasi-periodic matrix operators and a new version of the Coupling Lemma in the Fröhlich–Spencer techniques. *IMRN* **17**, 907–35.

Kuksin, S.B. (1987). Hamiltonian perturbations of infinite-dimensional linear systems with an imaginary spectrum. *Funkt. Anal. Prilozh.*, **21**(3), 22–37. [English translation in *Funct. Anal. Appl.*, **21**, 192–205 (1987).]

Kuksin, S.B. (1988). Perturbations of quasiperiodic solutions of infinite-dimensional Hamiltonian systems, *Izv. Akad. Nauk SSSR, Ser. Mat.*, **52**, 41–63. [English translation in *Math. USSR Izv.*, **32**(1), 39–62 (1989).]

Kuksin, S.B. (1989). The perturbation theory for the quasiperiodic solutions of infinite-dimensional Hamiltonian systems and its applications to the Korteweg – de Vries equation. [English translation in *Math. USSR Sbornik*, **64**, 397–413 (1989).]

Kuksin, S.B. (1989). An averaging theorem for distributed conservative systems and its application to the von Karman equations. *Prikl. Mat. Mekh.*, **53**(2), 196–205. [English translation in *J. Appl. Math. Mech.*, **53**(2), 150–7 (1989).]

Kuksin, S.B. (1992). An infinitesimal Liouville–Arnold theorem as criterion of reducibility for variational Hamiltonian equations. *Chaos, Solitons and Fractals*, **2**(2), 259–69.

Kuksin, S.B. (1993). *Nearly integrable infinite-dimensional Hamiltonian systems*, Lecture Notes Math. 1556, Springer-Verlag, Berlin.

Kuksin, S.B. (1994). KAM-theory for partial differential equations. In *Proceedings of the First European Congress of Mathematics*, (Paris, 6–10 July 1992) Vol. 2, Birkhäuser, 123–57.

Kuksin, S.B. (1998*a*). Elements of a qualitative theory of hamiltonian PDEs. In *Proceedings of the International Congress of Mathematicians* (Berlin, 1998), Vol. 2, pp. 819–29.

Kuksin, S.B. (1998*b*). A KAM-theorem for equations of the Korteweg–de Vries type. *Rev. Math. & Math. Phys.*, **10**(3), 1–64.

Kuksin, S.B. and Pöschel, J. (1996). Invariant Cantor manifolds of quasi-periodic oscillations for a nonlinear Schrödinger equation. *Ann. Math.*, **143**, 149–79.

Lang, S. (1972). *Differential manifolds*. Addison-Wesley, Reading.

Lax, P.D. (1964). Development of singularities of solutions of nonlinear hyperbolic partial differential equations, *J. Math. Phys.*, **5**, 611–13.

Lax, P.D. (1975). Periodic solutions of the KdV equations. *Commun. Pure Appl. Math.*, **28**, 141–88.

Lazutkin, V.F. (1993). *KAM-theory and semiclassical approximation to eigenfunctions*. Springer-Verlag, Berlin.

Lions, J.L. (1969). *Quelques méthodes de résolution des problèmes aux limites no linéares*, Dunod, Paris.

Lions, J.-L. and Magenes, E. (1968). *Problémes aux limites non homogénes et applications*. Dunod, Paris. [English translation, Springer-Verlag, Berlin (1972).]

Marčenko, V.A. (1977). *Sturm-Liouville operators and applications*. Naukova Dumka, Kiev, (Russian). [English translation, Birkhäuser, Basel (1986).]

McKean, H.P. (1981). The Sine–Gordon and Sinh–Gordon equation on the circle. *Commun. Pure Appl. Math.*, **34**, 197–257.

McKean, H.P. and Trubowitz, E. (1976). Hill's operator and hyperelliptic function theory in the presence of infinitely many branch points. *Commun. Pure Appl. Math.*, **29**, 143–226.

Melnikov, V.K. (1965). On some cases of conservation of conditionally periodic motions under a small change of the hamiltonian function. *Dokl. Akad. Nauk SSSR*, **165**(6), 1245–8. [English translation in *Sov. Math. Dokl.*, **6**, 1592–6 (1965).]

Melnikov, V.K. (1968). A family of conditionally periodic solutions of a Hamiltonian system. *Dokl. Akad. Nauk SSSR*, **181**(3), 546–9. [English translation in *Sov. Math. Dokl.*, **9**, 882–6 (1968).]

Milnor, J. (1968). *Singular points of complex hypersurfaces*. Princeton University Press.

Moser, J. (1973). *Stable and random motions in dynamical systems*. Princeton University Press.

Moser, J. Various aspects of integrable hamiltonian systems. In *Proc. CIME Conference*, Bressansone, Italien, 1978. Progr. Math. 8 Birkhäuser, 1980.

Nekhoroshev, N.N. (1994). The Poincaré-Lyapunov-Liouville-Arnold theorem. *Funkt. Analiz Prilozn.*, **28**(2). [English translation in *Funct. Anal. Appl.*, **28**, 128–9 (1994).]

Nirenberg, L. (1974). *Topics in non-linear functional analysis. Lect. Notes by R.A. Artino.*, Courant Inst. N.Y. University. [Russian translation, Mir, Moscow (1977).]

Novikov, S.P. (1974). A periodic problem for the Korteweg-de Vries equation, I. *Funkt. Anal. Prilozh.*, **8**(3), 54–66. [English translation in *Funct. Anal. Appl.*, **8**, 236–46 (1974).]

Pazy, A. (1983). *Semigroups of linear operators and applications to partial differential equations*. Springer-Verlag, Berlin.

Pesin, Ya. (1989). General theory of smooth hyperbolic dynamical systems. In *Encyclopaedia of mathematical sciences*, Vol. 2, Springer-Verlag, Berlin.

Pöschel, J. (1989). On elliptic lower dimensional tori in Hamiltonian systems. *Math. Z.*, **202**, 559–608.

Pöschel, J. (1996*a*). A KAM-theorem for some nonlinear PDEs. *Ann. Scuola Norm. Sup. Pisa, Cl. Sci.*, IV Ser. 15 **23**, 119–48.

Pöschel, J. (1996*b*). Quasi-periodic solutions for a nonlinear wave equation. *Comment. Math. Helv.*, **71**, 269–96.

Pöschel, J. and Trubowitz, E. (1987). *Inverse spectral theory*. Academic Press, New York–London.

Reed, M. and Simon, B. (1975). *Methods of modern mathematical physics*. Vol. 2, Academic Press, New York–London.

Siegel, C.L. and Moser J. (1971). *Lectures on celestial mechanics*. Springer-Verlag, Berlin.

Sigal, I.M. (1993). Nonlinear wave and Schrödinger equations I: Instability of periodic and quasiperiodic solutions. *Commun. Math. Phys.*, **153**, 297–320.

Springer, G. (1981). *Introduction to Riemann surfaces* (2nd edn). Chelsea Publishing Company, New York.

Treschev, D.V. (1998). *Introduction to the perturbation theory for Hamiltonian systems*. Fazis, Moscow, (Russian).

Veselov, A.P. and Novikov S.P. (1984). Poisson brackets and complex tori. *Trudy Matem. Inst. Steklov*, **165**, 49–61. [English translation in *Proc. Steklov Inst. Mathem.*, Issue 3, 53–65 (1985).]

Vey, J. (1978). Sur certain systèmes dynamiques séparables. *Am. J. Math.*, **100**, 591–614.

Vishik, M.I. and Fursikov A.V. (1980). *Mathematical problems of statistical hydrodynamics*. Nauka, Moscow. [English translation, Kluwer, Dordrecht (1988).]

Wayne, C.E. (1990). Periodic and quasi-periodic solutions of nonlinear wave equations via KAM theory. *Commun. Math. Phys.*, **127**, 479–528.

Weinstein, A. (1977). *Lectures on symplectic manifolds*. A.M.S. Conf. Board, Reg. Conf. in Math. **29**.

Zakharov, V.E., Manakov S.V., Novikov S.P., and Pitaevskij L.P. (1980). *Theory of solitons*. Nauka, Moscow. [English translation, Plenum Press, New York, (1984).]

# Index